VOLUME THREE

Developments in Neuroethics and Bioethics

Ethical Dimensions of Commercial and DIY Neurotechnologies

Serial Editor

Judy Illes, CM, PHD
Division of Neurology, Department of Medicine
The University of British Columbia
Neuroethics Canada
2211 Wesbrook Mall, Koerner S124
Vancouver BC
Canada
V6T 2B5

Katherine Bassil, Editorial Assistant to Judy Illes

VOLUME THREE

DEVELOPMENTS IN NEUROETHICS AND BIOETHICS

Ethical Dimensions of Commercial and DIY Neurotechnologies

Edited by

IMRE BÁRD
The London School of Economics and Political Science, United Kingdom

ELISABETH HILDT
Center for the Study of Ethics in the Professions, Illinois Institute of Technology, Chicago, United States

ELSEVIER

ACADEMIC PRESS
An imprint of Elsevier

Academic Press is an imprint of Elsevier
50 Hampshire Street, 5th Floor, Cambridge, MA 02139, United States
525 B Street, Suite 1650, San Diego, CA 92101, United States
The Boulevard, Langford Lane, Kidlington, Oxford OX5 1GB, United Kingdom
125 London Wall, London, EC2Y 5AS, United Kingdom

First edition 2020

ISBN: 978-0-12-816181-4
ISSN: 2589-2959

For information on all Academic Press publications
visit our website at https://www.elsevier.com/books-and-journals

Publisher: Zoe Kruze
Acquisitions Editor: Sam Mahfoudh
Editorial Project Manager: Peter Llewellyn
Production Project Manager: Denny Mansingh
Cover Designer: Alan Studholme

Typeset by SPi Global, India

Contents

Contributors

Imre Bárd
Methodology Department, London School of Economics and Political Science, London, United Kingdom

Kimberly Rose Clark
Department of Psychological & Brain Sciences, Dartmouth College, Hanover, NH, United States

Iris Coates McCall
Neuroethics Canada, Division of Neurology, Department of Medicine, University of British Columbia, Vancouver, BC, Canada

Ishan Dasgupta
Department of Philosophy; Center for Neurotechnology, University of Washington, Seattle, WA, United States

Elisabeth Hildt
Center for the Study of Ethics in the Professions, Illinois Institute of Technology, Chicago, IL, United States

Viorica Hrincu
Neuroethics Canada, Division of Neurology, Department of Medicine, University of British Columbia, Vancouver, BC, Canada

Marcello Ienca
Department of Health Sciences and Technology, ETH Zurich, Zurich, Switzerland

Judy Illes
Neuroethics Canada, Division of Neurology, Department of Medicine, University of British Columbia, Vancouver, BC, Canada

Ying-Tung Lin
Institute of Philosophy of Mind and Cognition, National Yang-Ming University, Taipei City, Taiwan

Nicole Martinez-Martin
Stanford Center for Biomedical Ethics, Stanford University, Stanford, CA, United States

Nicole Minielly
Neuroethics Canada, Division of Neurology, Department of Medicine, University of British Columbia, Vancouver, BC, Canada

James Scheibner
Department of Health Sciences and Technology, ETH Zurich, Zurich, Switzerland

Naznin Virji-Babul
University of British Columbia, Vancouver, BC, Canada

Anna Wexler
Department of Medical Ethics & Health Policy, Perelman School of Medicine, University of Pennsylvania, Philadelphia, PA, United States

Maya Willms
University of British Columbia, Vancouver, BC, Canada

Preface: Ethical aspects of DIY and commercial neurotechnologies

Imre Bárd[a], Elisabeth Hildt[b]
[a]Methodology Department, London School of Economics and Political Science, London, United Kingdom
[b]Center for the Study of Ethics in the Professions, Illinois Institute of Technology, Chicago, IL, United States

The term neurotechnology refers to technologies that "enable the monitoring and/or modulation of the function of the brain" (Eaton & Illes, 2007, p. 393). In research and clinical medicine, this includes neuroimaging techniques, such as functional magnetic resonance imaging (fMRI), brain stimulation methods such as deep brain stimulation (DBS) or transcranial direct current stimulation (tDCS), cochlear implants, and brain-computer interfaces. In recent years a variety of views have emerged with regard to how direct the monitoring or modulation of brain function needs to be in order to warrant the neurotechnology label (Ienca & Vayena, 2019). Some favor a rather narrow definition, which is restricted to technologies that directly read or upload brain data (Kellmeyer, 2018). On this account, neuromonitoring devices like electroencephalography (EEG) and neurostimulation interventions would exhaust the category of neurotechnologies. Others take a somewhat more expansive view and allow brain training games and mental health applications to be designated as neurotechnologies as well (Wexler & Reiner, 2019). Still others embrace an even more expanded view, arguing that this class of technologies is characterized by being personal, digital and mobile, which opens up the category to various other devices that use non-neural biomarkers to measure psychological properties, such as stress (Kreitmair, 2019a). Therefore, it has been suggested that the relationship between different devices in the broad category of neurotechnologies is best understood along the lines of family resemblance instead of a clear set of criteria (Kreitmair, 2019b).

During the past 2 decades, ethical issues related to monitoring and modifying the human brain with the help of neurotechnologies have been discussed intensively, with topics like autonomy, identity, privacy, safety, agency, and equality emerging as the most important considerations (Illes & Raffin, 2002; Farah, 2008; Jotterand & Giordano, 2011; Mecacci & Haselager, 2019; Schermer, 2011; Yuste, Goering, Arcas, et al., 2017). The enhancement uses of neurotechnologies, as well as their ethical implications and policy approaches have also been addressed

(Lavazza, 2019; Maslen, Douglas, Cohen Kadosh, Levy, & Savulescu, 2014, Fitz & Reiner, 2015). These issues are particularly salient outside research laboratories and clinics, as neurotechnologies have found their way into the workplace, into the offices of advertising and market research companies, and into the hands of citizen scientists and members of the Do-It-Yourself (DIY) community. While this process is rapidly picking up pace, facilitated by technological developments, a decrease in the size of devices, and a reduction of costs, such a trend in neuroscience is not new. The insights and tools of the field have been used to better understand consumer behavior since at least the late 1990s. By the early 2000s the concept of neuromarketing had been born—a marketing approach in which neuroscience methods are used to analyze consumer behavior and preferences in order to develop more successful marketing strategies (Ariely & Berns, 2010; Lee, Broderick, & Chamberlain, 2007; Morin, 2011). Scholars in the burgeoning field of neuroethics were quick to highlight the ethical challenges raised by the practice of using advanced technologies to probe the mind for commercial benefit, emphasizing potential threats to the autonomy of individuals, and the importance of protection from exploitation (Murphy, Illes, & Reiner, 2008; Ulman, Cakar, & Yildiz, 2015). In the late 2000s advances in microelectronics manufacturing started to make electroencephalography devices viable as consumer products and companies like NeuroSky and Emotiv released affordable brain sensing headsets marketed as the next interface for human-computer interaction and as powerful tools to take care of and improve one's mind. The first games and toys driven by brain-computer interfaces, such as Mattel's Mindflex appeared around this time as well, along with the vision of an app store for brain technologies.[1] The early 2010s saw the birth of the DIY brain stimulation movement, whose members construct simple electrical brain stimulation devices similar to those used in academic research, in order to self-treat certain medical conditions and to experiment with enhancing cognitive performance (Wexler, 2016). The practice of DIY brain stimulation has been situated at the interface of broader social and cultural contexts and trends, such as citizen science, the quantified-self movement, maker culture and the fascination for neuroenhancement (Wexler, 2017). Although levels of neuroscience knowledge may vary considerably among members of the DIY community, the benefits and risks of DIY applications such as tDCS are far from clear (Fitz & Reiner, 2014).

[1] https://web.archive.org/web/20091119214028/http://store.neurosky.com/collections/applications

The expansion of access to tools for the management of the mind may bring about great improvements in health, well-being and productivity, but it also presents a range of ethical and societal questions that are likely to become one of the primary concerns of neuroethics over the next decade. In fact, the Emerging Issues Task Force of the International Neuroethics Society recently identified growing consumerization as a key future topic for the field (Emerging Issues Task Force, International Neuroethics Society, 2019). At the center of the ethical and regulatory challenges related to DIY and commercial neurotechnologies is the fact that these exist outside established ethical frameworks, oversight and accountability mechanisms that guide the application of interventions in the medical domain. While the same technology, for example EEG or tDCS, may be used in various contexts—such as the clinical field, scientific research, DIY, or consumer products—the knowledge level of those applying the technology, the purpose of use, and the applicable regulations differ significantly. There is considerably less regulation in the context of nonmedical consumer neurotechnologies as compared to medical applications or institutional research practice, and there is an obvious lack of regulation in the DIY area. This divergence may facilitate an inadequate distinction between medical and purported lifestyle uses of neurotechnologies (Forlini, Lipworth, Carter, & Kerridge, 2019) and prompt companies to target their products primarily to nonmedical applications and to advertise accordingly. Labeling a device as a consumer product instead of a medical device may avoid restrictions and lead to disappointment if the device does not live up to its marketing, or even cause harm to the consumer. As editors of this volume, we believe that there is a clear need for responsible policies and oversight of consumer neurotechnologies and DIY applications. The democratization of neurotechnologies and their growing uptake in extraclinical spheres of use is only set to increase over the coming years, as governments and private investors continue to fund research into new ways of recording, understanding and influencing brain activity. Interest in all things *neuro* continues to grow as the brain is increasingly understood as a malleable organ, the resources and capacities of which should be actively managed and optimized (Rose & Abi-Rached, 2013).

The utilization of neuroscience tools and knowledge in the consumer marketing domain, the direct-to-consumer (DTC) sale of neuromonitoring and neurostimulation products, and the practice of DIY self-experimentation with brain stimulation represent the core of and motivate the present volume. Interest for neurotechnology products is on the rise, as evidenced by the fact that the largest online retailer of neurostimulation devices has recorded over 50% growth every quarter since 2014 (Waltz, 2019). In addition, new companies continue to emerge in the consumer neurotechnology space (Coates McCall et al., 2019; Wexler & Thibault, 2019) and some of these companies mobilize vast financial resources to expedite technology development. Kernel, a start-up founded by tech entrepreneur Brian Johnson, is developing a "non-invasive mind/body/machine interface (MBMI) to improve, evolve and extend human cognition"[2] and Silicon Valley magnate Elon Musk leads Neuralink, a company working on a high-bandwidth wireless, implanted brain-computer interface to connect humans and computers. Although Neuralink is initially developing therapeutic solutions for neurological conditions, the long-term vision is to create a consumer product that achieves a symbiotic relationship between humans and machines (Neuralink, 2019). Although such visions may appear fanciful, a recent report by the British Royal Society estimated that by 2040 several game-changing, nonmedical applications of neural interfaces may become feasible, such as hands-free control of computers. Invasive neurotechnologies are expected to improve drastically as well, evolving into miniature, intelligent, high-bandwidth implants (Royal Society, 2019). The efforts of social media giant Facebook to develop a novel non-invasive brain-computer interface technology also point in the direction of easing and speeding up communication between users and the extended network of digital tools including augmented and virtual reality devices, smartphones and other products (Dujmovic, 2019).

This interconnection serves to highlight the ways in which the boundaries of the neurotechnology category are becoming increasingly porous. On the one hand, neurotechnologies like brain-computer interfaces will be embedded in an expanded technological ecosystem that comprises the Internet of Things, online platforms and large data infrastructures. Scholars have also noted the growing convergence of neurotechnologies and artificial intelligence, which opens up entirely new possibilities in terms of increasing the capabilities, efficacy and precision of devices. However, a range of serious challenges around the use of algorithms are bound to emerge as well (Yuste et al., 2017). Questions related to data protection, privacy, transparency and accountability, which are fundamental aspects of the ethics of algorithms, are increasingly important for neurotechnologies

[2] https://www.kernel.co/

(Rainey, Bublitz, Maslen, & Thornton, 2019; Wolkenstein, Jox, & Friedrich, 2018). In addition, given to the rapid rise of big data and machine-learning capabilities, a wide range of other wearable technologies, from smart earphones and smart watches to smart rings and smart glasses can be used to make inferences about mood, behavior and mental state, as well as to offer suggestions and nudges to manage the mind.

The uncertainty around the exact definition of neurotechnologies also points to a more fundamental question about the proper scope of neuroethics as a discipline. People are increasingly surrounded by technologies that contribute to making us transparent and predictable in new ways, opening up possible channels for influencing our mental states, moods, behaviors, and well-being without directly interfacing with the brain. As a result, the breadth of topics legitimately falling within the purview of neuroethics seems to be growing. While some scholars have been arguing for the recognition of this extension for a number of years (Levy, 2007; Reiner & Nagel, 2017) the proliferation of technologies capable of influencing our cognitive and affective functioning in meaningful ways necessitates that neuroethicists engage more widely in discussions about technology development, design and deployment.

The present volume seeks to canvass the range of issues that emerge from these developments. Below we offer a brief summary of each contribution.

The collection opens with a piece by Iris Coates McCall and Anna Wexler, who offer a robust overview of the ethical issues related to consumer electroencephalography (EEG) devices for neuromonitoring and neurofeedback, which kickstarted the wave of direct-to-consumer neurotechnology products. The authors provide an accessible review of the workings of EEG neuromonitoring devices as well as an assessment of the current market and manufacturers' claims about the capabilities of their devices. Coates McCall and Wexler discuss current, mid-term and long-term ethical issues, arguing that widely shared concerns about brain data privacy with regard to consumer EEG devices are vastly overstated. Instead of focusing on such hypothetical concerns underpinned by a belief in the potency, accuracy and reliability of neuromonitoring devices, they suggest that steps should be taken to address misleading and unsubstantiated claims.

The chapter by Kimberly Rose Clark provides a unique look at the rapidly growing field of consumer neuroscience, written from the perspective of a pracademic who is involved in both academic research and commercial work. Clark gives a detailed description of the plethora of methods ranging from familiar brain imaging techniques through nudges and exploitations of psychological vulnerabilities, all the way to the most recent tools involving machine-learning methods for emotion detection, voice analysis, and other predictive analytics. Richly illustrated with real-world examples, the chapter revolves around the fundamental question whether neuroscience knowledge in the consumer domain allows for such a fine-grained profiling of human behavior and preferences that their use represents a threat to individual autonomy and freedom.

The piece by Nicole Martinez-Martin offers an in-depth look at the field of consumer digital mental health tools. Such tools hold out the promise of making mental health therapies more widely accessible, which is an urgent imperative given the large unmet medical need in this domain. However, as Martinez-Martin lays out, there are serious concerns with regard to privacy, transparency, accountability, safety and effectiveness. Mental health apps present challenges to established frameworks of the therapeutic encounter between patient and provider, and Martinez-Martin further demonstrates where regulatory oversight would be beneficial.

In his contribution, Imre Bárd looks at sensory enhancement, arguing that this class of neurotechnologies has largely been ignored by neuroethicists. He gives an overview of the variety of technologies for sensory enhancement methods that allow for the ongoing mediation of a user's experience of the world. Bárd's chapter describes the practices of the DIY sensory enhancement community and argues that these represent a hitherto untapped source for neuroethicists to study proactionary attitudes toward enhancement. In addition, the chapter looks at consumer augmented reality technologies, highlighting the risks with regard to autonomy and manipulation inherent to the expansion of algorithmically mediated experiences to the everyday perception of our environment.

The next three contributions focus more closely on the use of non-invasive electrical brain stimulation. In her chapter, Anna Wexler, who is arguably the world's foremost expert on the subject, gives an overview of the issues related to DIY and DTC neurostimulation outlining the history of the rise and development of the field and its associated ethical issues. She also highlights how the focus on these two domains of use has led to insufficient attention to the provision of brain stimulation services in various private clinics and presents pilot empirical data about this area. Wexler describes how brain stimulation is entering a third sphere of use besides DIY and DTC, which presents ethical and regulatory questions analogous to those related to Complementary and Alternative Medicine and the off-label uses of pharmaceuticals.

The contribution by Maya Willms and Naznin Virji-Babul offers an analysis of the ethical issues related to the use of brain stimulation in young people. The authors explain the ways in which the developmental characteristics of adolescence represent unique challenges with regard to managing brain stimulation technologies. Willms and Virji-Babul put forward a neurodevelopmental framework to respond to these challenges, which highlights the importance of multiple maturities with regard to brain-behavior interactions, the role of sensitive periods, and brain connectivity. The framework is then used to develop guidelines for practice, which cover the dosing of brain stimulation in adolescents, considerations around using brain stimulation in highly charged emotional contexts, the social responsibility of researchers to understand the long-term effects of brain stimulation, and mechanisms for involving young people in crafting frameworks for the responsible use of this type of neurotechnology.

The chapter by Ying-Tung Lin on brain stimulation analyzes the difficulties related to measuring the effectiveness of do-it-yourself brain stimulation practices. On the one hand, these difficulties are related to the lack of adequate controls in the home use of brain stimulation devices, but Lin also highlights the contribution of metacognitive monitoring to perceived and subjective effects. She argues that this uncertainty with regard to the real effects of stimulation and the existence of illusory effects raise concerns with regard to truthfulness and authenticity. On the other hand, for Lin, DIY brain stimulation also represents an opportunity for individuals to exercise agency over their mental states and she argues that the practice can contribute to users' sense of well-being.

The piece by Marcello Ienca and James Scheibner provides an exemplary analysis of the extensively used term *neurohacking*. The authors describe the results of a scoping review of the relevant literature and advance a taxonomy for differentiating between the various uses of neurohacking to create conceptual clarity where currently there is mostly confusion. Ienca and Scheibner demonstrate that neurohacking is employed in four distinct but partially overlapping ways in the popular and scientific literature, and they offer an analysis of the ethical and legal questions that accompany these uses.

In his chapter, Ishan Dasgupta focuses on one of the most crucial challenges related to DTC and DIY neurotechnologies, the regulatory uncertainty surrounding these developments. He offers an in-depth analysis of the existing regulatory frameworks for DTC neurotechnologies in the United States, considering neurostimulation, neuromonitoring, mental health apps and brain training games. Dasgupta describes the merits and limitations of various regulatory mechanisms, including regulation via federal agencies, legislative action, tort liability, mechanisms of self-regulation, and independent working groups.

The final chapter in the volume, by Nicole Minielly, Viorica Hrincu, and Judy Illes, presents the findings from an empirical study conducted among the representatives of companies in the DTC neuromonitoring and neurostimulation markets. The focus of their interview-based research was to explore practices for the handling of incidental findings and adverse events associated with consumer neurotechnology products. In addition to offering insights into the limited status quo, the authors also advance recommendations for how industry actors could better engage with these important responsibilities.

To the best of our knowledge, ours is the first themed volume devoted to the topic of DIY and commercial neurotechnologies. Although there are certainly important areas that we should have covered or addressed in a more in-depth manner, we do hope that the book will serve as a useful resource to students, researchers, entrepreneurs, and others interested in the development of this fascinating field.

As editors we declare that we have no conflicts of interest.

Acknowledgments

We would like to express our gratitude to Series Editor Judy Illes for her guidance during the editorial process, and to Peter Llewellyn and Sam Mahfoud from Elsevier for their professional support. I.B. would also like to thank Izabella Fekete and Alice Bárd.

References

Ariely, D., & Berns, G. S. (2010). Neuromarketing: The hope and hype of neuroimaging in business. *Nature Reviews Neuroscience*, *11*(4), 284–292.

Dujmovic, J. (2019). Facebook passes its first hurdle on a brain–computer interface. *MarketWatch*. https://www.marketwatch.com/story/facebook-passes-its-first-hurdle-on-a-brain-computer-interface-2019-08-08. (Last accessed 31 January 2020).

Eaton, M. L., & Illes, J. (2007). Commercializing cognitive neurotechnology—The ethical terrain. *Nature Biotechnology*, *25*(4), 393–397.

Emerging Issues Task Force, International Neuroethics Society. (2019). Neuroethics at 15: The current and future environment for neuroethics. *AJOB Neuroscience*, *10*(3), 104–110. https://doi.org/10.1080/21507740.2019.1632958.

Farah, M. J. (2008). An ethics toolbox for neurotechnology. *Neuron*, *86*(1), 34–37.

Fitz, N. S., & Reiner, P. B. (2014). The perils of using electrical stimulation to change human brains. In R. C. Kadosh (Ed.), *The stimulated brain* (pp. 61–83). Academic Press.

Fitz, N. S., & Reiner, P. B (2015). The challenge of crafting policy for do-it-yourself brain stimulation. *Journal of Medical Ethics*, *41*, 410–412.

Forlini, C., Lipworth, W., Carter, A., & Kerridge, I. (2019). Beyond flourishing: Intersecting uses and interests in the neurotechnology marketplace. *AJOB Neuroscience, 10*(4), 178–180.

Ienca, M., & Vayena, E. (2019). Direct-to-consumer neurotechnology: What is it and what is it for? *AJOB Neuroscience, 10*(4), 149–151.

Illes, J., & Raffin, T. R. (2002). Neuroethics: An emerging new discipline in the study of brain and cognition. *Brain and Cognition, 50*(3), 341–344.

Jotterand, F., & Giordano, J. (2011). Transcranial magnetic stimulation, deep brain stimulation and personal identity: Ethical questions, and neuroethical approaches for medical practice. *International Review of Psychiatry, 23*(5), 476–485.

Kellmeyer, P. (2018). Big brain data: On the responsible use of brain data from clinical and consumer-directed neurotechnological devices. *Neuroethics*, 1–16. https://doi.org/10.1007/s12152-018-9371-x.

Kreitmair, K. V. (2019a). Dimensions of ethical direct-to-consumer neurotechnologies. *AJOB Neuroscience, 10*(4), 152–166.

Kreitmair, K. V. (2019b). Response to open peer commentaries on "ethical dimensions of direct-to-consumer neurotechnologies" *AJOB Neuroscience, 10*(4), W1–W3.

Lavazza, A. (2019). Transcranial electrical stimulation for human enhancement and the risk of inequality: Prohibition or compensation? *Bioethics, 33*, 122–131.

Lee, N., Broderick, A. J., & Chamberlain, L. (2007). What is 'neuromarketing'? A discussion and agenda for future research. *International Journal of Psychophysiology, 63*, 199–204.

Levy, N. (2007). Rethinking neuroethics in the light of the extended mind thesis. *The American Journal of Bioethics*, 7(9), 3–11.

Maslen, H., Douglas, T., Cohen Kadosh, R., Levy, N., & Savulescu, N. (2014). *Mind machines. The regulation of cognitive enhancement devices.* Oxford Martin Policy Paper, Oxford. https://www.oxfordmartin.ox.ac.uk/downloads/briefings/Mind_Machines.pdf.

McCall, I. C., Lau, C., Minielly, N., & Illes, J. (2019). Owning ethical innovation: Claims about commercial wearable brain technologies. *Neuron, 102*(4), 728–731.

Mecacci, G., & Haselager, P. (2019). Identifying criteria for the evaluation of the implications of brain reading for mental privacy. *Science and Engineering Ethics, 25*, 443–461.

Morin, C. (2011). Neuromarketing: The new science of consumer behavior. *Society, 48*(2), 131–135.

Murphy, E. R., Illes, J., & Reiner, P. B. (2008). Neuroethics of neuromarketing. *Journal of Consumer Behaviour: An International Research Review*, 7(4–5), 293–302.

Neuralink. (2019). *Neuralink launch event [video file].* https://www.youtube.com/watch?v=r-vbh3t7WVI. (Last accessed 31 January 2020).

Rainey, S., Bublitz, J. C., Maslen, H., & Thornton, H. (2019). Data as a cross-cutting dimension of ethical importance in direct-to-consumer neurotechnologies. *AJOB Neuroscience, 10*(4), 180–182.

Reiner, P. B., & Nagel, S. K. (2017). Technologies of the extended mind: Defining the issues. In J. Illes (Ed.), *Neuroethics: Anticipating the future* (pp. 108–122). Oxford University Press.

Rose, N., & Abi-Rached, J. M. (2013). *Neuro: The new brain sciences and the management of the mind.* Princeton University Press.

Royal Society. (2019). *iHuman: Blurring lines between mind and machine.* Royal Society.

Schermer, M. (2011). Ethical issues in deep brain stimulation. *Frontiers in Integrative Neuroscience, 5*, 17.

Ulman, Y. I., Cakar, T., & Yildiz, G. (2015). Ethical issues in neuromarketing: "I consume, therefore I am!" *Science and Engineering Ethics, 21*(5), 1271–1284.

Waltz, E. (2019). The brain hackers. *Nature Biotechnology, 37*(9), 983.

Wexler, A. (2016). The practices of do-it-yourself brain stimulation: Implications for ethical considerations and regulatory proposals. *Journal of Medical Ethics*, *42*(4), 211–215. https://doi.org/10.1136/medethics-2015-102704.

Wexler, A. (2017). The social context of "do-it-yourself" brain stimulation: Neurohackers, biohackers, and lifehackers. *Frontiers in Human Neuroscience*, *11*, 224.

Wexler, A., & Reiner, P. B. (2019). Oversight of direct-to-consumer neurotechnologies. *Science*, *363*(6424), 234–235.

Wexler, A., & Thibault, R. (2019). Mind-reading or misleading? Assessing direct-to-consumer Electroencephalography (EEG) devices marketed for wellness and their ethical and regulatory implications. *Journal of Cognitive Enhancement*, *3*(1), 131–137.

Wolkenstein, A., Jox, R. J., & Friedrich, O. (2018). Brain–computer interfaces: Lessons to be learned from the ethics of algorithms. *Cambridge Quarterly of Healthcare Ethics*, *27*(4), 635–646.

Yuste, R., Goering, S., Arcas, B., et al. (2017). Four ethical priorities for neurotechnologies and AI. *Nature*, *551*, 159–163. https://doi.org/10.1038/551159a.

Introductory note from the series editor

This third volume in the *Developments in Neuroethics and Bioethics* series tackles a much newer and narrower topic than its predecessors on pain (Buchman & Davis, 2018) and global mental health (Cratsley & Radden, 2019). In *Ethical Dimensions of Commercial and DIY Neurotechnologies*, Editors Bard and Hildt have brought together a remarkable set of chapters that explore the state-of-the-art of neurowearables, technological development in neuroscience, and discourse in ethics that is merely 10 years old. Even in the early stages of this field in which both research and commercialization are still evolving rapidly, the ethical issues are salient: industry and consumer motivation, intention, evidence, clinical application, authenticity, enhancement, validity, placebo, trustworthiness, agency, reporting, and more, are all examined and challenged in this volume. The chapters further span considerations for neurowearables across the lifespan, with different implications for the developing, developed, and aging brain. The chapters fundamentally create a new landscape for neurolaw and neurosecurity.

As for each volume in the series, the goal of the editorial team is to gather and delivers the most robust, balanced, and up-to-date material to readers of this series, but we do not take a position on the content. Positional stances—positive or negative—belong to the authors of each chapter, as does responsibility for disclosures and acknowledgements of academic–industry relationships, funding, and partnerships. As Editors, we declare no financial or other conflict of interest with the overall content herein, an especially important point given the significant interactions authors, ethicists, and innovators must embrace to ensure that the highest ethical standards of this expanding industry are realized.

We hope you enjoy this stimulating volume.

Judy Illes, CM, PhD
DNB Series Editor

CHAPTER ONE

Peering into the mind? The ethics of consumer neuromonitoring devices

Iris Coates McCall[a,*], Anna Wexler[b]
[a]Neuroethics Canada, Division of Neurology, Department of Medicine, University of British Columbia, Vancouver, BC, Canada
[b]Department of Medical Ethics & Health Policy, Perelman School of Medicine, University of Pennsylvania, Philadelphia, PA, United States
*Corresponding author: e-mail address: iriscmccall@gmail.com

Contents

1. Introduction

Direct-to-consumer (DTC) neurotechnologies represent a growing market as companies vie to bring the promise of brain-based devices into consumers' homes (Coates McCall, Lau, Minielly, & Illes, 2019; SharpBrains, 2018; Wexler & Reiner, 2019). Brain stimulation headsets that hook up to smartphones are marketed for everything from treating disease to enhancing focus, while brain recording devices are sold with promises of optimizing users' cognitive abilities and moods. Here, we use the term *direct-to-consumer* to refer to products that are marketed to the public for personal use and that are available for purchase on the open marketplace

Developments in Neuroethics and Bioethics, Volume 3
ISSN 2589-2959
https://doi.org/10.1016/bs.dnb.2020.03.001

(i.e., without the need for physician involvement). We use the term *neurotechnology* to encompass products that claim to modulate, manipulate, or monitor brain function. In this chapter we focus specifically on one type of DTC neurotechnology, electroencephalography (EEG) devices, which represent a significant portion of the market.

Typically, neuroimaging seeks to elucidate the structures and functions of the brain that underpin our lived experiences. While not delivering pictures of the brain itself, EEG captures the activity of the brain and transforms it into visualized waveforms for analysis. The increasing ability to picture the brain on a finer and finer scale has brought immeasurable good both in the diagnosis and treatment of disease and in its contribution to our understanding of the brain and mind. Over time, brain imaging has also allowed us to picture the neural correlates of increasingly sophisticated behaviors and, some have claimed, even specific thoughts, leading to concerns about mind reading and neural privacy (Ienca & Andorno, 2017; Ienca, Haselager, & Emanuel, 2018; Pustilnik, 2013).

Indeed, the idea of being able to peer into the mind of others and read their thoughts has long captured the imaginations of authors, scientists, and philosophers. The idea that one could gain access to the most intimate minutiae of our conscious selves both excites and terrifies. In an increasingly interconnected and data-driven world, the mind may be seen as the last vestige of human privacy—the one remaining frontier that cannot, and should not, be trespassed upon. The transition of EEG from clinical and research settings into people's homes has reignited this debate over mental privacy and fears about mind reading. However, in this chapter, we suggest that the focus on privacy has drawn attention away from other, potentially more imminent concerns related to DTC EEG devices, such as misleading advertising claims and potential uses of EEG in workplace settings.

We begin this chapter with a brief background on EEG technology—the basic science upon which it is based and its clinical and research uses—and provide an overview of the present state of the DTC EEG market. Next, we address the short-term ethical concerns raised by these devices, particularly with regard to manufacturers' questionable marketing claims. We then discuss mid-term concerns, such as those related to the use of consumer devices in the workplace. Finally, we examine long-term concerns relating to mental privacy, specifically in the context of assumptions that would need to be upheld in order for these concerns to warrant further attention. We conclude that the source of most ethical concerns about DTC EEG technology lies not so much in the devices themselves, but in what people believe about these devices and their capabilities.

2. Review of EEG

EEG is an electrobiological monitoring system designed to record the electrical activity of the brain. The fluctuations recorded, often referred to as brainwaves, result from ionic currents within neurons. When neurons are activated, local currents are produced. In EEG, these are recorded by a series of metal electrodes placed directly on the scalp, either with or without conductive gel. Devices can vary in the number of electrodes used, ranging from a single channel up to 256 channels, and electrodes are typically placed in set locations according to the internationally recognized 10–20 placement system (Towle et al., 1993). As the local currents produced by individual neurons are weak, EEG measures the combined currents of many neurons to provide an averaged level of brain activity. Due to this population level recording, the spatial resolution of EEG is low. Further, the EEG field mostly reflects the behavior of certain kinds of cortical cells, such as vertically-oriented pyramidal neurons, and as such EEG is not well suited to recording the activity of deep brain structures (Olejniczak, 2006). However, despite these spatial limitations, EEG has high temporal resolution, recording on the scale of milliseconds (Olejniczak, 2006).

Research has identified five main universal patterns of EEG activity that occur in humans. Gamma waves are in the frequency range of 31 Hz and up, and while not well understood, are associated with arousal, alertness, and active information processing (Pfurtscheller & Da Silva, 1999). Beta waves are in the frequency range of 13–30 Hz, and have been associated with action, concentration, conscious thought, and external focus, and are dominant during normal states of wakefulness with open eyes (Teplan, 2002). Alpha waves, the best understood rhythm, are in the frequency range of 8–12 Hz, and reportedly reflect relaxed wakefulness, disengagement, and a spacey or dreamy state (Niedermeyer & Da Silva, 2005). Theta waves ranging from 4 to 7 Hz have been linked to inefficiency, daydreaming and subconscious activity (Cahn & Polich, 2006). Finally, delta waves ranging from 0.5 to 4 Hz are the slowest waves and typically occur when a user is in hypnotization or is deeply unconscious (Li, Ding, & Conti, 2015).

EEG is used in the lab for research purposes to better understand cognitive functions. In addition to the frequency bands described above, EEG can capture individual responses to events and stimuli. An event-related potential (ERP) is an EEG-measured brain response that is the direct result of a specific sensory, cognitive, or motor event (Luck, 2014). As such, it can be used to associate certain brain activity patterns with specific events or stimuli. For example, a subtype of the ERP, the evoked potential (EP), involves averaging the EEG activity timed to the presentation of a stimulus, allowing researchers to obtain a stereotyped brain response to a certain stimulus such as a particular sound, tone, or mental event. The particular response can then be compared to known properties of different kinds of neurons, thus identifying those groups of neurons responsible for the response (Regan & Lee, 1993).

ERPs are often used by neuroscientists and cognitive scientists to study mechanisms underlying cognitive functions such as perception and attention. A specific kind of ERP is the P300 waveform that indicates the salience of a stimulus, and has been shown to be produced when an individual observes something that she has previously seen. The presence of the P300 has been used to identify the subject's own name out of a list of random names (Rosenfeld, Biroschak, & Furedy, 2006), discriminate familiar faces from non-familiar faces (Marcel & Millán, 2007), and detect high-impact autobiographical information (Rosenfeld et al., 2006).

EEG forms the basis of some brain computer interfaces (BCIs), which utilize brain activity as the input for a system that can then be transformed into output that reflects or represents the user's thoughts or intentions (Pfurtscheller, Guger, Müller, Krausz, & Neuper, 2000). These outputs can be prosthetic devices or cursors that allow individuals with brain damage or locked-in states to communicate and perform basic tasks (Pfurtscheller et al., 2000; Rezeika et al., 2018). Communication often occurs through BCI spellers, which can be based on the P300, EPs, or motor imagery (MI) (Rezeika et al., 2018). MI is a mental process by which an individual rehearses or imagines performing a given action and the accompanying EEG data is used as a way to control the output of the device (Kosmyna, Lindgren, & Lécuyer, 2018).

Historically, EEG has been used in clinical contexts to inform decision-making and provide diagnostics related to neurological conditions. Clinical-grade EEG has been used to show sleep patterns (Abad & Guilleminault, 2003), help diagnose and manage epilepsy (Smith, 2005), and facilitate the diagnosis of brain death (Wijdicks, Varelas, Gronseth, & Greer, 2010). EEG is non-invasive and can be used repeatedly on adults, children, the healthy, and the ill with virtually no risk or limitation (Mantri, Dukare, Yeole, Patil, & Wadhai, 2013).

EEG is also used for a technique called neurofeedback, where individuals attempt to self-regulate their brainwaves based on EEG readings. Neurofeedback has been proposed for wellness applications in the form of concentration training or meditation augmentation. Proponents of neurofeedback claim that self-regulation can alter mental states and ultimately behavior (Thibault, Lifshitz, Birbaumer, & Raz, 2015). Neurofeedback also forms the basis of many BCIs by helping to train the user to provide clear brain data for use in MI techniques (Hwang, Kwon, & Im, 2009). EEG neurofeedback has also been proposed for clinical use to treat anxiety, stress, and chronic pain, as therapy for attention deficit hyperactivity disorder (ADHD), and even to help treat irritable bowel syndrome (Birbaumer, Murguialday, Weber, & Montoya, 2009; Hammond, 2007; NeuroSky, n.d.a). However, the clinical validity of neurofeedback is disputed—EEG neurofeedback is not a standard treatment/therapeutic for any condition and research suggests its effects can be attributed to little more than placebo (Schönenberg et al., 2017; Thibault & Raz, 2017).

3. Direct-to-consumer EEG devices

3.1 Assessing the market

The earliest consumer-grade EEG devices—inspired by the EEG systems used in research, but stripped to the bare elements necessary to record some degree of brain activity—appeared on the market roughly a decade ago, and were initially marketed for thought control of either real-world or virtual objects (Wexler & Thibault, 2019). In recent years, however, new companies have entered the consumer EEG market, with some conducting crowdfunding campaigns for their devices. As a whole, these companies have raised more than $7 million via outlets such as Kickstarter and Indiegogo (Roy, 2017). Currently, consumer EEG devices range in price from approximately $99 to $799, and can be purchased online from company websites and via retailers such as Amazon (Wexler & Thibault, 2019).

Until recently, consumer EEG devices marketed to the general public had escaped scholarly critique, as attention had largely been focused on the consumer use of neurostimulation devices. However, two recent studies, by Coates McCall et al. (2019) and Wexler and Thibault (2019), provide a critical assessment of the DTC EEG neurotechnology market and manufacturers' claims. Both sets of authors systematically mapped the claims made about DTC EEG devices. As of the time of writing, 26 available devices have been identified that, depending on definition and search criteria, can be considered DTC EEG devices (Table 1). Wexler & Thibault identified 18 such devices while Coates McCall et al. identified 22. Discrepancies are due to different sampling strategies and inclusion criteria (for example, the

Table 1 DTC EEG devices currently available as identified by Wexler & Thibault (*) and Coates McCall et al. (^).

Aurora Dreamband by iWinks*^	**Melomind by My Brain Technologies*^**
Aware by United Sciences*	Mindball by Interactive Productline*
B2v2 by BRAINtellect^	Mindset Headphones by Mindset*^
Brainlink Pro by Macrotellect^	MindWave Mobile 2 by NeuroSky*^
Cricket by Somaxis^	Muse by Interaxon*^
Dreem Headband by Rhythm*^	Naptime by Entertech^
EPOC+ by Emotiv^	Narbis Glass by Narbis^
FocusBand*^	NeuroOn Open by InteliClinic*^
iBand+ by Arenar B.V.*^	Neuroplus by Neuroplus Inc.*
Insight by Emotiv*^	Senzeband Neeuro*^
Kokoon by Kokoon Technology Limited*^	Sleep Shepherd Blue by Sleep Shepherd*^
Life by SmartCap Tech^	Super Brain II by REX*
Lowdown Focus by Smith^	Versus Headset by Neurotherapeutics*^

Emotiv EPOC+ was excluded by Wexler & Thibault because it is marketed for use in research settings, but the product was included by Coates McCall et al. as it can be purchased by consumers for personal use).

3.2 Mapping claims

Coates McCall et al. (2019) classified the advertised claims made for consumer EEG devices into four overarching categories: health, wellness, enhancement, and practical applications (Table 2). While acknowledging that enhancement, wellness, and health represent a continuum with blurred boundaries, Coates McCall et al. identified a distinct set of claims regarding enhancement as augmented performance beyond the user's standard level of functioning. Claims that fell into this category typically stated that enhancement could be achieved through self-regulation using neurofeedback for abilities such as concentration and memory, and even physical and athletic performance. Claims included in the wellness category included statements that a device could improve general wellness such as sleep quality or stress management. Similar classifications were made by Wexler & Thibault who found that the greatest number of claims made by consumer EEG companies were related to improving concentration, relaxation, meditation, and sleep (Wexler & Thibault, 2019).

Table 2 Claims made about DTC EEG products.

Category	Claim	Number of companies making the claim
Enhancement	Assist with self-regulation/ neurofeedback	20
	Increase concentration/focus	10
	Increase efficiency/productivity	8
	Improve general cognition	7
	Support learning/training	5
	Boost physical performance	3
	Enhance memory	2
Wellness	Improve general wellness	13
	Improve sleep quality	12
	Augment meditation/relaxation	10
	Increase self-awareness	9
	Relieve stress/anxiety	8
	Improve mood	2
Health	Promote autonomy after disability	3
	Improve a medical condition	1
Practical applications	Perform research	5
	Control technology	4
	Create art	2
	Promote safety	1

Adapted from data reported in Coates McCall, I., Lau, C., Minielly, N., & Illes, J. (2019). Owning ethical innovation: Claims about commercial wearable brain technologies. *Neuron, 102*(4), 728–731.

Both Wexler and Thibault (2019) and Coates McCall et al. (2019) found that a number of companies made health-related claims, defined as claims pertaining to an identified medical condition. As noted by Wexler and Thibault, several companies made these health claims more explicitly than others. For example, while some of NeuroSky's 100-plus third-party apps merely allude to medical benefits (e.g., Neuroplus references media coverage about how the device ameliorates the symptoms of ADHD), others make more explicit medical claims about treating chronic pain and ADHD.

Still other companies, like Versus and Mindset, do not make clear claims about their devices specifically but reference scientific literature regarding how neurofeedback devices in general can be used to treat a variety of clinical indications.

Beyond wellness, health, and enhancement, Coates McCall et al. (2019) found that devices were also marketed for practical applications. These included using the device to control virtual or real-world objects via brainwaves, making art, conducting do-it-yourself research, playing games, and/or detecting fatigue for safety purposes (Coates McCall et al., 2019). Gaming applications such as those marketed by companies such as Emotiv and NeuroSky use EEG data as a method of object control. These objects can be either physical, as in the case of the MindFlex by Mattel Games, where players use EEG signals to levitate a ball through a maze (Rettner, 2016), or virtual, as in the case of the game FlappyMind, where players use their level of concentration (as measured by EEG) to navigate a flying brain around obstacles (NeuroSky, n.d.b).

Many DTC EEG devices targeted a specific consumer group, either explicitly or implicitly. Targeted groups include older adults, athletes/trainers, children, independent researchers, health care providers/professionals, people with medical conditions, employees/employers, students, marketers, and gamers (Coates McCall et al., 2019).

4. Ethical considerations

Given the recent proliferation of DTC EEG devices—and indeed, DTC neurotechnology as a whole—it is important to carefully consider the ethical and social issues these products raise. Some scholars have expressed concerns about the potential for privacy and security breaches (Hallinan, Schütz, Friedewald, & de Hert, 2014; Ienca & Andorno, 2017; Pustilnik, 2013; Shen, 2013), arguing that EEG devices already yield information that is rich and personally revealing (Ienca et al., 2018). While we agree that privacy and security may represent potential long-term concerns, we do not believe that these devices are currently capable of revealing personal information, nor is there good evidence to suggest that this will occur in the near future (Wexler, 2019). Rather, we argue that the ethical concerns at present center on the misleading nature of some manufacturers' claims. In addition, several mid-term concerns relate to the use of DTC EEG devices in the workplace. Thus, this section is divided temporally, to distill present, mid-term, and potential future longer term concerns related to consumer

EEG devices. We have endeavored to acknowledge the underlying assumptions that would need to hold true for the mid-term and long-term ethical concerns to be realized.

4.1 Current considerations

Given the claims made about the capabilities of DTC EEG devices, it is unsurprising that there is concern surrounding what they can reveal about individuals' thoughts and preferences. However, these privacy concerns rely on the assumption that the information provided by these devices is valid and accurate. Here we argue that this is not the case, by showing that there is insufficient evidence to demonstrate that devices validly and reliably record brain activity. Thus, we suggest that the short-term ethical considerations are not ones related to privacy, but rather to misleading marketing practices and public deception. We highlight the potential for both psychological and physical harms stemming from receiving inaccurate information about brain function.

Consumer EEG devices differ from research-grade devices, both in their design and in the quality of signal they produce. For example, consumer EEG devices typically only include a few electrodes and utilize dry (rather than wet) electrodes, which have a higher impedance and noise level (Mathewson, Harrison, & Kizuk, 2017). Furthermore, the software used in artifact removal for consumer EEG devices is usually proprietary, as opposed to the well-validated programs used in research EEG. Finally, consumer EEG devices require the users themselves to place the device—rather than a trained technician. Though a few companies have attempted to establish the validity and reliability of their devices, overall there is little evidence to suggest that data captured from most consumer EEG devices is valid or reliable (Wexler & Thibault, 2019).

Even if devices do accurately record brainwave activity (i.e., the raw data), there remain questions regarding associations between brain activity and purported associated behavior or mental states. Although generalized signal frequency bands have been associated with specific mental states—for example, alpha is commonly thought to be correlated with increased concentration—the same pattern of activity can also arise during numerous other states (Schabus et al., 2017; Thibault & Raz, 2017). Furthermore, an individual can experience that state without producing the purportedly associated frequencies (Beyerstein, 1990), meaning there is no one-to-one correlation between EEG frequency band and mental state.

Thus, any claims of consumer EEG devices regarding reading mental states such as engagement or concentration should be interpreted with caution.

Companies that manufacture DTC EEG devices often go a step further, claiming, as we have shown in the previous section, that the devices can help increase attention or focus, or improve sleep, health, or general wellness. The implicit theoretical basis for these claims is related to neurofeedback: that providing individuals with their brainwave data can help them alter that behavior or mental state (Wexler & Thibault, 2019). However, neurofeedback has failed to gain widespread acceptance of efficacy beyond a placebo effect despite 60 years of research (Schönenberg et al., 2017; Thibault & Raz, 2017). Thus, the wellness and enhancement claims of consumer EEG devices are largely based on a body of literature that has yet to establish the benefit of receiving genuine brainwave data.

Furthermore, few DTC EEG device companies have conducted original research utilizing their own devices. Coates McCall et al. found that the majority of EEG device websites (16/22) cited or referred to general science research and scientific concepts, but did not cite peer-reviewed research utilizing the specific device itself (Coates McCall et al., 2019). For example, one website discussed how EEG in general has been demonstrated to improve cognitive functioning, but did not explicitly claim that their commercial device has been proven to do so. Indeed, Wexler & Thibault found that only three companies had made available research on their specific devices. The remaining supporting evidence consisted of links to user testimonials, gray literature, company/in-house research, or peer-reviewed literature that was simply irrelevant to the device or claim at hand.

Notably, the methods used by Coates McCall et al. (2019) and Wexler and Thibault (2019) were both proxy methods of assessing level of evidence: neither study involved testing the devices themselves, which would be the most accurate measure of whether the device's benefits aligned with the manufacturers' claims. Still, given the rather dubious and unspecific nature of the supporting science presented by these companies, there is little evidence to suggest that DTC EEG devices work as advertised.

In our view, the most salient ethical concern at present with regard to DTC EEG devices is misleading advertising. In other words, the current problem with consumer EEG devices is not that we need to worry about the privacy of the rich, accurate, and revealing data they collect; the problem is that consumer EEG companies are misleading consumers into thinking the data is rich, accurate, and revealing. To adhere to ethical business practices, companies should conduct research to support their products' advertising claims, practice honest marketing to ensure that such claims are not exaggerated or misleading, and adhere to robust regulatory oversight (Bubela & McCabe, 2013; Coates McCall et al., 2019).

In addition to misleading claims, there are ethical concerns surrounding psychological harms. Providing individuals with information about their mental states is not inconsequential: EEG devices may lead individuals to modify their behavior to align with their purported brain state. For example, if individuals are told by a consumer EEG device that their brain is in a stressed state (even if they are not), they may become stressed to align with this interpretation of their brain's activity. This raises ethical issues relating to non-maleficence as inducing stress or other negative symptoms could constitute a harm to the user. Moreover, issues relating to personhood also arise, as information taken from EEG devices may be incorporated into what the user believes about herself and could in turn alter her sense of self (Kreitmair & Cho, 2017).

In addition, there may be other harms, such as opportunity and financial costs. As mentioned above, some EEG devices are marketed for medical indications. For example, the NeeuroFIT website states: "While we can't cure conditions like Dementia or Alzheimer's, our solution could act as a tool for early detection and to slow down cognitive decline" (Neeuro, n.d.). This kind of language could cause consumers to make less than optimal medical decisions regarding their own health, such as possibly forgoing traditional medical interventions in favor of unsupported self-treatment with DTC EEG devices. Individuals might also be spending a significant financial sum on these endeavors, as consumer EEG devices can cost hundreds of dollars (Wexler & Thibault, 2019).

Despite the numerous ethical concerns surrounding DTC EEG devices, no regulatory enforcement action has been taken against EEG companies that claim their device can improve health. In the United States, where the majority of consumer EEG companies are located, products that make medical claims are considered medical devices and regulated by the Food and Drug Administration (FDA). Thus, consumer EEG products making explicit (and even implied) medical claims—such as treating or diagnosing ADHD or chronic pain—without prior FDA approval would be in violation of federal law. The vast majority of DTC EEG devices, however, make wellness and enhancement claims that lie just shy of health claims. While the Federal Trade Commission (FTC) has the authority to take action against unfair or deceptive business practices—and has disciplined several companies manufacturing brain-training *games* (Federal Trade Commission, 2016)—it has not taken any action in the domain of consumer EEG *devices* that make the same claims.

In sum, there is little evidence or compelling reason to believe that DTC EEG data is valid or reliable, nor that it provides accurate readings of individuals' emotional states, improves wellness, or enhances cognition. And yet, this does not mean that there are not potential negative outcomes associated with their use spanning misleading advertising, potential physical harm of forgoing evidence-based therapies, and psychological harms from receiving inaccurate information about oneself. Thus, the main problem with consumer EEG devices is not currently one of mental privacy; it is that these products are being sold to consumers as mind reading devices when they are not. Our main concern, therefore, is that consumers are likely being misled into buying neurotechnological snake oil.

4.2 Mid-term: Fatigue and attention monitoring

Although there is little evidence at present to support manufacturers' claims of enhancement, treatment, or mind reading, there is one domain where even consumer-grade EEG devices might prove effective: the ability to detect fatigue. Studies supporting the efficacy of EEG use in this capacity are based on machine learning that associates brain activity states with task performance, thereby identifying which brain states are associated with drowsiness-related poor performance. Studies have shown that EEG, either alone or in combination with other biosensor data, can accurately detect drowsiness in participants performing simulated driving tasks (Correa, Orosco, & Laciar, 2014; Dong, Hu, Uchimura, & Murayama, 2010; Mardi, Ashtiani, & Mikaili, 2011). One study found that a wireless, single-electrode EEG device accurately detected drowsiness in a practical environment, supporting the potential for EEG device use in the measurement of classroom mental engagement, evaluation of movies, and office environment evaluation (Ogino & Mitsukura, 2018).

EEG-based drowsiness detection has formed the basis of technology to detect fatigue in workers, either for safety or productivity purposes (Aryal, Ghahramani, & Becerik-Gerber, 2017; Rackspace, 2014; Wang et al., 2017). For example, a Chinese state-sponsored program has been reported to have employees ranging from train drivers to production line workers wearing helmets outfitted with EEG sensors in order to boost productivity, detect fatigue, and monitor emotional risk factors (Chen, 2018). Similarly, SmartCap Technologies has developed a trucker's cap that it claims can detect drowsiness for safety purposes—if a driver's EEG cap detects drowsiness behind the wheel, it sounds an alarm in the truck cab to alert the driver, and can even alert the dispatcher about a potentially unsafe situation (SmartCap Technologies, n.d.). The product has been used by Australian companies Rio Tunto's Hunter Valley Mines and Newcrest Mining, as well as British mining firm Anglo American Mining (Tractica, 2016). While the companies have attributed huge savings to these devices, it remains unclear whether, if real, the effect is due to the devices functioning as intended or simply acting as a deterrence to poor productivity.

Still, given that there are already a few small pilot projects, EEG drowsiness detection could be an appreciable consideration in the coming decade as the science, unlike that supporting claims of enhancement and wellness, is well established. While current usage data is limited, several market research reports indicate that wearable use in the workplace is steadily increasing (Tractica, 2016; Wild, 2017). Although the wearables provided to employees have mainly been used for the purposes of monitoring physiological measurements such as heart rate and footsteps, it is not inconceivable that drowsiness detection devices could also be adopted by companies. Indeed, there is a substantial body of research looking into EEG in office settings for the purpose of cognitive functioning assessment and monitoring (Blanco, Vanleer, Calibo, & Firebaugh, 2019; Nayak et al., 2017; Ogino & Mitsukura, 2018; Wascher et al., 2016), with proponents claiming that continuously monitoring levels of task engagement in a work environment could help identify more efficient methods for employees to interact with technology (Berka et al., 2007).

However, the potential use of this technology raises numerous ethical questions. As discussed above, the legitimacy of these devices rests on the assumption that they are validly and accurately measuring individuals' brain signals and mental states. However, the majority of studies supporting the efficacy of EEG use in the workplace are either conducted in tightly controlled laboratory settings or, if conducted in a naturalistic setting, involve very basic tasks meant to replicate one particular cognitive behavior, not complex multitasking. EEG is infamous for being highly susceptible to artifacts, and the seriousness of potential invalid signals depends on the purpose for which the EEG is being used. Constant monitoring of levels of engagement during work hours could be used to punish those who are distracted, tired, unfocused or unalert. A payment system that compensates employees only for the number of supposedly alert hours worked is not inconceivable (Dubljević, Coates McCall, & Illes, 2019). This is not a far-fetched idea, as similar applications for rewarding cognitive behavior are purportedly already occurring in Chinese classrooms where EEG devices monitor students' attention levels throughout the day in order to reward those most concentrated and punish those most distracted (Ye, 2019). If a device is not accurately reading its wearers' attention levels, it is conceivable that workers could be unfairly punished (or compensated).

Even if it becomes the case that EEG can reliably and validly provide employers with information on their employees' cognitive functioning, ethical issues still arise—not as an issue of device and signal quality but rather one of application. In other words, even if the information collected is accurate, there may be limitations to its interpretation. For example, with regard to proposals for attention-based compensation, different individuals may have different ways of accomplishing tasks: one individual may outperform another, even in a fatigued or drowsy state. Similarly, some individuals may have a higher performance level than others even when in a distracted state. Thus, in a hypothetical scenario where EEG data is used to determine culpability (e.g., a workplace accident where an employee's EEG is assessed for level of focus or distraction; see Dubljević et al., 2019), there would be questions about the analytic validity of these devices, or the extent to which the interpretation of their signals is warranted. Fatigue specifically, and brain states more generally, are not necessarily a sufficient proxy of performance or justification of blame given that we are inherently contextual beings grounded in an extended physical world.

Finally, ethical issues could arise in an employment or educational situation that mandates the use of EEG neurotechnology for monitoring purposes. As mentioned above, there is a growing movement for employers to collect physiological information about employees, often with little recourse for workers who object to such monitoring (Wild, 2017). Although requiring an employee to undergo an EEG assessment would likely be legal within the realm of current U.S. regulations (Tovino, 2007), mandatory EEG assessments raise similar issues to those put forth regarding mandatory drug testing in the workplace—namely, whether the benefits gained in terms of workplace safety justify the invasion of privacy (Christie, 2015; Raskin, 1993). Privacy with relation to fatigue or attention monitoring is distinct from privacy of thought, as the former relates to a general emotional or cognitive state, and not to specific factual information such as the content of thoughts. Still, one's emotional or cognitive state may be considered an aspect of one's life that ought not to be infringed upon without sufficient demonstrable evidence of benefit. In that sense, mandatory EEG monitoring for drowsiness detection for employees who operate heavy machinery may be considered ethically permissible if there is evidence that it improves physical safety, while mandatory EEG monitoring of attention for the purposes of enhancing worker productivity would be less ethically permissible.

In sum, given the solid scientific basis of fatigue monitoring using EEG in laboratory settings, the increasing trend of employers adopting wearable devices in the workplace, and the small number of EEG pilot projects that have already been launched, fatigue monitoring for productivity and performance tracking is a likely phenomenon on the mid-horizon. The associated mid-term ethical implications should not be left unattended.

4.3 Long-term: Neural privacy

Being prepared for the future—and anticipating harms before they can occur—is a cornerstone of bioethics. As such, it is important to interrogate the potential social and political impacts of new technologies as they arise. As companies such as Facebook, Neuralink and Kernel have in recent years invested hundreds of millions of dollars in efforts to create neural interfaces for consumer use (Levy, 2019; Mannes, 2016; Markoff, 2019), there has been increased scholarly attention toward neural privacy (Bonaci, Herron, & Chizeck, 2015; Ienca & Andorno, 2017; Ienca et al., 2018; Pustilnik, 2013). For example, some researchers have foreseen a world where wearable neuromonitoring devices are ubiquitous and worn regularly by everyone, and highlight associated concerns relating to efficacy, privacy, and security. These scholars have pointed to the potential for constant monitoring either by state or corporate entities, and the probability of maleficent agents hacking the programs used in combination with these devices to access personal information. Others have raised concerns regarding the collection and sale of neural data, which may become increasingly incorporated into data brokerage companies' business models (Hallinan et al., 2014; Ienca et al., 2018; Yuste et al., 2017).

However, as others have argued, anticipatory ethics requires underlying assumptions to be identified and supported in order to be valuable (Racine, Rubio, Chandler, Forlini, & Lucke, 2014). We identify two assumptions which, should they hold true, warrant further discussion of these concerns. First, the voiced concerns rest on the assumption that these devices can reveal personal information. As previously discussed in this chapter and argued by Wexler and Thibault (2019), there is good reason to believe the assumptions of efficacy are unfounded.

Scholars concerned about neural privacy related to EEG have pointed to research suggesting that personal thoughts and reflections are now accessible using EEG without the thinker's participation or cooperation. In support of such claims, they typically cite computer science papers that conduct side channel attacks or investigate leaks from EEG data. These studies are typically proof of principle demonstrations of how a malicious agent could extract EEG data from a device in the course of normal use without the user's knowledge. Examples of such studies include those that have claimed to use EEG to identify alcoholics (Matovu & Serwadda, 2016), extract users' preferences between brands (Bonaci et al., 2015), or reduce the uncertainty of guessing an unaware user's private information (Frank et al., 2013; Martinovic et al., 2012).

However, as one of us has previously argued (Wexler, 2019), these studies have often been misinterpreted. For example, Martinovic et al. (2012) is one of the most cited studies in support of the view that EEG can be used to reveal personal information such as PIN codes better than chance alone. However, better than chance does not mean accurate. In fact, for nearly all measures assessed, the algorithms used in Martinovic et al. were highly inaccurate, with the first guess being wrong 70–80% of the time (the one exception being birth month, where one algorithm guessed correctly 60% of the time on the first try). For these and other reasons, research revisiting the Martinovic et al. study concluded that while theoretically feasible, decoding PIN codes from EEG signals "may not be a very critical threat" (Lange, Massart, Mouraux, & Standaert, 2017). The literature on EEG side channel attacks should be read as a theoretical feasibility demonstration, which should not necessarily be equated with real-world potential for an attack.

Further, EEG information is not dissimilar to other kinds of data captured by technology. Eye gaze information provides similar mind reading potential in that with the provision of contextual information (such as what a subject is gazing at, either consciously or unconsciously), it is possible for private information to be revealed. Other side channel attack studies have shown that wrist movements captured by wearable devices like smart watches can be used to determine PIN codes with high accuracy (Wang, Guo, Wang, Chen, & Liu, 2016). Keystrokes and thus the content of messages can be inferred with a high degree of accuracy from the motion of a smartphone (Cai & Chen, 2011) and even partially revealed by video of hand movements (Shukla, Kumar, Serwadda, & Phoha, 2014). These studies, which often also use data that is directly captured from our bodies, have shown far greater accuracy than EEG side channel attack studies and use technologies that are already ubiquitous.

Given that gleaning EEG information from side channel attacks is not an imminent worry, then for this category of devices, what data might present concerns? If consumer neurotechnology data—or even research/clinical grade EEG data—had personal and revealing qualities, their concerns would merit further consideration. Although we have here argued that EEG devices cannot currently reveal much private information, we acknowledge that it is possible that EEG may 1 day attain this capability. However, unlike other authors, we do not take this to be an inevitability. All technologies have their limitations. EEG is an old technology that is susceptible to muscle artifacts, and by its very nature is noisy, as it is measuring subtle electrical signals from inside the skull. Still, if EEG were to 1 day be capable of revealing more personal information, then further ethical considerations would be warranted.

Many other concerns rest on a second assumption—that EEG devices will achieve widespread public uptake, allowing for widespread neuromonitoring. If this is assumption is unfounded and these devices are not widely adopted, then the flurry of ethical concerns regarding surveillance and mental privacy will also be unfounded. As of now, while some signs indicate that the personal DTC neurotechnology market may be blossoming (SharpBrains, 2018), uptake of wearable DTC neurotechnologies has been slow. In the realm of DTC neurostimulation, for example, devices have remained part of a subculture but have not been embraced by the mainstream (Wexler, 2017). It is also not a given that individuals will choose to adopt wearable EEG devices even if the device does provide some benefits and effectively performs its stated function. For example, individuals might not trust the company offering the device, might be too embarrassed to wear the device in public, or may find that the benefits of the device are not worth the trouble of wearing it. Google Glass, for example, partially failed because of the look of the device, not because it did not provide added functionality (Richardson, Mabbitt, Wilson, & Petriches, 2016). At present, therefore, it is not an inevitability that we will have widespread DTC EEG usage.

If both assumptions do hold, then we will indeed need to address the widespread privacy concerns raised by the prospect of the mind reading capabilities of these devices. However, until the technology improves sufficiently to ensure efficacy and adoption of these devices is widespread—ethicists should focus their energies on responsible science communication and educating the public on the limitations of these devices, so as to pre-empt these privacy concerns even in the absence of device reliability.

5. Conclusion

The fact that companies are marketing health and wellness products that lack a rigorous scientific basis to support their claims is not in itself novel. However, among brain improvement products, consumer EEG devices are unique in that their benefits supposedly come from providing information to individuals about their own brain states. No other consumer brain intervention marketed to improve health or cognition (not dietary supplements, brain-training games, nor neurostimulation devices) claim to do so by measuring or tracking a person's brain activity. Consumer EEG companies, on the other hand, claim to provide snapshots of mental states (e.g., concentration, fatigue, arousal) by reading electrical activity captured from the scalp—and it is this reading and interpretation of that activity that is claimed to provide the benefits of the device.

Presently, the most serious issues presented by DTC EEG devices are due to misleading or unsubstantiated claims being made in a climate of an enthused yet relatively neuro-illiterate marketplace. Concerns that are on the horizon center on issues regarding workplace use of EEG devices for fatigue monitoring. In the long-term future, if DTC EEG devices become more accurate and are widely adopted by the general public, then we will need to begin to seriously consider addressing concerns regarding neuromonitoring and the sanctity of mental privacy. In the meantime, however, the problem with DTC EEG devices is not so much what monitoring these devices can achieve, but rather what the public is led to believe about them.

References

Abad, V. C., & Guilleminault, C. (2003). Diagnosis and treatment of sleep disorders: A brief review for clinicians. *Dialogues in Clinical Neuroscience*, *5*(4), 371.

Aryal, A., Ghahramani, A., & Becerik-Gerber, B. (2017). Monitoring fatigue in construction workers using physiological measurements. *Automation in Construction*, *82*, 154–165.

Berka, C., Levendowski, D. J., Lumicao, M. N., Yau, A., Davis, G., Zivkovic, V. T., et al. (2007). EEG correlates of task engagement and mental workload in vigilance, learning, and memory tasks. *Aviation, Space, and Environmental Medicine*, *78*(5), B231–B244.

Beyerstein, B. L. (1990). Brainscams: Neuromythologies of the new age. *International Journal of Mental Health*, *19*, 27–36.

Birbaumer, N., Murguialday, A. R., Weber, C., & Montoya, P. (2009). Neurofeedback and brain–computer interface: Clinical applications. *International Review of Neurobiology*, *86*, 107–117.

Blanco, J. A., Vanleer, A. C., Calibo, T. K., & Firebaugh, S. L. (2019). Single-trial cognitive stress classification using portable wireless electroencephalography. *Sensors*, *19*(3), 499.

Bonaci, T. L. B. M. T., Herron, J., & Chizeck, H. J. (2015). How susceptible is the brain to the side-channel private information extraction. *American Journal of Bioethics Neuroscience*, *6*(4), 82–83.

Bubela, T., & McCabe, C. (2013). Value-engineered translation for regenerative medicine: Meeting the needs of health systems. *Stem Cells and Development*, *22*(S1), 89–93.

Cahn, B. R., & Polich, J. (2006). Meditation states and traits: EEG, ERP, and neuroimaging studies. *Psychological Bulletin*, *132*(2), 180.

Cai, L., & Chen, H. (2011). TouchLogger: Inferring keystrokes on touch screen from smartphone motion. HotSec, *11*(2011), 9.

Chen, S. (2018). *Chinese surveillance programme mines data from workers' brains*. Retrieved from https://www.scmp.com/news/china/society/article/2143899/forget-facebook-leak-china-mining-data-directly-workers-brains.

Christie, T. (2015). A discussion of the ethical implications of random drug testing in the workplace. *Healthcare Management Forum*, *28*(4), 172–174. https://doi.org/10.1177/0840470415581251.

Coates McCall, I., Lau, C., Minielly, N., & Illes, J. (2019). Owning ethical innovation: Claims about commercial wearable brain technologies. *Neuron*, *102*(4), 728–731.

Correa, A. G., Orosco, L., & Laciar, E. (2014). Automatic detection of drowsiness in EEG records based on multimodal analysis. *Medical Engineering & Physics*, *36*(2), 244–249.

Dong, Y., Hu, Z., Uchimura, K., & Murayama, N. (2010). Driver inattention monitoring system for intelligent vehicles: A review. *IEEE Transactions on Intelligent Transportation Systems*, *12*(2), 596–614.

Dubljević, V., Coates McCall, I., & Illes, J. (2019). Neuroenhancement at work: Addressing neuroenhancement at work: Addressing the ethical, legal, and social implications. In J. Trempe-Martineau & E. Racine (Eds.), *Organizational neuroethics* (pp. 87–103). Montreal, QC: Springer.

Federal Trade Commission. (2016). *Lumosity to Pay $2 Million to settle FTC deceptive advertising charges for its "Brain Training" program*. Press Release. Retrieved from https://www.ftc.gov/news-events/press-releases/2016/01/lumosity-pay-2-million-settle-ftc-deceptive-advertising-charges.

Frank, M., Hwu, T., Jain, S., Knight, R., Martinovic, I., Mittal, P., et al. (2013). Subliminal probing for private information via EEG-based BCI devices. *arXiv* preprint https://arxiv.org/abs/1312.6052v2.

Hallinan, D., Schütz, P., Friedewald, M., & de Hert, P. (2014). Neurodata and neuroprivacy: Data protection outdated? *Surveillance and Society*, *12*(1), 55–72.

Hammond, D. C. (2007). What is neurofeedback? *Journal of Neurotherapy*, *10*(4), 25–36.

Hwang, H. J., Kwon, K., & Im, C. H. (2009). Neurofeedback-based motor imagery training for brain–computer interface (BCI). *Journal of Neuroscience Methods*, *179*(1), 150–156.

Ienca, M., & Andorno, R. (2017). Towards new human rights in the age of neuroscience and neurotechnology. *Life Sciences, Society and Policy*, *13*(1), 5.

Ienca, M., Haselager, P., & Emanuel, E. J. (2018). Brain leaks and consumer neurotechnology. *Nature Biotechnology*, *36*(9), 805–810.

Kosmyna, N., Lindgren, J. T., & Lécuyer, A. (2018). Attending to visual stimuli versus performing visual imagery as a control strategy for EEG-based brain-computer interfaces. *Scientific Reports*, *8*(1), 13222.

Kreitmair, K. V., & Cho, M. K. (2017). The neuroethical future of wearable and mobile health technology. In J. Illes (Ed.), *Neuroethics: Anticipating the future* (pp. 80–107). Oxford, UK: Oxford University Press. https://doi.org/10.1093/oso/9780198786832.003.0005.

Lange, J., Massart, C., Mouraux, A., & Standaert, F. X. (2017). Side-channel attacks against the human brain: The PIN code case study. In *International workshop on constructive side-channel analysis and secure design* (pp. 171–189), Cham: Springer.

Levy, S. (2019). Facebook's latest purchase gets inside users' heads-literally. Retrieved November 26, 2019, from https://www.wired.com/story/facebooks-latest-purchase-gets-inside-users-headsliterally/.

Li, Q., Ding, D., & Conti, M. (2015). Brain-computer interface applications: Security and privacy challenges. In *2015 IEEE conference on communications and network security (CNS)* (pp. 663–666), IEEE.

Luck, S. J. (2014). *An introduction to the event-related potential technique*. MIT press.

Mannes, J. (2016). Bryan Johnson invests $100 million in Kernel to unlock the power of the human brain. Retrieved November 26, 2019, from https://techcrunch.com/2016/10/20/bryan-johnson-invests-100-million-in-kernel-to-unlock-the-power-of-the-human-brain/.

Mantri, S., Dukare, V., Yeole, S., Patil, D., & Wadhai, V. M. (2013). A survey: Fundamental of EEG. *International Journal of Advance Research in Computer Science and Management Studies*, *1*(4), 2321–7782.

Marcel, S., & Millán, J. D. R. (2007). Person authentication using brainwaves (EEG) and maximum a posteriori model adaptation. *IEEE Transactions on Pattern Analysis and Machine Intelligence*, *29*(4), 743–752.

Mardi, Z., Ashtiani, S. N. M., & Mikaili, M. (2011). EEG-based drowsiness detection for safe driving using chaotic features and statistical tests. *Journal of Medical Signals and Sensors*, *1*(2), 130.

Markoff, J. (2019). Elon Musk's Neuralink wants 'sewing machine-like' robots to wire brains to the internet. *The New York Times*. Retrieved from https://www.nytimes.com/2019/07/16/technology/neuralink-elon-musk.html.

Martinovic, I., Davies, D., Frank, M., Perito, D., Ros, T., & Song, D. (2012). On the feasibility of side-channel attacks with brain-computer interfaces. In *Presented as part of the 21st {USENIX} security symposium ({USENIX} Security 12)* (pp. 143–158).

Mathewson, K. E., Harrison, T. J., & Kizuk, S. A. (2017). High and dry? Comparing active dry EEG electrodes to active and passive wet electrodes. *Psychophysiology*, *54*(1), 74–82.

Matovu, R., & Serwadda, A. (2016, September). Your substance abuse disorder is an open secret! Gleaning sensitive personal information from templates in an EEG-based authentication system. In *2016 IEEE 8th international conference on biometrics theory, applications and systems (BTAS)* (pp. 1–7), IEEE.

Nayak, T., Zhang, T., Mao, Z., Xu, X., Pack, D. J., Dong, B., et al. (2017, July). Prediction of temperature induced office worker's performance during typing task using EEG. In *2017 39th annual international conference of the IEEE engineering in medicine and biology society (EMBC)* (pp. 1684–1687), IEEE.

Neeuro PTD LTD. (n.d.). Frequently asked questions—FAQs answered, Retrieved February 6, 2020, from https://www.neeuro.com/frequently-asked-questions-faqs-answered/.

NeuroSky. (n.d.a). IBS: Calm & control your gut, Retrieved November 26, 2019, from https://store.neurosky.com/products/calm-control-your-gut.

NeuroSky. (n.d.b). FlappyMind, Retrieved November 26, 2019, from https://store.neurosky.com/products/flappymind.

Niedermeyer, E., & Da Silva, F. L. (2005). The normal EEG of the waking adult. *Electroencephalography: Basic principles, clinical applications, and related fields*. Philadelphia, PA: Lippincott Williams & Wilkins., pp. 167–192. Retrieved from https://books.google.ca/books?id=tndqYGPHQdEC&printsec=frontcover&dq=isbn:9780781751261&hl=en&sa=X&ved=0ahUKEwjMw6yplYbmAhUprVkKHWQ1DyoQ6AEIKTAA#v=onepage&q&f=false.

Ogino, M., & Mitsukura, Y. (2018). Portable drowsiness detection through use of a prefrontal single-channel electroencephalogram. *Sensors*, *18*(12), 4477.

Olejniczak, P. (2006). Neurophysiologic basis of EEG. *Journal of Clinical Neurophysiology*, *23*(3), 186–189.

Pfurtscheller, G., & Da Silva, F. L. (1999). Event-related EEG/MEG synchronization and desynchronization: Basic principles. *Clinical Neurophysiology, 110*(11), 1842–1857.

Pfurtscheller, G., Guger, C., Müller, G., Krausz, G., & Neuper, C. (2000). Brain oscillations control hand orthosis in a tetraplegic. *Neuroscience Letters, 292*(3), 211–214.

Pustilnik, A. C. (2013). Neurotechnologies at the intersection of criminal procedure and constitutional law. In J. Parry & S. Richardson (Eds.), *The constitution and the future of the criminal law,* (pp. 2012–2061).

Racine, E., Rubio, T. M., Chandler, J., Forlini, C., & Lucke, J. (2014). The value and pitfalls of speculation about science and technology in bioethics: The case of cognitive enhancement. *Medicine, Health Care and Philosophy, 17*(3), 325–337.

Rackspace. (2014). *The human cloud at work: A study into the impact of wearable technologies in the workplace.* White Paper. Retrieved October 1, 2019 from Rackspace http://smoothmedia.com/project/the-human-cloud-at-work.

Raskin, C. (1993). Drug and alcohol testing in the workplace: Moral, ethical and legal issues. *Bulletin on Narcotics, 45*(2), 45–81.

Regan, D., & Lee, B. B. (1993). A comparison of the 40-Hz response in man, and the properties of macaque ganglion cells. *Visual Neuroscience, 10*(3), 439–445.

Rettner, R. (2016). Mind-tracking devices: Do 'brain wearables' really work? Retrieved November 26, 2019, from https://www.livescience.com/53840-do-brain-wearable-devices-really-work.html.

Rezeika, A., Benda, M., Stawicki, P., Gembler, F., Saboor, A., & Volosyak, I. (2018). Brain–computer interface spellers: A review. *Brain Sciences, 8*(4), 57.

Richardson, A., Mabbitt, B., Wilson, C., & Petriches, J. (2016). Google glass: A case study in design failure. Retrieved October 21, 2019, from https://www.appliedcommunication.org/?p=493.

Rosenfeld, J. P., Biroschak, J. R., & Furedy, J. J. (2006). P300-based detection of concealed autobiographical versus incidentally acquired information in target and non-target paradigms. *International Journal of Psychophysiology, 60*(3), 251–259.

Roy, Y. (2017). EEG & BCI crowdfunding landscape. Retrieved September 29, 2019, from https://medium.com/neurotechx/eeg-bci-crowdfunding-landscape-cfdb0da08937.

Schabus, M., Griessenberger, H., Gnjezda, M. T., Heib, D. P., Wislowska, M., & Hoedlmoser, K. (2017). Better than sham? A double-blind placebo-controlled neurofeedback study in primary insomnia. *Brain, 140*(4), 1041–1052.

Schönenberg, M., Wiedemann, E., Schneidt, A., Scheeff, J., Logemann, A., Keune, P. M., et al. (2017). Neurofeedback, sham neurofeedback, and cognitive-behavioural group therapy in adults with attention-deficit hyperactivity disorder: A triple-blind, randomised, controlled trial. *The Lancet Psychiatry, 4*(9), 673–684.

SharpBrains. (2018). Market report on pervasive neurotechnology: A groundbreaking analysis of 10,000+ patent filings transforming medicine, health, entertainment and business. https://sharpbrains.com/pervasive-neurotechnology/.

Shen, F. X. (2013). Neuroscience, mental privacy, and the law. *Harvard Journal of Law & Public Policy, 36,* 653.

Shukla, D., Kumar, R., Serwadda, A., & Phoha, V. V. (2014, November). Beware, your hands reveal your secrets!. In *Proceedings of the 2014 ACM SIGSAC conference on computer and communications security* (pp. 904–917), ACM.

SmartCap Technologies. (n.d.). SmartCap technologies: Our product, Retrieved October 1, 2019, from http://www.smartcaptech.com/life-smart-cap/.

Smith, S. J. M. (2005). EEG in the diagnosis, classification, and management of patients with epilepsy. *Journal of Neurology, Neurosurgery, and Psychiatry, 76*(Suppl. 2) ii2–ii7.

Teplan, M. (2002). Fundamentals of EEG measurement. *Measurement Science Review, 2*(2), 1–11.

Thibault, R. T., Lifshitz, M., Birbaumer, N., & Raz, A. (2015). Neurofeedback, self-regulation, and brain imaging: Clinical science and fad in the service of mental disorders. *Psychotherapy and Psychosomatics*, *84*, 193–207.

Thibault, R. T., & Raz, A. (2017). The psychology of neurofeedback: Clinical intervention even if applied placebo. *American Psychologist*, *72*(7), 679.

Tovino, S. A. (2007). Functional neuroimaging information: A case for neuro exceptionalism. *Florida State University Law Review*, *34*, 415.

Towle, V. L., Bolaños, J., Suarez, D., Tan, K., Grzeszczuk, R., Levin, D. N., et al. (1993). The spatial location of EEG electrodes: Locating the best-fitting sphere relative to cortical anatomy. *Electroencephalography and Clinical Neurophysiology*, *86*(1), 1–6.

Tractica. (2016). *Enterprise wearable technology case studies*. White Paper. Retrieved October 1, 2019 from Tractica https://www.tractica.com/resources/white-papers/enterprise-wearable-technology-case-studies-2016/.

Wang, D., Chen, J., Zhao, D., Dai, F., Zheng, C., & Wu, X. (2017). Monitoring workers' attention and vigilance in construction activities through a wireless and wearable electroencephalography system. *Automation in Construction*, *82*, 122–137.

Wang, C., Guo, X., Wang, Y., Chen, Y., & Liu, B. (2016, May). Friend or foe?: Your wearable devices reveal your personal pin. In *Proceedings of the 11th ACM on Asia conference on computer and communications security* (pp. 189–200), ACM.

Wascher, E., Heppner, H., Kobald, S. O., Arnau, S., Getzmann, S., & Möckel, T. (2016). Age-sensitive effects of enduring work with alternating cognitive and physical load. A study applying mobile EEG in a real life working scenario. *Frontiers in Human Neuroscience*, *9*, 711.

Wexler, A. (2017). The social context of "do-it-yourself" brain stimulation: Neurohackers, biohackers, and lifehackers. *Frontiers in Human Neuroscience*, *11*, 224.

Wexler, A. (2019). Separating neuroethics from neurohype. *Nature Biotechnology*, *37*(9), 988–990.

Wexler, A., & Reiner, P. B. (2019). Oversight of direct-to-consumer neurotechnologies. *Science*, *363*(6424), 234–235.

Wexler, A., & Thibault, R. (2019). Mind-reading or misleading? Assessing direct-to-consumer electroencephalography (EEG) devices marketed for wellness and their ethical and regulatory implications. *Journal of Cognitive Enhancement*, *3*(1), 131–137.

Wijdicks, E. F., Varelas, P. N., Gronseth, G. S., & Greer, D. M. (2010). Evidence-based guideline update: Determining brain death in adults: Report of the quality standards subcommittee of the American Academy of Neurology. *Neurology*, *74*(23), 1911–1918.

Wild, J. (2017). Wearables in the workplace and the dangers of staff surveillance. Financial Times. Retrieved from https://www.ft.com/content/089c0d00-d739-11e6-944b-e7eb37a6aa8e.

Ye, Y. (2019). Teachers snoop on Pupils' brains. *New Scientist*, *241*(3213), 13. https://doi.org/10.1016/S0262-4079(19)30101-0.

Yuste, R., Goering, S., Bi, G., Carmena, J. M., Carter, A., Fins, J. J., et al. (2017). Four ethical priorities for neurotechnologies and AI. *Nature News*, *551*(7679), 159.

CHAPTER TWO

A field with a view: Ethical considerations for the fields of consumer neuroscience and neuromarketing

Kimberly Rose Clark*,†
Department of Psychological & Brain Sciences, Dartmouth College, Hanover, NH, United States
*Corresponding author: e-mail address: kimberly.r.clark@dartmouth.edu

Contents

† Bellwether Citizen Response & Merchant Mechanics, Hanover, NH, United States.

Developments in Neuroethics and Bioethics, Volume 3
ISSN 2589-2959
https://doi.org/10.1016/bs.dnb.2020.03.002

1. Introduction

Commercial applications for the hardware and software used to measure the activity of the human nervous system have proliferated from 2005 to the present time. During the early rise of the field in the 2000s, neuromarketing was considered nothing more than a new trend in market research (Ariely & Berns, 2010; Botswick, 2010; Williams, 2010). Indeed, the utility and success of the field was once illustrated through the systematic rise and fall of a Gartner hype cycle (Linden & Fenn, 2003; Plassmann, Ramsøy, & Milosavljevic, 2012). Since this time, well over a decade of successful neuromarketing research has determined that in-moment measures of the brain and body (i.e., measurements made at the moment of consumer choice) are becoming standard tools to measure consumer behavior. For example, central and peripheral nervous system response measurements are increasingly in use by the research divisions of large scale marketing and advertising firms (Plassmann, Venkatraman, Huettel, & Yoon, 2015). The use of these applied neuroscience techniques in the commercial space has largely transcended the Gartner hype cycle and manifested into a reliable methods to further build on moving forward (Clark, 2017) (see Fig. 1).

At the turn of the millennium, following The Decade of the Brain (Baars, 2002), neuroscientists and philosophers pondered the potential ethical issues that might stem from advances in medical technology. Data streams garnered from the brain and body could uncover neural and physiological processes that dictate behaviors (Farah, 2012; Roskies, 2002). Such data are often

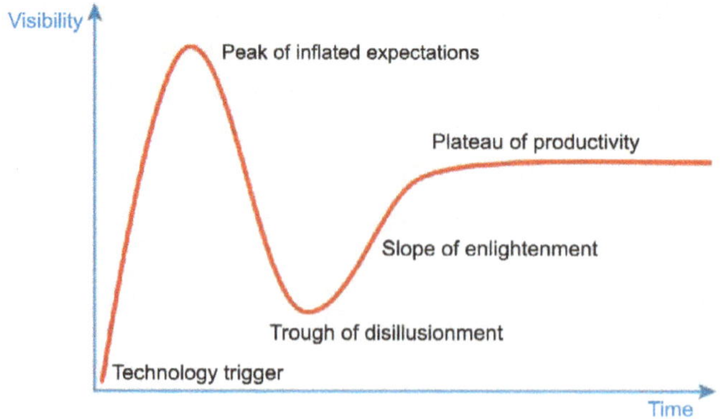

Fig. 1 Gartner-type cycle.

derived from biologic and automatic processes that can occur prior to, and outside of, conscious control. If commercial entities or governments had access to information (i.e., data) that increased their understanding of the mechanisms that precede behavior, it may is conceivable that these data could be used to manipulate the behavior of individuals without their knowledge or consent.

The term neuroethics was first defined in 2002 to describe a field that bridges the disciplines of the neurosciences, medical bioethics, cognitive science, public policy, law, and philosophy (Roskies, 2002; Iorga et al., 2017). The breadth of the field has provided for the consideration of complex ethical issues at an abstract and highly philosophical level, allowing for debate on the semantics of terminology as the field has crystallized (Baumeister & Bargh, 2014). As the neuromarketing industry continues to gain traction with a wider arena of clients, there is a growing need to consider a pragmatic focus on potential ethical issues that may arise from commercially applied neuroscience measures. The breadth of neuroethics must now comprise both academically oriented philosophical questions as well as application-centered considerations. Indeed, some academic institutions, practitioners, and client-side adopters have begun to understand this and put into place rules and language concerning the commercial use of applied measures of neuroscience (Al Pop, Dabija, & Iorga, 2014; Thomas, Pop, Iorga, & Ducu, 2016). In general, these rules speak to the need for the ethical treatment of human participants, focusing on informed consent and assurances of personal data protection to maintain participant privacy following their participation in neuromarketing research.

This chapter will first address the ethical considerations related to methodologies commonly used in commercially applied neuroscience, focusing on whether the measurements from the brain and body were collected via active consent or passive data capture (i.e., outside of the consumer's consent or awareness). The use and/or misuse of personal data, such as personality traits and emotional and self-regulatory states, are also discussed with respect to privacy rights and intentional manipulation of consumer emotions and choices. The final portion of this chapter discusses the ethical implications of inferences that are made possible by combining several data streams from brain, body and behavior to obtain information about the consumer that may allow for unethical manipulation of their future perceptions, emotions, and behaviors.

2. Ethical issues associated with common methods used in Neuromarketing

2.1 Functional magnetic resonance imaging (fMRI)

Functional magnetic resonance imaging (fMRI), measures brain activity by detecting changes associated with blood flow metabolism. The most common measurement technique used in fMRI is referred to as blood oxygen level dependent (BOLD) because it relies on the fact that oxygen extraction from the local cerebral blood supply and neuronal activation are coupled. Thus, the BOLD signal measures changes in localized metabolic activity in specific brain regions, with the assumption that increases in BOLD signal relate to increased activity within a local brain region that has a specific processing function.

While fMRI has proven instrumental in advancing the field of neuroscience, numerous logistical issues impede its application to consumer neuroscience. These include (a) the high cost of these scans, (b) the necessary institutional approval that must be granted by the site where the scan will occur, (c) the long turn-around times in data processing, and (d) a lack of generalizability of results outside of a scanner. Moreover, some neuroscience experts have voiced concern over the lack of reproducibility of fMRI results (Zuo, Biswal, & Poldrack, 2019). Taken together, these issues make routine commercial use of fMRI impractical.

Though infrequently used at the commercial level, fMRI studies can produce data that highlight ethical issues in consumer neuroscientific techniques. That is, fMRI has been used in academic neuroscience research to uncover the neural processing dynamics that may arise from different consumption experiences (Karmarkar, Yoon, & Plassmann, 2015). Here, I present two specific examples.

2.1.1 Price primacy

Recent research shows that fMRI responses are sensitive to the order in which price and product information is presented to a potential consumer. For example, brain activity differs in response to a given product presentation depending upon whether a price was presented before, rather than after, viewing the related product. To elaborate, BOLD signal in the medial prefrontal cortex (mPFC), a brain area known to be critical in decision making (Ly, Wang, Bhanji, & Delgado, 2019), increased immediately before making a purchase decision—but only if the price of the item was presented before viewing the product (Karmarkar, Shiv, & Knutson, 2015). This increase in mPFC activity did not occur in scenarios where a product was viewed before showing its related price; thus, the term price primacy. It should be clear how this information, derived from academic neuroscience, could be commercially applied to visual merchandizing and its effect on purchase behavior. That is, instead of having consumers evaluate whether a product is worth the initial price that is shown, simply showing the product before the price is displayed might diminish the importance consumers place on their purchase decisions. This would be a simple tactic that social media sites could easily implement in their product presentations. But whether it is ethical for Facebook to do this based on this neuroscience data remains an open question. Does the commercial application of these findings constitute an unethical persuasion tactic simply because of the neural origin of the findings?

2.1.2 Fandom

fMRI has been used to assess whether changes in neural processing of broadcast television show content vary as a function of an individual's fandom for a show. For example, a study commissioned by Viacom and Comedy Central was conducted at Harvard University to assess the influence of fandom on neural processes (The Neurological Effects of South Park, 2016). Fans, casual viewers and non-fans of the popular television show South Park viewed an episode of the show while changes in their BOLD signal were measured. For the routine musical introduction, referred to as "the title sequence," fans of the show experienced a significant increase in activation in the nucleus accumbens (NAcc), a region of the brain that has been widely implicated in reward processing (Knutson, Fong, Adams, Varner, & Hommer, 2001; Knutson, Fong, Bennett, Adams, & Hommer, 2003; Knutson, Wimmer, Kuhnen, & Winkielman, 2008). This research has not been published academically but was disseminated during an Interdisciplinary Symposia for Decision Neuroscience at Stanford University in 2017. Viacom did not give the researchers permission to publish explicit details of the study in an academic fashion and chose to maintain study details as brand assets. This use of fMRI to understand the impact that familiar music has on fans could be used to manipulate or unfairly persuade fans at a level outside their explicit awareness. By pairing all or part of the musical intro with a novel product or experience, a fan could associate a false reward with that product simply as a result of the halo effect the music has on its fan base. This may result in purchase behaviors that otherwise would not occur.

2.2 Electroencephalography (EEG)

Electroencephalography, or EEG, measures changes in the electrical activity of the brain via sensors attached to the scalp. EEG is referred to as temporally sensitive because it measures brain responses to a given stimulus presentation at the millisecond level (Hammond, 1981). The sensitivity of EEG allows for fine grain measures of brain responses to dynamic (i.e., quickly changing) stimuli, such as web and video-based advertisements and music.

For this reason, EEG has been adopted to study the effects of temporally sensitive marketing strategies on implicit brain responses. For example, frontal alpha asymmetry (FAA) is one example of EEG activity measured from the frontal lobe that has been shown to reflect motivation in attending to or avoiding a stimulus (Harmon-Jones, Gable, & Peterson, 2010). Neuromarketers often use the FAA metric to study "approach" motivation toward and/or an "avoidance" motivation to draw back from branded content. FAA is becoming an important implicit measure involved in temporally sensitive assessment of the consumer's willingness to engage with branded commercial content (Çakar & Gez, 2017).

2.2.1 Predicting market-level key performance indicators

Some researchers have correlated multiple data components of ongoing EEG to assess engagement with a stimulus (Dmochowski, Sajda, Dias, & Parra, 2012). In this research, engagement is defined as "emotionally laden attention" and can be used to predict the market-level success of stimuli. There are distinct advantages to having the means to understand the neural substrates of engagement in one consumer or in a specific population of interest. Quantitative assessments of entertainment are possible by this means (Gunter, 2000). Further, the prediction of future sales successes is possible by harnessing EEG data. Novel computational approaches for extracting neurophysiological data from EEG, along with eye-gaze based metrics, have been shown to predict population-wide behavior of movie goers (Christoforou, Papadopoulos, Constantinidou, & Theodorou, 2017). This type of neural metrics modeling was shown to predict close to 72% of the variance of a films' revenue performance at its premiere. This technique carries predictive information about a broader audiences' decision to watch a movie, above and beyond traditional methods. If a small groups' voluntary responses can predict a large groups' response to the same commercial stimulus, we must consider whether this is fair to citizens who do not wish their behaviors to be predicted. Those in the advertising and marketing industries have long attempted to predict future market level success, but the accuracy of those predictions was affected by the lack of having a sensitive enough measure. With the combination of neuroscience-based data and machine learning, it is now possible to predict certain behaviors at the market level.

2.2.2 Consent at market-level

Traditional self-report instruments used to measure future performance have historically been fraught with bias based upon the consumer's intentions. These intentions can differ widely from subsequent consumption behaviors. Alternatively, the in-moment responses of small groups can be used to accurately predict individual and temporal biases in the responses of a larger audience group. Thus, using neuro-derived computational models of prediction removes much of the uncertainty associated with traditional prediction techniques. That said, we must consider whether this increased accuracy in predicting consumer behavior results in an ethical concern for audience privacy if an entire population's consent is not possible. Policies may someday be put in place that establishes a prediction accuracy threshold that would define when the rights of a population are violated, especially if their future behaviors can be known at an extremely high level of certainty.

2.2.3 Integrating EEG with other neuro measures

Additional EEG metrics linked to mental workload and task engagement can be used to assess the dynamics of decision making. EEG indices of both engagement and workload have been recorded during cognitive tasks such as the forward/backward-digit-span, grid-recall, mental addition, image-learning, and memory tests. Whereas EEG-measured engagement has been found to reflect information-gathering, visual processing, and allocation of attention, EEG-measured workload has been associated with the increased demand experienced during activities that involve problem solving, integration of information, and analytical reasoning (Berka et al., 2007). These measures also have been used in conjunction with physiological metrics such as heart rate and heart-rate variability, both of which are believed to reflect the focusing of attention and increased cognitive load (Stikic, Johnson, Tan, & Berka, 2014). Heart rate and skin conductance also are known to be sensitive to certain aspects of emotion and may exhibit significantly different responses during arousal (Walla, Brenner, & Koller, 2011).

EEG studies have been commonly supplemented by measures of eye-tracking. These measures provide important information regarding where subjects fixate on a screen, the rate at which they fixate on different objects, and the period spent gazing at a given point in space. These data then can be used to determine where attention is focused while making a choice (Kamienkowski, Ison, Quiroga, & Sigman, 2012). Real-time facial expressions can be measured and affectively classified to measure emotional valence associated with motivation (Movellan et al., 2014; Ohme, Reykowska, Wiener, & Choromanska, 2009).

2.3 Biometrics

Biometrics provides a method for measuring physiological responses of the peripheral nervous system. These physiological responses to external stimuli are an effective means of assessing a consumer's in-moment responses to brand-oriented products or experiences. For example: (a) Changes in face-based emotions allow researchers to understand the extent to which a stimulus is perceived positively, neutrally, or negatively. Other emotions such as anger and surprise can also be resolved via changes in facial muscles. (b) Changes in heart rate and specifically the inter-beat-interval allow for an understanding of how exciting or mentally taxing a stimulus might be to engage with. (c) Minute changes in the sweat on the fingertips allow for an understanding of the strength of arousal that may arise from experiencing a stimulus.

All of these measures are lab-based where researchers can carefully assess consenting participants. However, technological advances have allowed for data capture from peripheral devices, such as mobile phones, and may be measured without the consent of a participant. Policies are not currently in place that address the extent to which biometric information can be collected and used for commercial purposes that the consumer user may never know exist. The following section will highlight some of the most commonly used biometric technologies.

2.3.1 Face-based data

Use of facial information has proliferated in the last decade. Identification of individuals and their emotions is commonplace in applied commercial neuroscience and done by resolving information gleaned from facial features. Two common means to measure changes in facial affect are electromyography (fEMG) and facial action unit coding (FACs). fEMG is a technique where sensors are placed on key muscle groups of the face to measure minute changes in the physiological properties of facial muscles and is an effective way to analyze voluntary and involuntary facial movements in real time. When compared to self-report measures, fEMG provides a more sensitive analysis of emotional reactions to TV advertisements and is related to brand recall (Ohme, 2008). Facial affect classification (FACs) measures emotional information of the face, but through classification of face-based emotions obtained from video that is processed by software trained to identify similarities in muscle groupings associated with positive and negative affect (Bartlett, Littlewort-Ford, Movellan, Fasel, & Frank, 2016; Ekman, 1989).

2.3.2 Cardiac measures

Changes in the interbeat interval of the cardiac signal is referred to as heart rate variability, or HRV. Once used as a measure to assess driver fatigue (Egelund, 1982; O'Hanlon, 1972), this signal is commonly used to assess changes in mental workload, or cognitive effort. Eye position information is also used in conjunction with biometrics in order to identify specific information being attended to at varying levels of arousal and cognitive effort (Clark, Leslie, Garcia-Garcia, & Tullman, 2018).

2.3.3 Galvanic skin conductance

Momentary increases in the skin conductance occur as a result of processing external or internal stimuli (Groeppel-Klein, 2005). Dublin-based Shimmer Research (Dublin, Ireland), for example, uses skin conductance to measure responses to advertising and political messaging, allowing for the assessment of particular moments of video and/or audio-based content that elicit strong affective arousal. The valence, or direction of this arousal is not measured, only the strength of the response. When used in conjunction with FACs, the valence of the arousal can be determined.

2.3.4 Combining measures

The psychophysiological and automatic behavioral responses detailed here have been validated in the academic arena to assess affective valence, arousal, cognitive effort and motivation. These measures now are available to study implicit responses commercially and continue to be improved in their footprint, memory storage and signal processing. The combination of these measures, in addition to traditional marketing techniques, provides a way of analyzing the multitude of facets that comprise the consumer response.

2.3.5 Integrating information across data sources

Data streams derived from these neuromarketing measures can be integrated with data from other sources, such as self-reported responses, social media shares, personality trait, emotional state and even genetic information. Linking these various consumer data sources offers a behavioral and emotional chain of causality to predict the decisions individuals will make related to commercial consumption. This ability for a massive cross-referencing of data is advancing more quickly than policy has been able to keep up with.

The ensuing sections of this chapter will give a pracademic (both academic and practitioner) perspective of the ethical implications of neuromarketing by presenting casework from the field. I will focus on data collection and information dissemination, opt-in versus opt-out data capture, unintended applications of this research, and the implications of inferential knowledge that can be gleaned from triangulating data streams.

3. Ethical issues associated with informed consent and dissemination of data

Two types of data acquisition are possible using measures of the brain and body: active opt-in acquisition, where individuals explicitly and actively agree to participate in a study. The other type of data acquisition is passive, typically using stand-off technologies that do not require the participant's knowledge or a need to place sensors on their person.

3.1 Active data acquisition

One of the most important aspects of neuromarketing studies is obtaining informed consent of the individuals who agree to participate. A written and verbal description of the study details, potential risks and benefits of participation, and how the information will be used are all part of obtaining informed consent from participants. Though not currently required outside of medical and academic research, most neuromarketing research firms obtain informed consent. Healthy individuals, who are most commonly studied, are at less risk than those in vulnerable populations, such as children, the elderly and any person with cognitive deficits. That said, it has been argued that true informed consent cannot be given by any consumer neuroscience study participant, because they cannot always be aware of the potential decision-based information (Iorga et al., 2017). In other words, if a consumer is persuaded to make a consumption choice through nudges, or persuasive techniques, derived from neuromarketing studies, they may never be aware that any manipulation of their behavior has occurred.

One consent strategy that might mitigate some of these concerns is the use of tiered levels of consent, meaning that a participant can approve of some uses of their data but not others. For example, a participant may be willing to have their data used for an academic publication, but may be less willing to share their personal data for commercial purposes, such as trade show presentation, webinars, and other outlets. Disclosing all known potential uses of participant data should be required of practitioners.

One relevant policy that is in place is known as the GDPR (General Data Protection Regulation) has been adopted in the United Kingdom. The GDPR gives individuals more control over how their personal data is used. If an organization processes personal data, the Regulation requires that organization to provide data participants with certain information. This typically takes the form of a data privacy statement or privacy notice. The GDPR sets out seven key principles: Lawfulness, fairness and transparency, purpose limitation, data minimization, accuracy, storage limitation, integrity and confidentiality, or security, accountability (Albrecht, 2016; Tankard, 2016; Voigt & von dem Bussche, 2017). The United Kingdom has set in place laws to help ensure compliance of the GDPR, and with misuse cases documented, a better understanding of the Regulation's effectiveness will be understood over time.

3.1.1 Incidental findings in non-diagnostic, commercial neuro data

For specific commercial studies using fMRI, it is imperative that any informed consent process include information about procedures to be followed in the case where an incidental neurological finding, such as a brain anomaly, is observed (see Box 1).

I have experienced such an issue in commercial work with fMRI. Prior to commencing with commercial fMRI work for a major television network, a protocol had been put in place that included a referral to a neurologist in the case of such an incidental finding. This protocol was reviewed and approved by a commercial Institutional Review Board (IRB) before the study launched. During the data collection phase of fMRI scans of one participant, a tumor was noticed behind the eye in the left prefrontal cortex (lPFC). In accordance with the study protocol, the participant was informed that they may have a neurological issue that would require further evaluation by a neurologist. The neurologist then described the anomaly to the participant and presented them with the choice as to whether they wanted to seek treatment.

3.1.2 Neuromarketers opt-in code of ethics

Though neuromarketing measures are designed to elucidate the brain and body processes that influence consumer decisions, the techniques are still

BOX 1 An example of consent form language.

The following language was approved by an Institutional Review Board for an fMRI study conducted by this author in 2015:

MRI is commonly used in medicine for the purpose of diagnosing abnormalities of the brain. The procedures that are to be used in this study are different from clinical MRI scanning. As researchers, we do not intend to make any medical diagnosis with the MRI as used in this research project, and we are not trained in medical diagnosis. However, if in the course of this research study we observe an anomaly in one or more of the MRI images, we feel ethically obligated to inform you of the observation. We believe it is important to inform you of such observations, because we cannot rule out the possibility that an anomaly may require medical attention. In this event, all information collected as part of this study will be made available to you for further examination by a medical professional. You will be fully responsible for the costs associated with a radiological examination and any further examinations or treatments that may be required for medical purposes. If you prefer not to be informed of an image anomaly, you must choose not to participate in the study.

This project is for research purposes only and is not directed toward, nor designed for, clinical diagnosis. The analyses performed in this study are not designed to find abnormalities and may not even include examination of structural images that would detect large abnormalities. The functional data, which are usually too blurry to detect abnormalities, are the primary focus of the study and what investigators examine. If you are participating in this study to obtain a clinical diagnosis or images that could be used for clinical diagnosis, you should not participate. If you believe you have a medical condition, now or in the future, you should seek care from a physician.

On occasion, a member of the research team may notice a finding on a scan that seems anomalous. We believe that it is our ethical obligation to inform you of such a finding. Being told about a finding may cause anxiety as well as suggest the need for additional clinical testing and its associated financial costs. The decision as to whether to proceed with further medical examination or treatment would remain entirely yours. If you do not want to be notified of this discovery, we ask that you not participate in this study.

grounded in diagnostic medical devices, and with that comes great ethical responsibilities for the researchers. Outside of fMRI research, the establishment of compliance with academic or commercial IRBs is uncommon in neuromarketing research. The majority of neuromarketing vendors today have subscribed to an ethical code established by members of the Neuromarketing Science and Business Association (NMSBA Code of

Ethics—NMSBA, 2019). Per this code, practitioners agree: (a) there will be no misrepresentation of data, (b) analyses and results that are delivered to clients must be backed by science and statistical evidence, (c) the research will not negatively impact the neuromarketing research profession, and (d) the research will comply with the highest research standards and accepted scientific principles. Moreover, the code outlines aspects of practitioner credibility and accounts for a need to disclose protocols for dealing with incidental findings during fMRI data capture. Despite these early promising efforts, there remains great variability in the protocols used across the industry with differing benefits to participants and neuromarketing firms.

In addition to core principles and credibility assurances, the NMSBA's Code of Ethics lays out the rights of individuals who actively consent to neuromarketing research participation. During consent, practitioners who follow the Code agree to explain the tools in use in lay terms, assure participants are able to explain back the procedures and rationale of the research, and fully inform participants about the intention of the study and their right to withdraw their participation at any time. Further, related to privacy, Code opting practitioners agree to having a privacy policy in place which is easily accessible to participants of a study, that participants' identities will be anonymized unless they provide explicit consent to waive that right, and that personal information will not be kept by the practitioner longer than is required to complete the study.

3.1.3 Considering the ethics of case examples when neuromarketing research is done for good

Several neuromarketing studies have been conducted with the betterment of user experiences in mind. In one example, the Advertising Research Foundation commissioned original research to understand, in a quantifiable manner, the extent to which in-moment brain and body responses change as a function of the type of placement and delivery vehicle used in advertising presentations (Clark et al., 2018). Their goal was to provide the advertising industry with an objective and statistically driven means by which to reduce the incidence of the most negative advertising experiences. Their work found that brain and body measures predicted what participants were willing to pay (in monetary terms) on a post-task self-report of their consumption experience. Unexpectedly, this research also uncovered the fact that that the most annoying and disruptive placement and delivery vehicles for ads negatively impacted the user's recall of the editorial content in which the advertisements were placed. Commercial entities could utilize this information

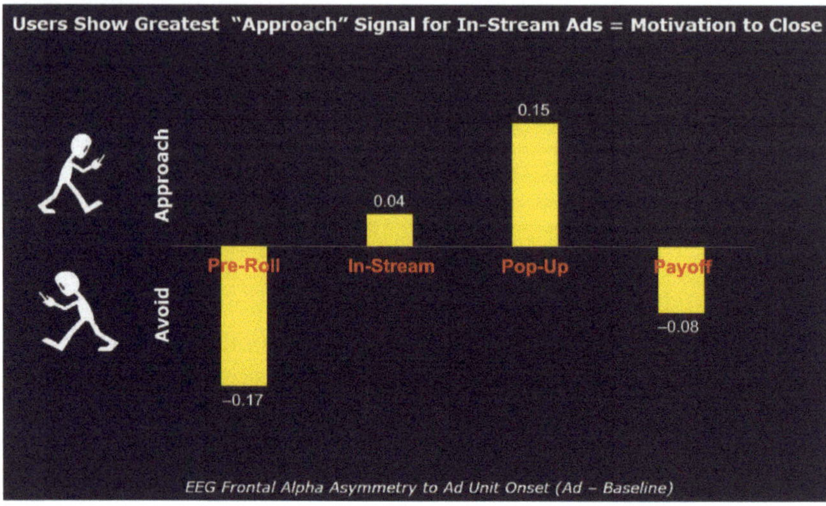

Fig. 2 EEG x Ad placement type.

for subversive means. For example, in highly contested election cycles, embedding highly annoying advertising delivery vehicles on a politician's website could subtly foster a negative association toward the politician through a nonconscious association with the ad's delivery (Fig. 2).

A second pro-bono study using neuromarketing measures for the benefit of society centered around identifying the most successful ads to hinder cigarette smoking behaviors. By combining traditional research methods with implicit reaction times, the study produced two findings demonstrating that the use of implicit measures strengthened the result of traditional research (Matukin, 2015). Study results suggested that pictorial health warning messages are more effective than text messages only, independent of culture. Generally speaking, warnings that contain images produced increased recall as well as recall confidence scores. In addition, even in cases where text-based warnings produced recall similar to image-based warnings, the use of images was more effective at eliciting emotional responses. Finally, the study found that warning oriented toward harm done to self and others were more effective than warnings focused only on smokers' health. This finding was also culturally independent.

These data provide a clear road map for the placement of warnings on cigarette packaging and in anti-smoking campaigns. However, this knowledge in the wrong hands could be used to purposely offer cigarette warnings that are less effective (i.e., text alone). Similarly, using warnings focused on only the smoker's health, when the research shows them to be less effective, would be equally unethical. In such a case, legislation should be enacted to govern advertising policy based on this scientific evidence.

3.2 Passive data acquisition

The second type of data acquisition I will consider here addresses physical and physiological information without consent. This type of data acquisition often occurs outside of individuals' explicit awareness. Passive data acquisition frequently occurs in public spaces, with the original impetus being one of public safety. For example, thermal cameras have been deployed at some airports. These cameras are used to measure the body temperature of deplaning passengers who arrive from locations known to have a high risk for certain communicable diseases (McBride, Buikstra, & FitzGerald, 2010; Ng, 2005). This measure can also be used to detect specific characteristics of emotional disposition in a crowd sourced fashion. A recent study has suggests that thermal imaging techniques can reliably classify basic emotions, including pleasure and stress (Merla & Romani, 2007). This type of technology could be deployed in commercial spaces to identify the predispositions of consumers as they consider purchases. In theory, products could be best positioned in the future for emotional merchandising, rather than visual merchandizing.

3.2.1 Video-based face recognition software

Cameras, in general, have proliferated in recent years in public spaces. Pilot tests using facial recognition software to identify students in some public schools in the United States are currently being conducted (Alba, 2019). Facial recognition software has increasingly improved at identifying individuals from various cultures. This is ethically questionable as racial profiling in public spaces can and has unfairly impacted citizens of certain minorities. China has been an early and active adopter of face recognition software based on video feeds. The Chinese government has actively used video-based recognition technology to socially shame its citizens for rule infractions (Ng, 2019). Their government has also used this technology to identify and then to randomly test individuals who have previously been charged with drug possession (Goldkorn, 2017). Shaming citizens and proactively seeking out former felons is not commensurate with ethical treatment on a humanitarian level. The implications for human rights are clear in this misuse of passive data capture technology. At the writing of this chapter, no available facial recognition software claims 100% accuracy. In fact, it is well known in academic and commercial use that this classification software yields high levels of inaccuracy, especially regarding women and people of color. The ethical implications of misidentifying a given citizens are clear. In the near future, it will not be difficult to connect data streams to a recognized face with purchase histories, location, medical records, or other personal information garnered from numerous sources of consumer data.

3.2.2 Video-based emotion detection

Video-based data capture is now being used for more than the identification of individuals based on facial features. Discerning the emotional disposition of citizens is now possible through the classification of different muscle groupings, or facial action units. Advertisers have begun to apply this method to better understand the consumer audience (Schwartz, 2019). Emotion detection technologies employ two techniques: computer vision, to identify facial expressions, and machine learning algorithms to analyze, interpret and then classify the emotional content of those facial features. While emotion detection has grown into a 20 billion dollar industry (Schwartz, 2019), there is a need to understand ethical issues inherent in this tool's commercial applications. Before considering the ethical implications of facial affect detection software, it will be useful to first consider the background of the origin of the tools.

Facial affect algorithms have historically been derived using supervised learning, which is an artificial intelligence approach that learns by being trained to recognize content. Software used in neuromarketing to measure and classify consumer emotions relies on databases that contain thousands of faces which have been classified by a system based on Paul Ekman's classic work in facial affect identification (Ekman, 1989, 1992; Ekman & Friesen, 2003). When a novel face is presented, specific muscle groupings are identified (e.g., the corrugator supercilii muscles near the eye) based on shadow or point detection and then compared against the groupings of faces in the seven emotions Ekman identified as universal. During the training process of the emotion classification model, patterns are extracted from the muscle grouping data and a model created. Ultimately, the model is used to classify previously unseen faces. The algorithm then computes an evidence value, which is a probability that the novel face is eliciting a given discrete emotion based on the trained model.

The software classifications are based on Ekman's classification of seven emotions based on changes in facial muscle group configurations particular to each emotion. The assumption that there exist seven discrete and universal emotions has been challenged (Barrett, 1997, 2006, 2017). In a typical study, Ekman had participants sort images of faces eliciting different emotions into pre-defined categories of emotions: joy, anger, disgust, contempt, sadness, surprise, fear. When the study was reconducted by Feldman Barret and colleagues, the pre-defined emotion categories were not used and participants were allowed to sort the images into their own subjective emotion categories. This modification resulted in participants sorting emotions into more categories than the initial seven from Ekman's classic study. Feldman Barrett does not feel that it is appropriate or accurate to generalize emotions that are derived from facial expressions across individuals, or from different cultures. When angry, one person might scowl, while another may smile while quietly plotting revenge against the instigator of their anger. Instead of capturing actual nuanced emotions, facial affect software captures stereotypical emotional constructs that are well known in specific cultures but are not universal across cultures.

3.2.3 Three ethical issues associated with face recognition software

The first ethical issue surrounding commercial application of facial affect software comes in the potential for its overuse and the issue of acquiring the consent of an individual in a public space or in a crowd of people. For example, employment recruiting firms may adopt this technique to gauge the emotional disposition of job candidates, which should require consent, but it is unclear if there exists legal language to enforce a consent protocol. If the analysis is not accurate or misclassifies the emotion being elicited, this has potential implications for our citizens' occupation acquisition and wellbeing. In a similar vein, schools may adopt this tool to identify student engagement, or airports could use this technology to decipher anger in faces that may be associated with a terrorist threat, both relying on the accuracy in the software's classification. If treatment of citizens depends on the accuracy of software to reliably identify emotions that may be bellwethers of imminent threat, regulations must be considered that include an understanding of how base rate accuracy can impact human lives.

A second issue faced by practitioners using facial affect software relates to its inherent cultural bias. A popular open source dataset known as Labeled Faces in the Wild is commonly used to train computer detection models. This dataset contains 77.5% male faces and 83.5% Caucasian faces (Huang & Learned-Miller, 2014; Learned-Miller, Huang, RoyChowdhury, Li, & Hua, 2016). A lack cultural diversity in the faces used to in the acquisition of the training data has led to facial recognition model overfitting and built-in racial biases. Additionally, items of culturally specific clothing (e.g., niqabs), can block parts of the forehead that are used by traditional facial affect classification algorithms.

A third issue in the commercial over-adoption of facial affect classification software relates to contextual drivers that work to elicit emotions. I have worked commercially with social media sites as well as large online retailers to identify attention and emotion dynamics along the digital consumer journey. In an academic consideration of some consumer journey data, expressions classified by software were systematically reviewed by human researcher coders to assess the software's accuracy. In this assessment, the coders were instructed to review all classified expressions and review the consumer journey context 3–5 s prior to the classified expression. In the findings, the software's classification accuracy compared to human coders was extremely poor when not considering the context in which the classified expression was made (Clark, Leslie, Tullman, Eglington, & Wolford, 2016). Across 118 participants, the software's classification of joy aligned with human coders only 9.5% of the time, disgust only 1.6% and contempt 2.4%. For example, when an individual clicked a product to purchase, and was subsequently informed that their chosen product was out of stock, the participant may grimace and laugh in a frustrated way. In context, this becomes evident. Out of context, it is classified as joy, when in fact it is frustration that is felt by the consumer. Reliance on this type of software in online purchase decisions is also problematic for a different reason. While movies and advertisements are meant to elicit strong emotions, online purchases simply are not as strongly emotionally oriented. As such, consumers spend the bulk of their shopping time eliciting *no* facial affect. Brand strategists that rely too heavily on automated facial affect classification throughout the online purchase journey may come to misleading conclusions based on the low sampling rate of emotions expressed while shopping online.

3.2.4 Ethical concerns over emotion recognition patents

Measuring emotional information online in social networking contexts has been shown to be an effective means of aligning content based on affect. In a best-case scenario, a brand's knowledge of users' emotional dispositions presents opportunities to enhance their online experiences by targeting communications through emotional alignments. However, the knowledge of a users' current or prevalent emotions can also be a tool that can intentionally or unintentionally be used against them. Three Facebook patents have been filed to create algorithms designed to predict, and then adapt to its users' emotional states. The first patent defines parameters for formatting messages based on users' emotions while typing. The second describes the displaying of content based on identifying changes in their facial expression responses to prior content, along with an Application Programming Interface, or API for third-party use. APIs are a set of functions and procedures allowing the creation of applications that access the features or data of an operating system, application, or other service. This third patent is designed to map facial expressions on to the most relevant emojis for its users to post. Facebook has shown interest in detecting faces and emotions for an extended time—from acquiring a service for matching novel photos with user accounts to introducing automatic alt text, which describes photo content with overlaid text (including the facial affect expressed in persons within content) to assist its blind users. Facebook has the ability to share this type of walled garden data to third party companies. Having knowledge of users' current emotional disposition can unfairly target them in a vulnerable moment. Users who are found to have negative emotions via these algorithms could potentially be presented with a series of mood enhancing pharmaceuticals they may otherwise not consider for purchase.

A lawsuit has been filed regarding a patent for a Facebook software algorithm that automatically identifies and tags friends in photos. This patent raises ethical concerns because of its ability to identify and classify the emotions based on the classification of the identified individuals. The identification of happy or sad individuals based on posted photos could unintentionally be used to create emotional alignments to sell products or to create unconscious brand associations. For example, placing a Coca∼Cola advertisement near a photo of happy individuals could serve to enhance positive associations to the Coca∼Cola brand. Conversely, presenting another brand near posted photos of unhappy individuals could create a negative halo for the brand. Moreover, with this technology, it is possible to identify individuals and to subsequently ascertain the proportion of specific emotions they elicit. This capability could allow for an understanding of an individual's emotional predispositions and speak to likely personality traits, without any consent or awareness by an individual. The promise of these user interface improvements comes at the cost of privacy rights if Facebook is not able to ensure third parties' ethical use of this data.

3.3 Vulnerable populations

In both active and passive data acquisition, it is important to consider the susceptibility of vulnerable populations. This issue is an especially prevalent concern in studying youth populations. A 2010 review article examined research focused on understanding motivational systems in adolescence and how they differ from other age groups (Doremus-Fitzwater, Varlinskaya, & Spear, 2010). Adolescents were found to be significantly more sensitive than adults to the reward properties of a variety of drugs and natural stimuli. These findings were corroborated with fMRI research, which showed a strengthening within the reward network [e.g. increased connectivity between the prefrontal cortices (PFC) and nucleus accumbens (NAcc)] from adolescence to adulthood. Adolescents' increased sensitivity and approach motivation toward positively rewarding stimuli makes them more vulnerable to brand communications that might seek to exploit this consumer age group, many of which have a discretionary income. One must consider whether it is ethical for any company to target advertising toward adolescents when this consumer group is known to have less inhibitory control over reward circuitry. Moreover, alcohol and drug related companies, as well as the rise of firms promoting the popular nicotine e-devices, often target youth. Companies should be monitored through evidence-based policies for potential exploitation of adolescents' heightened reward systems sensitivity and predilection to approach positively rewarding stimuli that may be harmful to their mental or physical health.

3.3.1 Cultural biases in neuro technologies

Another considerations to be addressed regards the unintentional bias of some of the measures used in neuromarketing. Specifically, it has been well documented that electroencephalography (EEG) is less effective in measuring response to stimuli in elderly populations (Loewenstein, Argüelles, Argüelles, & Linn-Fuentes, 1994). In addition, facial affect recognition software has been found to maintain different levels of accuracy in certain cultures compared to others (Dailey et al., 2010; Kilbride & Yarczower, 1983; Li & Deng, 2019).

4. Ethical issues associated with unintended applications of academic research

Academic social and affective neuroscience research has proliferated into applied commercial applications with many digital data sources now available for mining. In decades past, it was considered taboo for academic researchers to look outside of the ivory tower to field studies and instead focus concentration on experiments in lab settings. The blend of private funding, open data sources on users and use of artificial intelligence prediction algorithms has created a perfect storm within which to understand human emotion and behavior at the societal level. But this new age of academic inquiry lends itself to creating theoretical findings than can be applied in unethical ways. The next section will discuss several such examples of theoretical knowledge that can be or has been mapped onto commercial principles and tactics.

4.1 Facial information in the social media age

Facial Information from social media is a rich source of personal information. Neuroscience research has shown that the fusiform face area (FFA) is activated when exposed to partial or whole faces (Flournoy et al., 2016; McCarthy, Puce, Gore, & Allison, 1997). Familiar faces have been found to be more readily detected compared to faces of strangers. Consistent with this finding, familiar faces are processed with reduced attentional resources and even in the absence of awareness (Gobbini et al., 2013; Ramon & Gobbini, 2018; Taylor et al., 2009). Our own faces and other familiar faces require less effort to process, exhibiting our preference for the familiar over the unfamiliar.

4.1.1 Face familiarity and voter persuasion

The power of familiar faces also has been shown to sway important voter preferences and decisions. In a clever study, a group of undecided voters were asked to view photos of candidate faces and select which candidate they would vote for. Participants viewed images of two unfamiliar candidates morphed at various degrees with either the participant's own likeness or other voters. Results demonstrated a strong preference for facially similar candidates, despite no conscious awareness of the similarity manipulation (Bailenson, Iyengar, Yee, & Collins, 2008). The individuals who saw themselves literally in the candidate showed a preference for that face. This unfairly influenced their preference for candidates based on their implicit preference for their familiar face.

The current age of selfie photographs which highlight closeups of individuals faces has provided a wealth of facial images (Wickel, 2015). It would not be difficult for a social media giant to utilize Bailenson's academic research for commercial advantage. Facebook could create doppelgänger advertisements for brands by subtly blending a user's facial features with a model in imagery promoting the product. Given the known influence and advantage familiar faces have, it is likely that this would increase sales of the promoted product and increase client-side return on ad spend (ROAS). Clearly, there are several ethical issues embedded in this hypothetical advertising tactic, including a lack of clear consent, as well as the known and unfair advantage integrating users' likeness into promoting products with which they may otherwise have no interest.

4.1.2 Personality trait and emotional state information

In 2019, Facebook shut down over 400 third party vendors from accessing user information. Some of these third party apps were disguised as games, when in fact they were collecting personal information. Some apps came in the form of entertaining personality quizzes, with a potential for user responses being harnessed to tailor specific information in their newsfeeds to ultimately change voter behavior (Kelly, 2018).

4.1.3 Traits and political party affiliation

The creators of these apps can connect the dots between Facebook user's quiz responses and their predisposition for particular personality traits. One study assessed correlations between specific personality traits with political party affiliation. The study found significant correlations between trait-based risk taking and Republican Party affiliation. Further, individuals scoring high on trait-based creativity were significantly more likely to affiliate with the Democratic Party (Tyagi, Hanoch, Choma, & Denham, 2018). A political party or individual politician's campaign could use this type of academic finding and create campaign messaging framed to best resonate with the personality types of particular party members, creating an unfair means of communication that may intentionally and unfairly harness an individual's own personality against them.

Facebook has, since the 2016 Presidential election, reportedly stopped allowing third party access of such information gathering apps, games and quizzes. Their crackdown underscores the strong influence that strategic use of personality trait information can have. Advertisements purposely aligned with certain personality traits can unfairly persuade consumption behaviors, be the products consumer goods, political candidates, or even government policies. Our perception and resonance with content begins with these internal or endogenous lenses with which we take in and interpret information.

4.1.4 The Big Five

Other research has shown differences in individuals' Big Five personality traits, including outgoing, conscientious, extraversion, agreeableness, and neurotocism are predictive of whether an individual dislikes the opposing party and the degree to which they express this hostility. Modeling negative affect toward the opposing party as a two-stage process, extraverted individuals are found to be less likely to have negative affective evaluations of the opposing party (Webster, 2018). Facebook users taking quizzes to determine their spirit animal or the Disney character that most defines them were unwittingly providing ground truth information about their predispositions, which can inform marketing campaigns on the most effective communications tactics to unethically guide behavior change (Brown, 2019).

There have been similar leaps in the use of artificial intelligence related to the identification and classification of personality traits that will be integrated into marketing soon. An algorithm has been recently developed that links individuals' eye movements to their personality traits. By tracking eye movements, the software accurately deciphered participants' scores on four of the Big Five personality traits (neuroticism, extroversion, agreeableness, and conscientiousness) (Hoppe, Loetscher, Morey, & Bulling, 2018).

4.1.5 Predicting traits through visual search patterns

In a recent study, participants' eye movements were tracked throughout their everyday behaviors. The individuals' personality traits were then assessed through self-report measures which were compared with their visual gaze behavior patterns in an attempt to accurately classify personality traits. A marketer's ability to capitalize on understanding the personality traits of a consumer audience would allow for means of targeted communication without explicit consent. The impetus of the research was aimed at improving robot-human interactions to increase the personalization of the users' experience. However, with visual gaze data routinely available to third party vendors from wearable technology, built-in computer cameras and public security video, individual personality assets may be learned from this tech through involuntary data acquisition.

4.1.6 Covert detection

Marketers can capture gaze and facial affect information while an individual uses their computer or mobile phone via the built-in camera. These data can be used to assess emotion while consuming web content and can also determine where on the screen a user is most and least engaged. Firms such as RealEyes (New York, NY) measured the attention, emotions and sentiment of a sample audience as they watched a brand's content on their own device at home. While this firm reported testing a voluntary sample, other user agreements embed language that allow for apps to use personal data such as the web camera and the geospatial location of the user. Given the onslaught of user agreements presented to consumers, the vast majority do not take the time to read which rights they are voluntarily giving to brand researchers. Further, eye tracking data from a voluntary group can be paired with artificial intelligence to create inferences about the visual attention and engagement of consumers who did not consent to the research and may be unfairly targeted through smart communications.

4.1.7 Engineering social emotion contagion

Many commercial clients are using sentiment analysis to identify the attitudes consumers express toward given products (Kregor, Gomez, Kelly, & Stevenson, 2018; van den Broek-Altenburg & Atherly, 2019). By understanding the origin of the sentiments that drive consumers, enhanced advertisements could include messages that will best resonate with their emotional dispositions. But there is a dark side to this means of communication calibration. A 2014 study collaboration between Facebook and researchers at Cornell University used sentiment analysis to identify positive and negative emotional content in users' feeds (Flick, 2016). In order to test the extent to which social contagion of shared emotions spreads through the friend network, users' feeds were altered so that some users experienced fewer positive sentiments from friends and others experienced fewer negative sentiments. If a depressive individual was shown only negative emotional content, their depression may have been unfairly heightened without their knowledge of any manipulation of their newsfeed. The key ethical violation of this study is that the users were not informed that it was occurring. The researchers found that participants who experienced their Facebook friends' positive or negative emotions led to their sharing of that emotion—deemed emotional contagion. This study showed that the emotions of Facebook users in the study were unfairly influenced outside of their awareness or consent. With the walled gardens of proprietary data housed by the social media site, there is currently no outside regulation of this type of emotional manipulation. Given the Cornell research findings, Facebook can easily harness sentiment scraping algorithms to identify user's emotions and create content to cater to those emotions, without the Facebook user's awareness.

The active manipulation of a user's content to measure the extent to which the platform can spread emotion is alarming. Beyond the fact that no informed consent was established in this massive study of 689,003 users, actively tinkering with human emotions in this way is precarious for some of these individuals who may be predisposed to depression or suicidal tendencies. Overloading their Facebook experiences with bad news could have had the capacity to push someone over the edge in doing harm to self or others in the absence of positive emotional communications. Even more disturbing is the notion that depression has already been linked to social media use (Jelenchick, Eickhoff, & Moreno, 2013; Lin et al., 2016), which suggests that the proportion of users who may have been vulnerable to negative emotional content is likely higher than the general population.

4.2 The commercial advantage of high cognitive load & low self-regulatory mechanisms

The amount of mental effort, or cognitive load required by the consumer is another variable influencing consumer choices. Specifically, the amount of mental effort required to make a choice can decrease the consumer's ability to make rational decisions. Branders could take unfair advantage of consumers by taxing their mental effort at key choices along the consumer journey.

4.2.1 Self-regulation

Research has suggested that tasks requiring greater self-control deplete subsequent available cognitive resources to execute self-control in future tasks. Self-control is believed to be a limited and depletable resource (Baumeister, Bratslavsky, Muraven, & Tice, 2018; Baumeister, Vohs, & Tice, 2007). FMRI research has identified brain regions that are involved in self-control; recognizing a need for self-control engages the anterior cingulate cortex, while regions responsible for executing self-control are the dorsolateral PFC and dorsal frontal median cortex (Hedgcock, Vohs, & Rao, 2012). This study suggested that participants who had prior self-control tasks experienced more difficulty in subsequent control tasks, but self-control increased when individuals were presented with health reminders. This research can easily be mapped onto potential commercial uses. Navigating the consumer journey in physical brick-and-mortar retail stores as well as through online retailers requires deliberate cognitive processing. Advertisers and marketers have the potential to create scenarios that further increase cognitive load just prior to, or at the point of purchase for items meant for impulse purchase. For example, creating a confusing visual merchandising array on a shelf, or having difficult to understand pricing can increase a consumer's mental effort to make a reasonable choice. Advertisers can also present advertisements online in ways that tax viewers mental resources just prior to presenting them with impulse purchase options.

4.2.2 *Mental effort*

Increasing cognitive load has also been shown to have a positive impact on impulse purchases (Shiv & Fedorikhin, 1999). In this study, a manipulation of participants' cognitive resources occurred via task-instructions. Half of participants had to maintain a 7-digit number in working memory, while the other half were asked to maintain a 2-digit number. While keeping these respective number combinations in mind, participants were provided with both a healthy and a hedonic food snack choice. Participants who maintained a larger number set in working memory, and therefore a higher cognitive load, more often chose the hedonic over the healthy food choice. Snack aisle marketing, advertising and visual merchandizing could easily implement in-category tactics to increase cognitive load, which would act to increase the impulse purchase of brands with hedonically oriented brand assets.

4.3 Contextual drivers of consumer choice

Products are not considered or purchased in a vacuum devoid of contextual cues. Other proximal goods for sale, environmental cues, and product attributes such as price, weight, ingredients, and pictorial information work together to sway attention and relative salience of some product attributes over others in consumer decision making. Academic research has isolated specific components of the commercial environment and identified neural processes associated with product perception, valuation, and preference. An earlier reference to price primacy was discussed under the fMRI section of this chapter (Karmarkar, Shiv, & Knutson, 2015), however several other choice drivers, or nudges, have been researched and will be discussed in this chapter section.

4.3.1 *Gaze cascade and perceived preference*

The consideration sets presented in different visual merchandising strategies can also work to unfairly bias consumers' perceptions of value. For example, eye tracking research has shown that in-category product preference is influenced by how the products are presented in a visual merchandising array. Specifically, consumers tend to choose a product that is visually located in the center of an array over peripheral products (Atalay, Bodur, & Rasolofoarison, 2012). Products presented in the horizontal center of an array receive significantly more visual attention relative to other products and are subsequently more likely to be selected. This study found that consumers have an initial central fixation bias, which is a tendency to look first at the central option before other choices, and a central gaze cascade effect, where participant's gaze is directed toward a choice and increases steadily until the moment of choosing. Though this central gaze cascade effect was related to choice, in post task self-report, memory-based attention measures were not correlated with this attentional gaze effect. This research suggests that products in a central location have an automatic advantage in consumer's predisposed attention, which may serve as an unfair advantage in persuading consumer decisions, especially if the central product has fewer health benefits than the products flanking it.

4.3.2 Better-for-you nudging of choreographed visual merchandising

Another questionable issue in visual merchandising strategy relates to what is known in the consumer-packaged goods (CPG) industry as better-for-you (BFY) or alternatives. This technique is widely used by brands focused on the snacking category, which grew $3.4 billion globally in 2017 (CPG, FMCG, & RETAIL, 2018). Consumer demand in this category has risen for foods and beverages that are not only perceived as healthier, but also those that maintain a nutritional halo. Convenience stores that call attention to such products in a concerted and direct way can set themselves up for higher sales and an incremental consumer base, especially among the younger cohort of Millennials and Gen Z. But BFY visual merchandising strategies can often be unethical in terms of the overall product array to be chosen from. For example, positioning real fruit gummies in with overtly sugary snacks may lead consumers to believe that the real fruit choice is better than the alternatives in a set. Still, the actual nutritional values across products may not differ.

4.3.3 The decoy effect

Another visual merchandising strategy that may be deployed with negative consequences to the consumer involves the decoy effect. The decoy effect is a phenomenon whereby consumers will tend to have a specific change in preferences between two options when also presented a third option that is asymmetrically dominated (Huber, Payne, & Puto, 1982). What asymmetric domination means is that the decoy is priced to make one of the other options more attractive. It is said to be dominated in terms of perceived value, quantity, quality, extra features, among other product attributes. The decoy is not intended to be purchased and instead is meant to nudge consumers away from the competitor and toward the target, which is typically a more expensive or profitable option. For example, if a consumer has to consider the purchase of a $1400 computer and a $1700 computer, they may opt for the less expensive model. However, if the choice set also included a $2000 computer, the $1700 model now is framed as a more reasonable choice. fMRI work has shown that valuation processes change as a function of the decoy effect. A cognitive neuroscientific study was conducted to determine the extent to which a trade-off aversion works to explain the attraction effect (Hedgcock & Rao, 2009). This neuroimaging study measured changes in the cerebral activity of participants engaging in various choice tasks. The study found that the presence of a third, irrelevant, choice alternative yielded changed activation in areas of the brain associated with negative emotion that differed when compared to the activation during choice tasks involving two equal options. The results support the claim that choice sets including a decoy require consumers to consider trade-offs. These trade-offs are associated with relatively greater negative emotion. This increase in negative emotion likely stems from the increased deliberation required to evaluate products on a particular attribute, working to unfairly frame a particular product that otherwise would not have been selected. In other words, consumers can be unfairly swayed to make choices they would not otherwise consider when visual merchandizers harness academic research to knowingly persuade consumer decisions that may be financial or medically harmful.

4.3.4 *In-store slack*

Mental budgeting is the itemized portion of budget that the consumer has allocated to spend on items planned at the brand or product level and the in-store slack is the portion of the mental budget that is not assigned to be spent on any particular product but remains available for in-store decisions (Stilley, 2010). Much research has been conducted on where in the consumer journey, in what categories and at what price points do consumers tend to spend this mental, financial windfall. Brick-and-Mortar and online retailers spend substantial resources on understanding in-store slack so that they may position the appropriate products at the optimal temporal and spatial location relative to the common shopper routes in order to capitalize on this cognitive bias. Capitalizing on this understanding of an individual's mental discretionary budgets can lead to unintentional overspending.

4.4 Implications of inferential knowledge from triangulating data streams

In isolation, some data streams do not present issues with regard to consumers' privacy. However, increases in the ability to capture, manage and merge data from several sources has opened the door to unanticipated inferential learning of the consumer that may hurt the privacy of their personal data. Triangulation, i.e., using multiple data streams to transitively make inferences about other information about the consumer, is a threat to consumer privacy. Creating a framework that lays out all possible permutations of inferential information that may be learned though data triangulation would be a daunting but important step to learning where privacy might be violated through inferential knowledge.

4.4.1 Layered voice analysis

Smartphones, sensors and machine learning have been recently combined to develop real-time prediction and interventions for suicide prevention (Torous et al., 2018; Vahabzadeh, Sahin, & Kalali, 2016). While this is beneficial for at risk soldiers suffering from PTSD, harnessing such rich data sources of emotional sentiment captured from voice and conversations could ultimately create context alignments for marketing based on emotional state. Amazon Alexa's AI unit is now using layered voice analysis (LVA) to assess frustration levels in its users (Johnson, 2019). LVA has historically been employed as part of lie detecting polygraphs (Adler, 2009; Harnsberger, Hollien, Martin, & Hollien, 2009; Horvath, McCloughan, Weatherman, & Slowik, 2013). With the capability Amazon has in establishing correlations between changes in LVA response, sentiment scraping, and purchase behaviors, they maintain a unique and powerful ability to infer affective contexts surrounding certain categories of interest and, more generally purchase volume. Integrating these data streams would allow their platform algorithms to tactically present specific products to users and frame marketing communications to align with the user's affective state. While it is possible to override impulse purchases with motivation and deliberation, aligning messaging with tailored communication to the current mood state is a questionable tactic that may target vulnerable populations.

4.4.2 Wearables and fitness trackers

Another example of the consequences of the triangulation of data in terms of potential inferences on consumers relates to fitness tracking data. Bluetooth data from fitness trackers is akin to bio-signatures that can be used to identify individuals like fingerprints (Das, Pathak, Chuah, & Mohapatra, 2016a, 2016b). The lack of security surrounding personal biometrics raises ethical concerns. Bluetooth signals are used to wirelessly communicate from smart watches and fitness trackers. Their lack of security opens the door for hackers to hijack the information. An individual can be identified, and their biometrics sold to third party firms for any application. For example, fitness trackers now have the capacity to measure heart rate variability, or HRV, which is a measure of the heart's inter-beat intervals. HRV has been used historically to measure the cognitive load or stress that an individual maintains in mental processing (Barraza, Alexander, Beavin, Terris, & Zak, 2015; O'Hanlon, 1972; Thayer & Sternberg, 2006). In understanding which consumers are experiencing high cognitive load together with the geocaching of overall locations, and in-store locations identified through comprehensive geofencing, the possibility has arisen for advertisers and brands to capitalize on the endogenous and exogenous connections between time, space and the consumers' predisposition.

4.4.3 Wearable data security breaches

Several recent studies have assessed the potential health and security related issues that may arise from the emergence of consumer wearable devices (Kerr, Butler-Henderson, & Sahama, 2018; Piwek, Ellis, Andrews, & Joinson, 2016; Tana, Forss, & Hellstén, 2017). While many positive health oriented aspects exist, such as the wearable sensor's application to assess standing balance and walking stability in patients with Parkinson's disease (Hubble, Naughton, Silburn, & Cole, 2015), both positive and negative implications exist that are related to the potential unwanted commercial applications of these devices, as well as their data security.

Recent research has addressed encryption protocols embedded in data transfer (Tahir, Tahir, & McDonald-Maier, 2018; Tahir, Tahir, Sajjad, & McDonald-Maier, 2017). For example, ICMetric is believed to offer security provisions in wearable devices by providing sufficient cryptographic services to protect consumer data. Tahir's research involved the use of five identical Shimmer sensors (Burns et al., 2010) that utilize the device's accelerometer and gyroscope to illustrate a secure generation of devices that use the ICMetric which also acts as a cryptographic key theft deterrent. Even so, the bulk of commercial wearables do not utilize such security and data breaches are still possible across myriad devices, opening the door for the unethical, if not illegal use of private data.

4.5 Architecting preference

A series of classic studies by Libet suggest that conscious free will is determined by a build-up of neural activity seconds before a conscious decision is made (Libet & Libet, 1993). Many have argued that these neural processes, which occur before, and often outside of our conscious awareness, steer our behavior in ways we cannot resolve. These neural processes are influenced by our environmental contexts. Contextual cues, such as smells, sounds, and changes in color impact our perceptions. Social contexts steer neural processes related to physiological arousal that precede conscious emotions. Branders could shape consumer contexts, and thereby shape the neural processes that precede an outcome behavior. Branders can learn consumers' preferences by understanding their past purchase behaviors, their online sentiments, their geospatial locations, and a trove of other personal data. In understanding these bits of information, branders can tailor marketing and advertising content to align with such preferences early in the consumer journey. This could have the effect of driving consumption behavior, or choice, in a direction that favors the brand. Having this understanding of the neuroscience behind the timing of consciousness and volition magnifies the vulnerability of consumers as they navigate a world filled with branded content. The consumer journey no longer begins at a store entry or the homepage of an online retailer but is ever present in subtle messaging and social listening. Brands and brand strategists may be removing more of the agency in consumers' free will in a way that stretches beyond what is ethically appropriate.

5. Conclusions

Philosopher John Bargh and Psychologist Roy Baumeister have famously debated about whether individuals have free will (Baumeister & Bargh, 2014). They have come to agree that most human behaviors manifest from an integration of conscious and unconscious processes that work together to meet the critical needs of an individual and facilitate the pursuit of meaningful goals. Though they do not speak directly to neuromarketing, their considerations have direct implications on consumer choice. This is especially of concern when it comes to the efficacy of advertising campaigns and marketing initiatives. Conscious processing is often required to limit impulsive decisions, and this type of processing is effortful. Consumers can actively deliberate on every consumption decision made during each waking moment, but that is highly improbable given the cognitive resources that would require. Automatic decisions are made every moment for cognitive efficiency, and these types of decisions are predominantly driven by emotion. Neuromarketers have an unprecedented ability to quantify and categorize affective responses toward test stimuli which can then be connected to predictions of consumption behaviors at a market level. This ability calls for greater responsibility in the use of implicit brain and body measures. On the client side, what is learned through the use of these novel tools must be treated in a way that assures that any principles and tactics developed will not affect consumers unethically. The NMSBA's Code of Conduct outlines issues of consent, ethical treatment of participants, and privacy rights that should be the basis of discussion and consideration for each study that utilizes measures of the brain and body to explain and to predict consumption behaviors.

While neuromarketing practitioners do not have the power to make consumers purchase specific goods, from a pracademic's point of view, there are several important ethical considerations related to the commercial use of passive and active measures of the brain and body associated with emotional response. Privacy rights, disclosure of information collection and dissemination, the ability to extrapolate information about individuals through a triangulation of data streams, the unintended disclosure of personality predispositions and the implications of false positive classifications on human lives are some of my primary concerns based on my academic and client-side experience. As the applied commercial field of neuromarketing and the academic discipline of consumer neuroscience continue to develop in sophistication, policy makers will continue to understand the power of these measures and increase discussions on the types of regulation that should occur. Different countries have different tolerance for this type of data to be used commercially. In general, Europe's GDPR contains assurances for the protection of personal information, but it is not yet a standard across all countries. For example, French law prohibits use of fMRI for commercial purposes, but the United Kingdom does not.

The United States does not regulate commercial use of fMRI or other brain and body measures, though this is changing with respect to facial recognition technology (Langleben & Moriarty, 2013). Currently, most policy focuses on limiting use of facial recognition technology, but this is at the state and not the federal level. At the time of writing, three states have banned use of facial recognition technology, though this is difficult to regulate in the smart phone era (Stegall, 2019). With the existence of firms such as Clearview, a New York based company that claims to have a massive database of face photos and has an ability to use it to identify novel faces, privacy activists are speaking out against the artificial intelligence company (Hill, 2020).

The power to influence consumers' perceptions, emotions and consumption behaviors should be met with robust knowledge of potential ethical violations. Misuse of the technologies, the sensitive data measured through these devices, and the inferences that can be made all have the real potential to harm citizens without their awareness of any manipulation. While neuromarketing does not carry with it an ability to overtly control minds, human behaviors can be influenced outside of conscious control in certain contexts with specific cues.

William Gibson wrote "We have no idea, now, of who or what the inhabitants of our future might be. In that sense, we have no future... We have no future because our present is too volatile. We have only risk management. The spinning of the given moment's scenarios. Pattern recognition" (Gibson, 2003, p. 40). We cannot foresee all the ways in which applied commercial neuroscience will impact our society moving forward. Therefore, it is critical to identify current patterns of both ethical use and misuse of these measures to best guide future applications and the policies that help to regulate their use.

References

Adler, M. J. (2009). *Detecting deceptive responses in sex offenders: A comparison of layered voice analysis (LVA) and the polygraph.* Working paper. Limestone, TN: Counseling and Consultation Services, Inc.

Al Pop, N., Dabija, D.-C., & Iorga, A. M. (2014). Ethical responsibility of neuromarketing companies in harnessing the market research—A global exploratory approach. *Amfiteatru Economic, 16*(35), 26.

Albrecht, J. P. (2016). How the GDPR will change the world. Eur. Data Prot. L. Rev., 2, 287.n of the GDPR. In *The EU General Data Protection Regulation (GDPR)* (pp. 9–30). Cham: Springer.

Alba, D. (2019). *Lockport public schools will be the first in the country to use facial recognition on June 3. May.* https://www.buzzfeednews.com/article/daveyalba/lockport-schools-facial-recognition-pilot-aegis.

Ariely, D., & Berns, G. S. (2010). Neuromarketing: The hope and hype of neuroimaging in business. *Nature Reviews Neuroscience, 11*(4), 284–292.

Atalay, A. S., Bodur, H. O., & Rasolofoarison, D. (2012). Shining in the center: Central gaze cascade effect on product choice. *Journal of Consumer Research, 39*(4), 848–866.

Baars, B. J. (2002). The conscious access hypothesis: Origins and recent evidence. *Trends in Cognitive Sciences, 6*(1), 47–52.

Bailenson, J. N., Iyengar, S., Yee, N., & Collins, N. A. (2008). Facial similarity between voters and candidates causes influence. *Public Opinion Quarterly, 72*(5), 935–961.

Barraza, J. A., Alexander, V., Beavin, L. E., Terris, E. T., & Zak, P. J. (2015). The heart of the story: Peripheral physiology during narrative exposure predicts charitable giving. *Biological Psychology, 105*, 138–143.

Barrett, L. F. (1997). Descriptions, and retrospective ratings of emotion. *Personality & Social Psychology Bulletin, 23*(i10), 1100.

Barrett, L. F. (2006). Are emotions natural kinds? *Perspectives on Psychological Science, 1*(1), 28–58.

Barrett, L. F. (2017). *How emotions are made: The secret life of the brain.* Houghton Mifflin Harcourt.
Bartlett, M. S., Littlewort-Ford, G., Movellan, J., Fasel, I., & Frank, M. (2016). *U.S. Patent No. 9,530,048.* Washington, DC: U.S. Patent and Trademark Office.
Baumeister, R., & Bargh, J. (2014). Conscious and unconscious. Toward an integrative understanding of human mental life and action. In J. W. Sherman, B. Gawronski, & Y. Trope (Eds.), *Dual-process theories of the social mind* (pp. 35–49). The Guilford Press.
Baumeister, R. F., Bratslavsky, E., Muraven, M., & Tice, D. M. (2018). Ego depletion: Is the active self a limited resource? In *Self-regulation and self-control* (pp. 16–44). Routledge.
Baumeister, R. F., Vohs, K. D., & Tice, D. M. (2007). The strength model of self-control. *Current Directions in Psychological Science, 16*(6), 351–355.
Berka, C., Levendowski, D. J., Lumicao, M. N., Yau, A., Davis, G., Zivkovic, V. T., et al. (2007). EEG correlates of task engagement and mental workload in vigilance, learning, and memory tasks. *Aviation, Space, and Environmental Medicine, 78*(5), B231–B244.
Botswick, W. (2010, January 19). *Waiter, there's pseudoscience in my soup: Campbell's chooses "neuromarketing" over consumer feedback in rebranding it's iconic soup cans.* Fast Company. https://www.fastcompany.com/90184518/waiter-theres-pseudo-science-in-my-soup.
Brown, D. (2019). Facebook cracks down on personality quizzes to protect your personal info. *USA Today* https://www.usatoday.com/story/tech/2019/04/26/facebook-bans-personality-quizzes-could-access-your-personal-info/3585286002/.
Burns, A., Doheny, E. P., Greene, B. R., Foran, T., Leahy, D., O'Donovan, K., et al. (2010). SHIMMER™: An extensible platform for physiological signal capture. In *2010 annual international conference of the IEEE engineering in medicine and biology* (pp. 3759–3762), IEEE. August.
Çakar, T., & Gez, K. (2017). Neuroscience applications on the assessments of TV ads. In *Applications of neuroscience: Breakthroughs in research and practice* (pp. 231–256). IGI Global.
Christoforou, C., Papadopoulos, T. C., Constantinidou, F., & Theodorou, M. (2017). Your brain on the movies: A computational approach for predicting box-office performance from viewer's brain responses to movie trailers. *Frontiers in Neuroinformatics, 11,* 72.
Clark, K. R. (2017). The future of consumer neuroscience. In P. Kotler (Ed.), *Consumer neuroscience* (p. 321). MIT Press.
Clark, K. R., Leslie, K. R., Garcia-Garcia, M., & Tullman, M. L. (2018). How advertisers can keep Mobile users engaged and reduce video-ad blocking: Best practices for video-ad placement and delivery based on consumer neuroscience measures. *Journal of Advertising Research, 58*(3), 311–325.
Clark, K. R., Leslie, K. L., Tullman, M. L., Eglington, L., & Wolford, G. L. (2016). Using consumer neuroscience to understand online shopping: Predicting impulse-driven purchase decisions across e-channel and platform through integrated self-report and non-verbal data streams, *Vol. 1.* Neuromarketing Science and Business Association (pp. 18–21).
CPG, FMCG & RETAIL. (2018, December 3). Booming snack sales highlight a growth opportunity in emerging markets [research consultancy]. https://www.nielsen.com/eu/en/insights/article/2018/booming-snack-sales-highlight-a-growth-opportunity-in-emerging-markets
Dailey, M. N., Joyce, C., Lyons, M. J., Kamachi, M., Ishi, H., Gyoba, J., et al. (2010). Evidence and a computational explanation of cultural differences in facial expression recognition. *Emotion, 10*(6), 874.
Das, A. K., Pathak, P. H., Chuah, C.-N., & Mohapatra, P. (2016a). Characterization of wireless multidevice users. *ACM Transactions on Internet Technology (TOIT), 16*(4), 29.

Das, A. K., Pathak, P. H., Chuah, C.-N., & Mohapatra, P. (2016b). Uncovering privacy leakage in BLE network traffic of wearable fitness trackers. In *Proceedings of the 17th international workshop on mobile computing systems and applications* (pp. 99–104).

Dmochowski, J. P., Sajda, P., Dias, J., & Parra, L. C. (2012). Correlated components of ongoing EEG point to emotionally laden attention—A possible marker of engagement? *Frontiers in Human Neuroscience*, *6*, 112.

Doremus-Fitzwater, T. L., Varlinskaya, E. I., & Spear, L. P. (2010). Motivational systems in adolescence: Possible implications for age differences in substance abuse and other risk-taking behaviors. *Brain and Cognition*, *72*(1), 114–123. https://doi.org/10.1016/j.bandc.2009.08.008.

Egelund, N. (1982). Spectral analysis of heart rate variability as an indicator of driver fatigue. *Ergonomics*, *25*(7), 663–672.

Ekman, P. (1989). The argument and evidence about universals in facial expressions. In H. Wagner & A. Manstead (Eds.), *Handbook of social psychophysiology* (pp. 143–164). John Wiley & Sons.

Ekman, P. (1992). An argument for basic emotions. *Cognition & Emotion*, *6*(3–4), 169–200.

Ekman, P., & Friesen, W. V. (2003). *Unmasking the face: A guide to recognizing emotions from facial clues*. Ishk.

Farah, M. J. (2012). Neuroethics: The ethical, legal, and societal impact of neuroscience. *Annual Review of Psychology*, *63*, 571–591.

Flick, C. (2016). Informed consent and the Facebook emotional manipulation study. *Research Ethics*, *12*(1), 14–28.

Flournoy, J. C., Pfeifer, J. H., Moore, W. E., Tackman, A. M., Masten, C. L., Mazziotta, J. C., et al. (2016). Neural reactivity to emotional faces may mediate the relationship between childhood empathy and adolescent prosocial behavior. *Child Development*, *87*(6), 1691–1702.

Gibson, W. (2003). *Pattern recognition*. Retrieved August 30, 2019, from https://www.goodreads.com/work/best_book/2455062-pattern-recognition.

Gobbini, M. I., Gors, J. D., Halchenko, Y. O., Rogers, C., Guntupalli, J. S., Hughes, H., et al. (2013). Prioritized detection of personally familiar faces. *PLoS One*, *8*(6), e66620.

Goldkorn, J. (2017). *Drug users nabbed by facial recognition system at beer festival—China's latest top news*. SupChina. September 1, https://supchina.com/2017/09/01/drug-users-nabbed-facial-recognition-system-beer-festival-chinas-latest-top-news/.

Groeppel-Klein, A. (2005). Arousal and consumer in-store behavior. *Brain Research Bulletin*, *67*(5), 428–437.

Gunter, B. (2000). *Media research methods: Measuring audiences, reactions and impact*. Sage.

Hammond, G. R. (1981). Finer temporal acuity for stimuli applied to the preferred hand. *Neuropsychologia*, *19*(2), 325–329.

Harmon-Jones, E., Gable, P. A., & Peterson, C. K. (2010). The role of asymmetric frontal cortical activity in emotion-related phenomena: A review and update. *Biological Psychology*, *84*(3), 451–462.

Harnsberger, J. D., Hollien, H., Martin, C. A., & Hollien, K. A. (2009). Stress and deception in speech: Evaluating layered voice analysis. *Journal of Forensic Sciences*, *54*(3), 642–650.

Hedgcock, W., & Rao, A. R. (2009). Trade-off aversion as an explanation for the attraction effect: A functional magnetic resonance imaging study. *Journal of Marketing Research*, *46*(1), 1–13.

Hedgcock, W. M., Vohs, K. D., & Rao, A. R. (2012). Reducing self-control depletion effects through enhanced sensitivity to implementation: Evidence from fMRI and behavioral studies. *Journal of Consumer Psychology*, *22*(4), 486–495.

Hill, K. (2020, January 18). The secretive company that might end privacy as we know it. *The New York Times* https://www.nytimes.com/2020/01/18/technology/clearview-privacy-facial-recognition.html.

Hoppe, S., Loetscher, T., Morey, S. A., & Bulling, A. (2018). Eye movements during everyday behavior predict personality traits. *Frontiers in Human Neuroscience, 12*, 105. https://doi.org/10.3389/fnhum.2018.00105.

Horvath, F., McCloughan, J., Weatherman, D., & Slowik, S. (2013). The accuracy of auditors' and layered voice analysis (LVA) operators' judgments of truth and deception during police questioning. *Journal of Forensic Sciences, 58*(2), 385–392.

Huang, G. B., & Learned-Miller, E. (2014). *Labeled faces in the wild: Updates and new reporting procedures*. Amherst, MA: Department of Computer Science, University Massachusetts Amherst. Technical Report, 14-003.

Hubble, R. P., Naughton, G. A., Silburn, P. A., & Cole, M. H. (2015). Wearable sensor use for assessing standing balance and walking stability in people with Parkinson's disease: A systematic review. *PLoS One, 10*(4), e0123705.

Huber, J., Payne, J. W., & Puto, C. (1982). Adding asymmetrically dominated alternatives: Violations of regularity and the similarity hypothesis. *Journal of Consumer Research, 9*(1), 90–98. https://doi.org/10.1086/208899.

Iorga, A., Hsu, M., Shapira-Lichter, I., Nieuwenhuis, I. L., Marci, C., Murray, B., et al. (2017). *Consumer neuroscience*. MIT Press.

Jelenchick, L. A., Eickhoff, J. C., & Moreno, M. A. (2013). "Facebook depression?" Social networking site use and depression in older adolescents. *Journal of Adolescent Health, 52*(1), 128–130.

Johnson, K. (2019). *Amazon's Alexa may soon know if you're happy or sad*. VentureBeat. July 8, https://venturebeat.com/2019/07/08/amazons-alexa-may-soon-know-if-youre-happy-or-sad/.

Kamienkowski, J. E., Ison, M. J., Quiroga, R. Q., & Sigman, M. (2012). Fixation-related potentials in visual search: A combined EEG and eye tracking study. *Journal of Vision, 12*(7), 4.

Karmarkar, U. R., Shiv, B., & Knutson, B. (2015). Cost conscious? The neural and behavioral impact of price primacy on decision making. *Journal of Marketing Research, 52*(4), 467–481.

Karmarkar, U., Yoon, C., & Plassmann, H. (2015). Marketers should pay attention to fMRI. *Harvard Business Review*.

Kelly, H. (2018, August 23). *Facebook has suspended over 400 apps after Cambridge analytica scandal*. CNNMoney. https://money.cnn.com/2018/08/23/technology/facebook-mypersonality-cambridge-analytica/index.html.

Kerr, D., Butler-Henderson, K., & Sahama, T. (2018). Security, privacy, and ownership issues with the use of wearable health technologies. In IGI Global (Ed.), *Wearable technologies: Concepts, methodologies, tools, and applications* (pp. 1068–1083). IGI Global. https://doi.org/10.4018/978-1-5225-5484-4.ch048.

Kilbride, J. E., & Yarczower, M. (1983). Ethnic bias in the recognition of facial expressions. *Journal of Nonverbal Behavior, 8*(1), 27–41.

Knutson, B., Fong, G. W., Adams, C. M., Varner, J. L., & Hommer, D. (2001). Dissociation of reward anticipation and outcome with event-related fMRI. *Neuroreport, 12*(17), 3683–3687.

Knutson, B., Fong, G. W., Bennett, S. M., Adams, C. M., & Hommer, D. (2003). A region of mesial prefrontal cortex tracks monetarily rewarding outcomes: Characterization with rapid event-related fMRI. *NeuroImage, 18*(2), 263–272.

Knutson, B., Wimmer, G. E., Kuhnen, C. M., & Winkielman, P. (2008). Nucleus accumbens activation mediates the influence of reward cues on financial risk taking. *Neuroreport, 19*(5), 509–513.

Kregor, J., Gomez, B., Kelly, J. S., & Stevenson, K. (2018). Grassroots market research on grass: Predicting Cannabis brand performance using social media scraping. In *2nd international conference on advanced research methods and analytics (CARMA 2018)* (pp. 201–208).

Langleben, D. D., & Moriarty, J. C. (2013). Using brain imaging for lie detection: Where science, law and research policy collide. *Psychology, Public Policy, and Law: An Official Law Review of the University of Arizona College of Law and the University of Miami School of Law, 19*(2), 222–234. https://doi.org/10.1037/a0028841.

Learned-Miller, E., Huang, G. B., RoyChowdhury, A., Li, H., & Hua, G. (2016). Labeled faces in the wild: A survey. In *Advances in face detection and facial image analysis* (pp. 189–248). Cham: Springer.

Li, S., & Deng, W. (2019). A deeper look at facial expression dataset bias. *arXiv*. 1904.11150.

Libet, B., & Libet, B. (1993). *Neurophysiology of consciousness*. Springer.

Lin, L. Y., Sidani, J. E., Shensa, A., Radovic, A., Miller, E., Colditz, J. B., et al. (2016). Association between social media use and depression among US young adults. *Depression and Anxiety, 33*(4), 323–331.

Linden, A., & Fenn, J. (2003). *Understanding Gartner's hype cycles*. Strategic analysis report N R-20-1971. Gartner, Inc..

Loewenstein, D. A., Argüelles, T., Argüelles, S., & Linn-Fuentes, P. (1994). Potential cultural bias in the neuropsychological assessment of the older adult. *Journal of Clinical and Experimental Neuropsychology, 16*(4), 623–629.

Ly, V., Wang, S., Bhanji, J. P., & Delgado, M. R. (2019). A reward-based framework of perceived control. *Frontiers in Neuroscience, 13*, 65.

Matukin, M. (2015). *Neuro against smoking: The full report*. http://www.nmsba.com/news/3303317.

McBride, W. J., Buikstra, E., & FitzGerald, M. (2010). Investigation of febrile passengers detected by infrared thermal scanning at an international airport. *Australian and New Zealand Journal of Public Health, 34*(1), 5–10.

McCarthy, G., Puce, A., Gore, J. C., & Allison, T. (1997). Face-specific processing in the human fusiform gyrus. *Journal of Cognitive Neuroscience, 9*(5), 605–610.

Merla, A., & Romani, G. L. (2007). Thermal signatures of emotional arousal: A functional infrared imaging study. In *2007 29th annual international conference of the IEEE engineering in medicine and biology society* (pp. 247–249).

Movellan, J., Bartlett, M. S., Fasel, I., Littlewort, G. F., Susskind, J., & Whitehill, J. (2014). *U.S. Patent Application No. 14/180,352*.

Ng, E. Y.-K. (2005). Is thermal scanner losing its bite in mass screening of fever due to SARS? *Medical Physics, 32*(1), 93–97.

Ng, D. (2019). *From dispensing toilet paper to shaming jaywalkers, China powers up on facial recognition*. CNA. https://www.channelnewsasia.com/news/cnainsider/shaming-jaywalkers-china-facial-recognition-technology-privacy-11196684.

NMSBA Code of Ethics—NMSBA. (2019). *Buying-neuromarketing/code-of-ethics*. https://www.nmsba.com/.

O'Hanlon, J. F. (1972). *Heart rate variability: A new index of driver alertness/fatigue*. SAE technical paper.

Ohme, R. (2008). TV commercial second by second: applying EEG and EMG to track down consumers' emotions and attention. In *American academy of advertising. Conference. Proceedings (Online)* (p. 31).

Ohme, R., Reykowska, D., Wiener, D., & Choromanska, A. (2009). Analysis of neurophysiological reactions to advertising stimuli by means of EEG and galvanic skin response measures. *Journal of Neuroscience, Psychology, and Economics, 2*(1), 21.

Piwek, L., Ellis, D. A., Andrews, S., & Joinson, A. (2016). The rise of consumer health wearables: Promises and barriers. *PLoS Medicine, 13*(2), e1001953.

Plassmann, H., Ramsøy, T. Z., & Milosavljevic, M. (2012). Branding the brain: A critical review and outlook. *Journal of Consumer Psychology, 22*(1), 18–36.

Plassmann, H., Venkatraman, V., Huettel, S., & Yoon, C. (2015). Consumer neuroscience: Applications, challenges, and possible solutions. *Journal of Marketing Research, 52*(4), 427–435.

Ramon, M., & Gobbini, M. I. (2018). Familiarity matters: A review on prioritized processing of personally familiar faces. *Visual Cognition, 26*(3), 179–195.
Roskies, A. (2002). Neuroethics for the new millenium. *Neuron, 35*(1), 21–23.
Schwartz, O. (2019, March 6). Don't look now: Why you should be worried about machines reading your emotions. *The Guardian,* https://www.theguardian.com/technology/2019/mar/06/facial-recognition-software-emotional-science.
Shiv, B., & Fedorikhin, A. (1999). Heart and mind in conflict: The interplay of affect and cognition in consumer decision making. *Journal of Consumer Research, 26*(3), 278–292.
Stegall, K. (2019). 3-year ban on police use of facial recognition technology in California to start in the new year. (2019a, December 20). *San Diego Union-Tribune,* https://www.sandiegouniontribune.com/news/public-safety/story/2019-12-20/3-year-ban-on-police-use-of-facial-recognition-technology-in-california-to-start-in-the-new-year.
Stikic, M., Johnson, R. R., Tan, V., & Berka, C. (2014). EEG-based classification of positive and negative affective states. *Brain-Computer Interfaces, 1*(2), 99–112.
Stilley, K. M. (2010). *The role of in-store slack and mental budgets in shopper marketing.* University of Pittsburgh ETD. February 1. http://d-scholarship.pitt.edu/10330/.
Tahir, H., Tahir, R., & McDonald-Maier, K. (2018). On the security of consumer wearable devices in the Internet of Things. *PLoS One, 13*(4), e0195487.
Tahir, R., Tahir, H., Sajjad, A., & McDonald-Maier, K. (2017). A secure cloud framework for ICMetric based IoT health devices. In *Proceedings of the second international conference on internet of things, data and cloud computing* (p. 171).
Tana, J., Forss, M., & Hellstén, T. (2017). *The use of wearables in healthcare—Challenges and opportunities.* https://www.theseus.fi/handle/10024/140584.
Tankard, C. (2016). What the GDPR means for businesses. *Network Security, 2016*(6), 5–8.
Taylor, M. J., Arsalidou, M., Bayless, S. J., Morris, D., Evans, J. W., & Barbeau, E. J. (2009). Neural correlates of personally familiar faces: Parents, partner and own faces. *Human Brain Mapping, 30*(7), 2008–2020.
Thayer, J. F., & Sternberg, E. (2006). Beyond heart rate variability. *Annals of the New York Academy of Sciences, 1088*(1), 361–372.
The Neurological Effects of South Park. (2016). *Viacom fan theory.* http://fantheory.viacom.com/inside-the-brain/index.html.
Thomas, A. R., Pop, N. A., Iorga, A. M., & Ducu, C. (2016). *Ethics and neuromarketing: Implications for market research and business practice.* Springer.
Torous, J., Larsen, M. E., Depp, C., Cosco, T. D., Barnett, I., Nock, M. K., et al. (2018). Smartphones, sensors, and machine learning to advance real-time prediction and interventions for suicide prevention: A review of current progress and next steps. *Current Psychiatry Reports, 20*(7), 51.
Tyagi, V., Hanoch, Y., Choma, B., & Denham, S. L. (2018). The 'right' side of creativity: Creative personality and social risk-taking predict political party affiliation. *Creativity Research Journal, 30*(4), 451–460.
Vahabzadeh, A., Sahin, N., & Kalali, A. (2016). Digital suicide prevention: Can technology become a game-changer? *Innovations in Clinical Neuroscience, 13*(5–6), 16.
van den Broek-Altenburg, E. M., & Atherly, A. J. (2019). Using social media to identify consumers' sentiments towards attributes of health insurance during enrollment season. *Applied Sciences, 9*(10), 2035.
Voigt, P., & von dem Bussche, A. (2017). Scope of application of the GDPR. In *The EU General Data Protection Regulation (GDPR)* (pp. 9–30). Cham: Springer.
Walla, P., Brenner, G., & Koller, M. (2011). Objective measures of emotion related to brand attitude: A new way to quantify emotion-related aspects relevant to marketing. *PLoS One, 6*(11), e26782.
Webster, S. W. (2018). It's personal: The big five personality traits and negative partisan affect in polarized US politics. *American Behavioral Scientist, 62*(1), 127–145.

Wickel, T. M. (2015). Narcissism and social networking sites: The act of taking selfies. *Elon Journal of Undergraduate Research in Communications*, *6*(1), Retrieved from http://www.inquiriesjournal.com/a?id=1138.

Williams, J. (2010, February 22). *Campbell's soup Neuromarketing redux: There's chunks of real science in that recipe.* Fast Company. https://www.fastcompany.com/1558477/campbells-soup-neuromarketing-redux-theres-chunks-real-science-recipe.

Zuo, X.-N., Biswal, B. B., & Poldrack, R. A. (2019). Reliability and reproducibility in functional Connectomics. *Frontiers in Neuroscience*, *13*, 117.

CHAPTER THREE

Trusting the bot: Addressing the ethical challenges of consumer digital mental health therapy

Nicole Martinez-Martin*
Stanford Center for Biomedical Ethics, Stanford University, Stanford, CA, United States
*Corresponding author: e-mail address: ms.nicoleangela@gmail.com

Contents

1. Introduction

Digital technology, including apps, wearables, machine learning, and chatbots, is expected to revolutionize mental health care, by providing more effective diagnosis and treatment, as well as expanding access to mental health care (Mohr, Riper, & Schueller, 2017). Smartphones and other mobile technology can be used to reach users, wherever they are located, deliver services and, importantly, collect data that can be used to generate behavioral insights (East & Havard, 2015mHealthIntelligence, 2017; Morganstein, 2016; Payne, Lister, West, & Bernhardt, 2015; Reynolds, 2017). For that reason, mental health apps, in particular, have been a significant focus of technology

Developments in Neuroethics and Bioethics, Volume 3
ISSN 2589-2959
https://doi.org/10.1016/bs.dnb.2020.03.003

investment and have rapidly proliferated in recent years (Brodwin, 2019; Carlo, Ghomi, Renn, & Areán, 2019; Neary & Schueller, 2018; Research2guidance, 2017). Considering that only 41% of people in the United States with mental illness receive treatment each year, due in part to the need for more resources and therapists, consumer mental health technology has potential to play an important role in providing mental health services (Hedden, Kennet, Lipari, Medley, & Tice, 2015). Furthermore, gathering behavioral data through smartphones and wearables can lead to new understandings regarding the development and progression of mental illness that improve diagnostics as well as treatment (Areàn, Hoa Ly, & Andersson, 2016; Price et al., 2014). The use of mobile mental health could also encourage access to mental health services for those who normally would avoid traditional modes of service or are physically distant from support services (Nebeker, Murray, Holub, Haughton, & Arredondo, 2017; Vis et al., 2018).

The enthusiasm for this digital revolution in mental health has been tempered by concern over ethical issues involved, with privacy and the difficulties in regulating apps for safety and effectiveness receiving particular attention (Capon, Hall, Fry, & Carter, 2016; Kobayashi, 2015; Nebeker, Harlow, et al., 2017). The ethical issues raised by digital mental health technology broadly overlap with some other technologies, such as genetic testing, in terms of privacy and difficulties in communicating the risks and repercussions of the technology for consent purposes (Shoenbill, Fost, Tachinardi, & Mendonca, 2014; Udesky, 2010). At the same time, digital mental health technology raises novel ethical issues. Digital tools and information span clinical and commercial domains and some of the lines of accountability and oversight are not yet clearly delineated (Martinez-Martin & Kreitmair, 2018; Nebeker, Torous, & Bartlett Ellis, 2019). The sheer number of digital mental health devices and apps, as well as technological features have presented significant challenges for oversight of safety and efficacy (Berg, 2014; Martinez-Martin, 2019; Torous & Nebeker, 2017; Weisel et al., 2019). In particular, the use of artificial intelligence (AI) for development of some digital tools raises the potential for bias in the development and implementation of the algorithms and the impact on fiduciary relationships (Towards trustable machine learning, 2018).

Moreover, behavioral analytics can be applied to the data collected through apps and mobile technology to generate behavioral inferences, such as the likelihood of an episode of anxiety or risk of suicide (Kleiman et al., 2018; Saeb, Lattie, Schueller, Kording, & Mohr, 2016). Users of consumer mental health technologies are unlikely to realize the extent to which they are providing personal data to apps from which sensitive behavioral inferences can be drawn or the potential repercussions of sharing that data. For example, if someone is using a consumer app for mental health wellness, outside of the confidentiality or privacy provisions that would apply to an interaction with a professional therapist, the data could potentially be sold to third parties, analyzed for additional behavioral information, such as depression, and used toward marketing or insurance purposes in ways the consumer would not be expecting (Huckvale, Torous, & Larsen, 2019; Singer, 2018).

The ethical and regulatory frameworks that traditionally have been applied to mental health care services are generally rooted in expectations that these services take place between a human professional and client and/or through a healthcare provider. This chapter highlights key ethical challenges that arise because the use of consumer digital mental health technology entails shifts outside of those traditional paradigms. Trust, data protection, safety, oversight, and transparency are ethical challenges at the heart of integrating digital mental health technology into mental health care. When technology is inserted into mental health service—when part or even all of the therapeutic interaction is between person and device rather than person and person—what does trust in a device mean and how does that fit into the ethical obligations? As a recent Wired article on trust in technology asked, "How can you trust someone you'll never see? How can you trust an algorithm that is making thousands of decisions a second of which you aren't even aware? How can you trust a company that tracks your movement every day?" (Dobrygowski & Hoffman, 2019). The consumer domain presents particularly vexing issues for trust, because the frameworks for accountability and oversight, as well as the mechanisms for data protection and assuring safety and effectiveness, are still evolving (Hsin & Torous, 2018).

2. Consumer digital mental health

The landscape for consumer mental health apps includes self-guided therapy and education tools, platforms that can be used to connect the user to different talk therapy options, as well as tools that incorporate chatbots or other predictive analytic tools (Magee, Adut, Brazill, & Warnick, 2018). Despite the need for improvements in access and availability of mental health services, there has been slower than expected uptake of mental health apps (Ng, Firth, Minen, & Torous, 2019). Questions about the safety and effectiveness of the app, what will happen to the user's data and confidence that the app has a role to play in improving the person's health appear to have a significant role in people's hesitation to use consumer mental health apps (Meng, Hussain, Mohr, Czerwinski, & Zhang, 2018; Rendina & Mustanski, 2018; Schueller, Neary, O'Loughlin, & Adkins, 2018; Torous, Nicholas, Larsen, Firth, & Christensen, 2018) (Table 1).

While some mental health apps are designed for use in connection with therapy from a licensed professional, many are meant to offer mental health assistance outside of the scope of an interaction with a clinician or licensed professional (Huang & Bashir, 2017; Kao & Liebovitz, 2017).

Table 1 Types of digital mental health technology.
Types of digital mental health technology devices include:

Bridge to talk therapist (via text or voice)	These services may utilize people with a range of training for providing talk therapy, from people who are trained in specific approaches, such as cognitive behavioral therapy, to coaches who have minimal to no training in mental health issues or therapy (Jacobsen, 2014; Tozzi, 2019)
Interactive platform to provide therapy	Services include chatbot or conversational agent to guide the user (Nutt, 2017; Vaidyam, Wisniewski, Halamka, Kashavan, & Torous, 2019). An example of this approach is the Woebot, an automated conversational agent that provides tools for a user to monitor their mood and other symptoms (Fitzpatrick, Darcy, & Vierhile, 2017)
Informational	Digital platform that provides educational materials on mental health issues that can allow people to learn about symptoms, the disorder and/or describe methods of treatment (Christensen, 2010)
Goal setting	Provides tools to assist in achieving specific components of setting and reaching behavioral goals, such as tasks that need to be done that day (getting out of bed; taking medications), often utilizing gamification and reward approaches (receiving points or stars for achieving goals) (Cheng, Davenport, Johnson, Vella, & Hickie, 2019)
Symptom tracking	Provides tools to assist in self-monitoring of symptoms and may also involve people recording voice diaries or writing descriptions of their mood (Campbell, 2018)

Table 1 Types of digital mental health technology.—cont'd
Types of digital mental health technology devices include:

Behavioral analysis	Device collects a range of information from a user, including passively collected information such as screen taps or location, as well as other text or voice content, in order to analyze the information for cognitive and mood states of the individual. Digital phenotyping, for example, uses smartphones or mobile sensors to gather personal data from users, which is then analyzed in order to assess the user's cognitive and mental state, as well as make predictions (Corcoran & Cecchi, 2018; Insel, 2017; Martinez-Martin, Dunn, & Roberts, 2018; Martinez-Martin, Insel, Dagum, Greely, & Cho, 2018; Onnela & Rauch, 2016). In the consumer space, there has also been interest from companies such as Facebook in using these kinds of analytics drawn from user content to predict mood or even suicide risk (NPR, 2018)

There are different models of care into which mental health apps can be incorporated, in terms of how and if professional oversight is involved (Lipschitz et al., 2019; Malhotra, Chakrabarti, & Shah, 2019; Powell, Yue, Shan, & Torous, 2019). Some mental health apps are both designed and intentionally used as a part of care, such as mental health apps that have been created and utilized by the Department of Veteran's Affairs as part of treatment plans for veterans living in rural areas (Connolly et al., 2018). These apps can be beneficial in providing care in areas that have limited resources and access to psychiatrists, as well as for offering options for patients who may find interaction with a device preferable at times to an interaction with a physician (Owen et al., 2018). Mental health professionals are still determining best practices for integration of digital mental health, including which populations may benefit most. The use of mental health apps as part of psychiatric service has been found to provide effective care and patient satisfaction (Ben-Zeev et al., 2018), although the success depends on the type of apps and how they are integrated. When a consumer mental health app or device is used without some involvement of a mental health professional (Gravenhorst et al., 2015), the ethical concerns, such as data protection, accountability, safety and privacy, become more pressing because of the gaps in oversight and accountability that are discussed in more detail below.

2.1 Business models

The ethical challenges in consumer digital mental health can be exacerbated by the business models of the companies. Consumer data is valuable financially as well as for the variety of uses to which it can be applied (Botella, 2019). With the availability of databases of personal information, as well as advances in predictive analytics, consumer data can be translated into behavioral and health information that is useful for purposes such as targeted marketing, insurance, or public health (Glenn & Monteith, 2014; Singer, 2014). Mental health start-ups may use different models for generating income, such as subscription or ad-based services, but monetizing the data that is collected from their users is a common practice (Horn, 2019). Insurance companies are examples of common partnerships that mental health start-ups seek, which can mean implications for insurance rates (Allen, 2018). Particularly when it comes to direct-to-consumer apps, users may be unaware of the potential for their data to be sold, as well as the kinds of inferences that can be drawn from their data. Furthermore, the data gathered by mental health technology companies is of scientific value, with the potential to be used to improve diagnosis and assessment of mental illness, as well as providing data for purposes such as natural language analysis. The business landscape for data also presents impediments to sharing and benefiting from the collected data, as individual companies seek to protect their share of a valuable resource—an issue that has also been seen in the area of consumer genomics. Because of the value of personal data and potential repercussions for consumers, the business models employed by consumer digital mental health impact the ethical challenges for privacy and data protection.

3. Trust in consumer digital mental health

Trust is a complex and multidimensional concept, with dimensions that vary according to context (Gaebel, Muijen, Baumann, et al., 2014). Trust is invoked as an important issue in the domains of healthcare and technology (Chakravorti, 2018), but remains a challenging concept to define. In general terms, the concept of trust involves the willingness of a person in a position of vulnerability to take a risk despite uncertainty, placing confidence in the intentions and competence of another (Acquisti, Smith, & Sadeghi, 2010; Andras et al., 2018; Hall, Dugan, Zheng, & Mishra, 2001). In healthcare, trust generally includes perceptions of provider competence, confidentiality and privacy, dignity, safety, and absence of coercion or stigma (Goudge & Gilson, 2005). Trust is often described as fundamental to the therapeutic relationship (Birkhäuer et al., 2017; Langley & Klopper, 2005; Lee & Lin, 2008). Trust in healthcare is often grounded in the quality of the interpersonal relationship between patient and provider, as well as whether there is precedent to place trust in a professional or an institution (LoCurto & Berg, 2016). Trust in digital technology involves perceptions of privacy, transparency, safety, and perception of the competence of the vendor and technology (Asan, Perchonok, & Montague, 2012; O'Loughlin, Neary, Adkins, & Schueller, 2019; Ozawa & Sripad, 2013). Thus, the components of trust in technology and in healthcare providers overlap in terms of several key ethical obligations, particularly safety, competence/effectiveness, and privacy.

The ethical obligations of privacy, accountability, and competence are embedded in the trust relationship (Kraft et al., 2018). There are specific features of mental health apps and devices that could impact the trust relationship in digital mental health care, raising challenges for existing frameworks for addressing ethical obligations of privacy and confidentiality, as well as informed consent (Fisher & Appelbaum, 2017; Martinez-Martin, Dunn, et al., 2018; Martinez-Martin, Insel, et al., 2018; Mey et al., 2013; Torous & Roberts, 2017a). For example, the collection of large amounts of fine-grained personal information by some technology introduces a level of surveillance to a therapeutic interaction that could impact trust. When an app or device is inserted into or even replaces aspects of a mental health care interaction, it raises questions about what it means to place trust in a device or who, or what, exactly is party to the trust relationship. For devices, trustworthiness, the characteristics and actions that allow technology developers and digital mental health tools to be perceived as worthy of trust, is important (Gilson, 2003). While trust is a willingness to undertake risk, trustworthiness encompasses qualities, such as integrity and competence, that make it appropriate for someone to place their trust in another (Pavlou, 2003).

Accountability and oversight are important to establishing trustworthiness. If a mental health app is providing informational resources, one would need confidence in the quality and reliability of the information. With chatbots and conversational agents, it will be important to ensure they are constructed to interact in beneficial ways with individuals with specific mental health issues. More needs to be understood regarding how trust figures into consumer use of mental health apps and tools, including awareness of potential risks, whether people avoid use of some potentially effective apps because of general mistrust, or conversely, what might make people place trust in apps that are not safe or otherwise worthy of trust. Furthermore, people may have emotional attachments to their technology and devices (Vincent, 2006), such as smartphones, which is relevant to questions of how people may experience trust in relation to digital mental health care technology (Anderson & Agarwal, 2011). This is an area where additional empirical research can be helpful in providing clearer dimensions to the trust relationship in the new context of digital mental health technology.

Digital interventions are a part of a paradigm shift in healthcare, in line with precision medicine and consumer health that increasingly moves health information and responsibilities for health management outside of the primary control of clinicians (Bloss et al., 2016). Because behavioral health inferences can be drawn from data which users of mental health technology may not know is potentially sensitive, there is added need to assure that data is adequately protected and users are informed of these risks. The traditional dyadic physician-patient relationship has long been the construct through which the therapeutic relationship, as well as trust and associated ethical obligations, has been understood. The increasing use of digital mental health tools for mental health care will necessarily entail shifts in how these ethical obligations are delineated and engaged (Soni et al., 2018). In order to provide an environment that supports trust in digital mental health care, these issues of data protection, accountability and oversight, transparency, consent and algorithmic bias are thus particularly important to address.

4. Accountability and oversight

Oversight of consumer digital mental health technology contains a number of challenges. For direct-to-consumer technology, not overseen by a health care provider, there is the basic issue that the consumer does not have the benefit of guidance from a professional, and may not be aware that the protections for privacy or confidentiality that apply in a professional therapeutic interaction likely do not apply to the individual's usage of the direct-to-consumer application. The mechanisms validating safety and effectiveness of mental health apps are a critical area of concern (Torous et al., 2019). There are thousands of mental health apps available, but the mechanisms for directing consumers to the small percentage that are backed by scientific evidence (Chandrashekar, 2018; NHS Digital, 2014), and screening out apps with potentially harmful information of features, are limited. Larsen, Huckvale, Nicholas, et al. (2019) found that of 1435 apps for addressing depression, 38% advertised themselves as being effective for improving depression, but only 2.6% provided evidence to support their claims. In some cases, it has been found that the mental health app includes misleading or even wrong information about the disorder (Leigh & Flatt, 2015), such as suicide prevention tools that encouraged risky behavior (Larsen, Nicholas, & Christensen, 2016) or a bipolar app that advised drinking alcohol to address symptoms of mania (Nicholas, Larsen, Proudfoot, & Christensen, 2015). When mental health advice is incorrect, or there is ineffective diagnosis or treatment offered through an app, people can be directly harmed or be diverted from receiving effective care (Anthes, 2016). Direct-to-consumer mental health apps that are used without oversight from mental health professionals currently appear to be best suited for assisting with less serious mental health problems, such as anxiety or stress management (Hsin, Torous, & Roberts, 2016; Possemato, Kuhn, Johnson, et al., 2016). While many mental health wellness apps contain disclaimers that they should not be used to address a serious mental health crisis, it remains a challenge to identify best practices or suitable technological mechanisms for assuring that people with more severe crises who engage a direct-to-consumer mental health app are directed to more suitable services.

The American Psychiatric Association has issued guidance for psychiatrists on evaluating mental health apps (APA, 2019). The APA model sets forth the main areas for evaluation of mental health apps as: risks presented by use of the app, privacy and security, the evidence bases for the app, ease of use and interoperability (Torous, Chan, et al., 2018). These criteria are meant to be used by psychiatrists in recommending apps. For direct-to-consumer apps, however, consumers may mainly have the information and ratings at app stores, such as Google or Apple, that they use in selecting an app (Torous & Roberts, 2017b). On the consumer side, there are a few organizations such as PsyberGuide (PsyberGuide, 2019) which post reviews evaluating mental health apps, but consumer uptake of these kinds of guides is not clear and needs to be further studied.

In the United States, the FDA regulates the safety and effectiveness of computer software and digital health tools that meet the definition of "medical device" (Cortez, Cohen, & Kesselheim, 2014) including software that may be "intended for use in the diagnosis of disease or other conditions, or in the cure, mitigation, treatment, or prevention of disease." (FDA, 2019). Generally, the FDA uses its enforcement discretion to focus on digital health devices that it deems as high risk of causing harm. The FDA's record of enforcement indicates that it focuses on apps that may cause physical harm, rather than harm to mental health (Armontrout, Torous, Cohen, McNiel, & Binder, 2018). Products that promote "general wellness" rather than addressing specific diagnoses or health conditions are generally excluded from FDA regulation (Enriquez, 2016). For example, the FDA has identified apps that help management of stress or promotion of relaxation as "wellness" apps (FDA, 2019). Obtaining FDA approval is an expensive and time-consuming process and it is unsurprising that many consumer mental health apps present themselves as wellness apps. Some have argued that the FDA has an ethical obligation to do more to regulate wellness apps, in large part because of the potential harms that such apps still pose (Kasperbauer & Wright, 2019). Additionally, although some consumer mental health apps may not meet the FDA definition of medical device, they may still be reasonably perceived by consumers as offering a substitute for professional guidance, and thus necessitate some mechanism for assuring that they do not harm users (Duggal, Brindle, & Bagenal, 2018).

The FDA has faced challenges in determining its approach to regulating digital health technology, in part due to the sheer number of different digital tools that have emerged, as well as the difficulties in identifying and implementing appropriate evaluation standards for digital mental health technology (Price, 2017). For example, software functions based on a trained neural network, multivariate regressions, or fuzzy logic can make it difficult, if not impossible, for clinicians to review or evaluate the reasoning behind the recommendations because the calculations are trained on millions of data points (AMIA, 2018). Furthermore, algorithms are not static products, and their predictive performance may change over time as the algorithm is exposed to more data (Parikh, Obermeyer, & Navathe, 2019). The FDA has created new approaches for evaluating digital health devices, such as the Software Precertification Program, which certifies companies that make digital health technologies rather than the individual products (Shuren, Patel, & Gottlieb, 2018). The FDA's precertification process offers flexibility and streamlining of the approval process for Software as a Medical Device (SaMD), but there remain questions regarding evaluation and explainability of algorithms in health devices. A variety of stakeholders, including computing organizations and healthcare organizations, have advocated for improving the explainability of such algorithms (Holzinger, Biemann, Pattichis, & Kell, 2017), as well as improving the training of clinicians regarding the limitations and applications of such devices.

In addition to safety and effectiveness, accountability for digital mental health technology involves questions regarding liability for problems with the technology, and who, if anyone, may have the ethical duties and obligations expected as part of the trust relationship in mental health care (Yang & Silverman, 2014). In the U.S. mental health apps that connect people to professional therapists have been subject to regulations that require the therapist to be licensed in the same state as the client receiving care (Zur Institute, 2018). When the mental health technology provides care from non-professionals, not only is there an absence of professional oversight, but in some instances the services being provided may actually meet the definition of care that requires a license to provide (Barras, 2019).

For the companies and individuals involved in developing and implementing digital mental health interventions, there are questions and opportunities to incorporate ethical values into the design of apps. This can include early consideration of oversight issues such as what mechanisms will be available for reporting harms and identifying where accountability resides for mistakes or harms arising from use of digital mental health interventions. One approach is to encourage the participation of relevant stakeholders (such as consumers, patients, clinicians or mental health advocates) during the development of frameworks for accountability that will apply to professional or fiduciary obligations in using digital health tools (Cohen, Amarasingham, Shah, Xie, & Lo, 2014). For example, the Oxford Neuroscience, Ethics and Society Young People's Advisory Group held a series of meetings through which young people developed an ethical framework prioritizing privacy, confidentiality and efficacy for the use of conversational agents in therapy (Kretzschmar, Tyroll, Pavarini, Manzini, & Singh, 2019). Furthermore, as digital mental health applications produce increasingly powerful tools for assessing behavioral and mental health risk, there will be a need to consider more broadly how such risk assessments may be used and what obligations might arise when risk of harm is identified. There needs to be evaluation of whether and how ethical obligations that have become a part of professional practice in mental health care, like the "duty to warn," should be transferred to the domain of digital mental health devices.

5. Bias and fairness

AI is increasingly incorporated into digital mental health technology. AI refers to the use of machines to perform the kinds of tasks that resemble cognitive functions that we associate with human intelligence (Yu, Beam, & Kohane, 2018). Examples of AI applications for digital consumer mental health technology include prediction of cognitive and mood states, expert systems, chatbots and other conversational agents, analysis of texts and conversation for mental health content, and diagnosis (Luxton, 2016). A type of AI, machine learning (ML) is particularly useful for data mining, image recognition, and natural language processing among other applications (Bzdok & Meyer-Lindenberg, 2017).

ML uses a group of data inputs in order to produce a desired output by finding patterns in the data (Gavish, 2017). A mathematical model is built to map inputs to desired outputs. Then the model is trained with large amounts of data (known inputs and outputs) in order to identify patterns and create rules that adjust and improve the parameters of the model (Marcus, 2018; Ngiam & Khor, 2019). Subsequently, the resulting model is validated and tested on fresh data in order to determine the actual performance of the model (Pasquale, 2016). Most AI modifies its algorithms during this process (Gonfalonieri, 2019). More importantly, because of the large datasets involved, it can be difficult to pinpoint which data was weighted in which way that led to specific outputs. Thus currently, the specific reasoning underlying the outputs of ML are often not available for scrutiny, even to those who programmed the ML system (Knight, 2017; Mols, 2019). Furthermore, there are different stages at which ML systems can unintentionally produce bias, and the lack of explainability for the ML findings can create problems for addressing algorithmic bias (Challen et al., 2019; Gianfrancesco, Tamang, Yazdany, & Schmajuk, 2018; Glymour & Herington, 2019).

There are different types of algorithmic bias. First, there is the potential for bias in the training data. The accuracy of the performance of the resulting algorithm depends greatly on the quality of the training and validation datasets (Siau & Wang, 2018). One concern is that if the dataset does not reflect the necessary attributes of the population to which the algorithm is applied, then there can be bias in the outcomes generated by the algorithm. In the area of mental health, for example, the symptoms and trajectory of mental illness can be affected by social factors such as class or race. If the dataset used to construct the algorithm for the digital mental health application is not sufficiently inclusive of people of different race or socioeconomic background, there could be bias in the results. Furthermore, it can be difficult to evaluate how the algorithm takes such factors into consideration in the results. This issue has raised concerns that the findings of ML systems can reflect and reinforce biases in society. Additionally, ML systems may conflate simple statistics with more complex social interactions in the data (Lipton & Steinhardt, 2018). When algorithms used for a particular purpose or context are transferred to a new context, bias can result, such as when an algorithm used in an urban context is transferred to a rural context (Danks & London, 2017). Finally, algorithmic bias can stem from a lack of alignment between the type of information produced by a system and the type of information actually expected or required by the user. This can lead to misinterpretation of the algorithm that produces bias in the outcomes (Danks & London, 2017). Additionally, algorithms themselves may be designed to perform in unethical ways, such as algorithms that may be designed to identify behavioral risk for purposes, such as allocation of resources that would discriminate against specific populations (Mittelstadt, Allo, Taddeo, Wachter, & Floridi, 2016; Tramèr et al., 2015). Finally, individuals may be predisposed to automation bias in which the findings that come from an AI tool are perceived as more accurate or trustworthy than those coming from a human.

There are a number of solutions that are being advanced to address these types of bias. In relation to bias in the training data, efforts to increase the diversity of populations included in the datasets are key. Even when the device's algorithm has been validated for a particular use, there may be a need to do further evaluation for its use in specific contexts and populations, such as with elderly or adolescent users, in order to mitigate context transfer bias. Recently, there have been calls for the development of ML systems the results of which are explainable (Boire, 2018; Vellido, Martín-Guerrero, & Lisboa, 2012). At the same time, developers are often reluctant to share information about the workings of their ML systems for proprietary reasons (Dickson, 2017). There is a need for further empirical research in order to help determine effective ways to convey the limitations of AI systems to a consumer population. Individuals using mental health tools that draw on ML may need training or educational materials that explain issues like the nature of the data sets, and limitations, and potential areas of bias (Char, Shah, & Magnus, 2018).

6. Privacy

Concerns over privacy and data protection are a key factor impacting trust in consumer digital mental health therapy applications. Mental health care involves the sharing of sensitive information by a patient (Lambert & Barley, 2001). In mental health therapy, a patient must be able to trust that the sensitive and sometimes painful personal experiences they may share with a therapist will remain confidential and private (Miller & Thelen, 1986). In the United States, the Health Insurance and Portability Accountability Act (HIPAA) provides for the privacy and security of personal health information (HIPAA, 1996). This approach to protecting health information is rooted in a paradigm in which health information is primarily generated and used within a physician-patient interaction associated with a healthcare provider (Schairer, Rubanovich, & Bloss, 2018). However, in the current digital environment, and more specifically with digital mental health technology, behavioral health information may be generated, shared, stored and potentially sold to third parties in the consumer domain, outside of HIPAA or the healthcare institutions that traditionally protect health information (Glenn & Monteith, 2014; Lee, 2017; Rothstein, 2007). Huckvale et al. (2019) found that there were discrepancies between what some health apps stated in their privacy disclosures and their actual data practices, with some apps sending identifiable data to Facebook and Google even when they had not disclosed those practices. While many apps may have a primary goal of providing mental health support for users, sharing or selling user data for marketing or to other institutions can be lucrative, and is particularly problematic because the consumer may not understand or be able to foresee third-party uses of their data. Digital mental health apps that are provided through employer wellness programs may leave users vulnerable to invasion of privacy and employment discrimination (Ajunwa, 2017). It may be that people are less concerned about sharing some personal information, but when it comes to sensitive behavioral data, users may be much less inclined to have that information shared (O'Loughlin et al., 2019). For example, when Facebook was found to have allowed marketers to target anxious and depressed teenagers with ads (Reilly, 2017), and Google was revealed to offer a platform for referrals to substandard substance abuse centers, there was a public outcry against such practices (CBS, 2018). These incidents underline how digital behavioral data is sometimes sold in ways that can harm individual users.

Furthermore, when it comes to consumer digital mental health applications, users may not be informed or aware of the various ways the platform collects and analyses their data. Apps can potentially utilize computational behavioral analyses and machine learning to then analyze user information (Dagum, 2018; Torous, Onnela, & Keshavan, 2017). In turn, companies may be able to draw behavioral inferences from that data. For example, location data or changes in voice can be used to predict risk of depression or psychotic episodes (Bedi, Carrillo, Cecchi, et al., 2015). These types of practices can make it difficult for users to foresee the potential repercussions of their data being shared. The predictive features that can enhance mental health diagnostics and treatment also can lead to negative repercussions for users if they are shared or sold to third parties.

In Europe, the General Data Protection Regulation offers increased protections for consumers' data, such as requirements to inform and obtain consent from individuals about the specific data uses, providing information in clear terms, and not conditioning access to sites on consent to data use (GDPR, 2018). In the United States, California has passed similar legislation (CCPA, 2018). These steps are useful, although there is concern that these protections do not go far enough to protect individuals from the inferences that can be drawn from their data (Rocher, Hendrickx, & de Montjoye, 2019; Wachter & Mittelstadt, 2019). At the same time, efforts to regulate data practices and privacy in digital health technology need to be formulated to balance privacy with practices that allow data sharing for scientific research. Efforts to address privacy and data sharing practices for digital mental health information also may need to grapple with the practices of companies in the consumer domain that are not strictly concerned with providing digital mental health but analyze individual behavior, including predictive analyses that incorporate mental health behaviors and may sell data or reports to additional companies, such as for insurance (Bauer et al., 2017). Google, Facebook and online gaming companies have amassed enormous amounts of behavioral data, raising issues regarding potential repercussions for users, as well as for scientific research in behavioral health. One potential solution that has been discussed is the creation of a "data trust" approach to data collection that offers individuals greater control over their information, sets terms for companies to ethically use that data, and over which an independent agency could serve as a "data steward." (Lakshmanan, 2019).

One of the underlying problems with current approaches to privacy is that Big Data practices have made it increasingly hard to definitively de-identify data (Dankar, Ptitsyn, & Dankar, 2018). Thus, some have suggested that there is a need to increase emphasis on regulation that can protect individuals from misuse or harmful practices arising from their data, in line with previous efforts such as the Genetic Information Nondiscrimination Act to protect individuals from discriminatory practices in areas like employment or insurance (Alpert, 2003). Data protection is a key issue for forming trust in consumer digital mental health therapy applications, in order to allow people to access the potential benefits from these devices.

6.1 Surveillance

Surveillance—the use of digital mental health for government surveillance, as well as the discomfort that users and patients may feel at having health providers or a therapeutic device constantly monitor them—is a subset of privacy concerns (Martinez-Martin & Char, 2018). A range of organizations and government institutions potentially have an interest in monitoring individual behavior and conducting predictive analysis of individual's mental state, from criminal justice to education and onward (Azofeifa, 2018; Wicken, Nevin, & Ritchie, 2015). For example, there have been proposals in the United States for monitoring data from consumer wearables and apps to identify individuals for their potential to conduct workplace violence and mass shootings (Nixon, 2018; Rahim, 2019). People who have been diagnosed with mental illness may become particular targets of this type of monitoring. While such a program has not been implemented, there are government agencies, such as criminal justice or national security, that have an interest in monitoring the population for violent behavior and could potential apply digital phenotyping and related smartphone technologies to monitor people with mental illness. There are serious questions regarding the ethics of targeting specific populations for surveillance and which institutions, if any, should have access to personal data for behavioral surveillance purposes (Friedland, 2015). Conducting these kinds of surveillance activities can undermine the trust that people have in mental health technologies, the use of their data, and the healthcare system in general, and should therefore be addressed through public discussion and engagement of relevant stakeholders.

7. Transparency

In digital health discussions, transparency and trust are terms that are often linked. Clear and open communication regarding the nature of the technology, its uses and implications are seen as necessary, so that users can be appropriately informed of relevant benefits and burdens, and so that duties and obligations between the developers, institutions, clinicians, and individuals are set forth and understood. At the same time, in consumer digital mental health applications, there are tensions between transparency and the need of developers and companies to protect intellectual property. Additionally, because the business models of some applications may rely on having latitude to share user data, companies may have incentives to be less transparent with consumers.

Broadly speaking, transparency requires accurate information regarding how, what and when data are collected from the individual, and how that data will be stored and shared. Depending upon the type of digital mental health technology, there also may need to be explanation of how the data collected from users is being analyzed—whether the data may be used to draw additional inferences regarding the users' cognitive and mental state. These are best practices for consumer health apps in general. However, when it comes to digital mental health technology in particular, consumers may not be aware of how data that would ordinarily seem unrelated or inconsequential to mental health can be used to draw inferences about users' behavior and mental state. For example, location data is commonly collected by many apps and can be used to track a user's daily activities, and also analyzed for inferences regarding mental health and mood (Harari et al., 2016). In turn, these inferences can have repercussions for individuals that range from the pricing of items they search for online to employment and insurance. Transparency should include efforts to educate consumers about such inferences.

Digital health conducted for clinical or research purposes involves formal informed consent procedures. In contrast, in the consumer domain, information about data usage, protection, liability and other ethical concerns are often conveyed through Terms and Conditions statements. These documents are rarely read by consumers and often utilize dense and difficult language (Spector-Bagdady, 2015). The GDPR provides a model for requiring consumer information to be provided in language that is accessible (Chassang, 2017). However additional empirical research will be useful for understanding what information consumers may most need to make informed decisions about using consumer digital mental health applications in order to improve efforts at transparency. Moreover, such efforts to improve transparency may draw more upon recent efforts to use digital technology options, such as video tutorials or staged information screens, to better convey necessary information to consumers.

8. Digital divide

As with any technology, there are potential justice considerations in consumer digital mental health. Smartphones are widely used and that has buoyed hopes that digital mental health can help broaden access to mental health care. Will there be a digital divide that affects access to digital mental health therapy as a tool for intervention, and are there ways to mitigate that divide? The quality of the mental health care that may be accessed through consumer digital mental health apps still needs to be better understood. A digital divide could mean that lower income users do not have sufficient access to the more effective, cutting-edge digital mental health tools. A digital divide could also end up relegating low income consumers to low-tier, less effective types of digital mental health services. Concerns regarding accessibility and fairness factor into perceptions of trust and trustworthiness in mental health care (Adjekum, Blasimme, & Vayena, 2018). In the current data environment, some people may also feel that they are put in a situation in which they have to decide whether to ignore risks for their personal data through an app in exchange for the access to some mental health tools. This is another area for empirical research to help identify, and potentially mitigate, any digital divide in this area.

9. Conclusion

As digital mental health technology becomes a part of the landscape of mental health services, frameworks for addressing ethical obligations in mental health care need to be examined in light of the resulting shifts in the paradigmatic relationships that underlie those ethical obligations. Digital mental health devices may be used in ways that bring them in as stand-ins for human mental health providers or as an additional party to the therapeutic relationship. In doing so, there is an impact on traditional frameworks that were rooted in the idea that the therapeutic relationship is between therapist and patient, or that mental health information is mainly generated and maintained within a doctor-patient relationship. In particular, digital mental health technology may involve user interactions outside of the traditional health care domain, and data may be analyzed and shared in ways that a patient or consumer would not anticipate or easily understand. For those reasons, trust, data protection and privacy, accountability and transparency pose core ethical challenges in the use of digital mental health technology. One key issue in accountability is the need for reliable oversight and mechanisms to ensure the safety and effectiveness of digital mental health devices. Trust, as a concept, also provides a useful way of considering how relational obligations may be impacted in this emerging digital mental health landscape—does one have the therapeutic relationship with a device? Trustworthiness, as a concept, engages questions of what relational obligations and mechanisms need to be in place in order to support the ethical use of consumer digital technology tools to meet mental health needs.

Digital mental health offers promising ways to address the need for mental health care, but for direct-to-consumer mental health apps, there is a need for further research into approaches that support trust, and further investigate how associated ethical obligations are impacted by having digital technology as therapist. The risks to consumers' personal data in using apps remain considerable. There have been a number of efforts to formulate recommendations for the appropriate use of mental health apps and standards for how the technology is developed and employed. The APA has put forth guidance on recommending appropriate digital mental health care, and computer science organizations, such as the Association of Computing Machinery, have put forth guidance to address data protection, bias and transparency in digital health (ACM Code of Ethics, 2018). More empirical research is needed regarding the dimensions of trust in relation to the use of consumer digital mental health tools. It is clear, though, that oversight, accountability and data protection are critical components that need to be addressed further in order to support trustworthy consumer digital mental health technology.

References

ACM Code of Ethics (2018). *ACM code of ethics and professional conduct.* Association for Computing Machinery. Adopted by ACM Council on June 22, 2018. https://ethics.acm.org/.

Acquisti, A., Smith, S. W., & Sadeghi, A.-R. (2010). Trust and trustworthy computing. In *Third international conference, TRUST 2010, Berlin, Germany, June 21-23, 2010, proceedings*, Springer.

Adjekum, A., Blasimme, A., & Vayena, E. (2018). Elements of trust in digital health systems: Scoping review. *Journal of Medical Internet Research, 20*(12), e11254. https://doi.org/10.2196/11254.

Ajunwa, I. (2017, January). *Workplace wellness programs could be putting your health data at risk.* Harvard Business Review. https://hbr.org/2017/01/workplace-wellness-programs-could-be-putting-your-health-data-at-risk. Accessed 27 September 2017.

Allen, M. (2018, July 18). *Health insurers are gathering personal data—That could raise your rates. STAT.* Retrieved from https://www.statnews.com/2018/07/18/health-insurers-personal-details-raise-rates/.

Alpert, S. A. (2003). Protecting medical privacy: Challenges in the age of genetic information. *Journal of Social Issues, 59*(2), 301–322. https://doi.org/10.1111/1540-4560.00066.

American Medical Informatics Association. (February 6, 2018). *AMIA response to FDA draft guidance on clinical and patient support software,* AMIA. Retrieved from https://www.amia.org/public-policy/amia-response-fda-draft-guidance-clinical-and-patient-decision-support-software.

Anderson, C. L., & Agarwal, R. (2011). The digitization of healthcare: Boundary risks, emotion, and consumer willingness to disclose personal health information. *Information Systems Research, 22*(3), 469–490.

Andras, P., Esterle, L., Guckert, M., Han, T. A., Lewis, P. R., Milanovic, K., et al. (2018). Trusting intelligent machines: Deepening trust within socio-technical systems. *IEEE Technology and Society Magazine, 37*(4), 76–83. https://doi.org/10.1109/MTS.2018.2876107.

Anthes, E. (2016). Mental health: There's an app for that. *Nature, 532*(7597), 20. https://doi.org/10.1038/532020a.

APA. App evaluation model. 2019. https://www.psychiatry.org/psychiatrists/practice/mental-health-apps/app-evaluation-model. Accessed May 1, 2019.

Areàn, P. A., Hoa Ly, K., & Andersson, G. (2016). Mobile technology for mental health assessment. *Dialogues in Clinical Neuroscience, 18*(2), 163–169.

Armontrout, J. A., Torous, J., Cohen, M., McNiel, D. E., & Binder, R. (2018). Current regulation of mobile mental health applications. *The Journal of the American Academy of Psychiatry and the Law, 46*(2), 204–211, Online. https://doi.org/10.29158/JAAPL.003748-18.

Asan, O., Perchonok, J., & Montague, E. (2012). Contextual differences in the dynamic measurement of trust in websites. *International Journal of Cyber Society and Education, 5*(2), 91–110. https://doi.org/10.7903/ijcse.969.

Azofeifa, A. (2018). Evaluating behavioral health surveillance systems. *Preventing Chronic Disease, 15*, E53. https://doi.org/10.5888/pcd15.170459.

Barras, C. (2019). Mental health apps lean on bots and unlicensed therapists. *Nature Medicine* https://doi.org/10.1038/d41591-019-00009-6.

Bauer, M., Glenn, T., Monteith, S., Bauer, R., Whybrow, P. C., & Geddes, J. (2017). Ethical perspectives on recommending digital technology for patients with mental illness. *International Journal of Bipolar Disorders, 5*, 6. https://doi.org/10.1186/s40345-017-0073-9.

Bedi, G., Carrillo, F., Cecchi, G. A., et al. (2015). Automated analysis of free speech predicts psychosis onset in high-risk youths. *NPJ Schizophrenia, 1*, 15030. https://doi.org/10.1038/npjschz.2015.30.

Ben-Zeev, D., Brian, R., Jonathan, G., Razzano, L., Pashka, N., Carpenter-Song, E., et al. (2018). Mobile health (mHealth) versus clinic-based group intervention for people with serious mental illness: A randomized controlled trial. *Psychiatric Services, 69*(9), 978–985. https://doi.org/10.1176/appi.ps.201800063.

Berg, J. (2014). The E-health revolution and the necessary evolution of informed consent. *Indiana Health Law Review, 11*(2). http://journals.iupui.edu/index.php/ihlr/article/view/18889. Accessed 9 December 2018.

Birkhäuer, J., Gaab, J., Kossowsky, J., Hasler, S., Krummenacher, P., Werner, C., et al. (2017). Trust in the health care professional and health outcome: A meta-analysis. *PLoS One, 12*(2), e0170988. https://doi.org/10.1371/journal.pone.0170988.

Bloss, C., et al. (2016). Reimagining human research protections for 21st century science. *Journal of Medical Internet Research, 18*, e329.

Boire, R. (2018, January 15). *Explaining deep learning by making AI transparent—Predictive analytics times—Machine learning & data science news*. Retrieved August 3, 2018, from Predictive Analytics Times website: https://www.predictiveanalyticsworld.com/patimes/explaining-artificial-intelligence-ai-solutions/9175/.

Botella, E. (2019, November 16). *Facebook's Six4Three Pikinis lawsuit emails explain why your data is so valuable. Slate*. Retrieved from https://slate.com/technology/2019/11/facebook-six4three-pikinis-lawsuit-emails-data.html.

Brodwin, E. (2019, July 8). *VCs in the hottest part of healthcare explain why the time is ripe for a mental health funding boom*. Retrieved October 23, 2019, from Business Insider. https://www.businessinsider.com/mental-health-barriers-startups-venture-evidence-2019-6.

Bzdok, D., & Meyer-Lindenberg, A. (2017). *Machine learning for precision psychiatry*. ArXiv:1705.10553 [Stat]. Retrieved from http://arxiv.org/abs/1705.10553.

Campbell, D. (2018, November 22). *UK teenagers turn to mobile apps to help with mental ill health. The Guardian*. Retrieved from https://www.theguardian.com/society/2018/nov/21/uk-teenagers-turn-to-mobile-apps-to-help-with-mental-ill-health.

Capon, H., Hall, W., Fry, C., & Carter, A. (2016). Realising the technological promise of smartphones in addiction research and treatment: An ethical review. *The International Journal on Drug Policy, 36*, 47–57. https://doi.org/10.1016/j.drugpo.2016.05.013.

Carlo, A. D., Ghomi, R. H., Renn, B. N., & Areán, P. A. (2019). By the numbers: Ratings and utilization of behavioral health mobile applications. *NPJ Digital Medicine*, *2*(1), 1–8. https://doi.org/10.1038/s41746-019-0129-6.

CBS. (2018). *Google blocking addiction center ads amid dubious referrals*. http://newyork.cbslocal.com/2018/01/16/google-addiction-centre-ads/New York. Accessed February 6, 2018.

CCPA. (2018). *California Consumer Privacy Act of 2018*.

Chakravorti, B. (2018, January 4). *Trust in digital technology will be the internet's next frontier, for 2018 and beyond*. Scientific American. Retrieved from https://www.scientificamerican.com/article/trust-in-digital-technology-will-be-the-internet-rsquo-s-next-frontier-for-2018-and-beyond/.

Challen, R., Denny, J., Pitt, M., Gompels, L., Edwards, T., & Tsaneva-Atanasova, K. (2019). Artificial intelligence, bias and clinical safety. *BMJ Quality and Safety*, *28*(3), 231–237. https://doi.org/10.1136/bmjqs-2018-008370.

Chandrashekar, P. (2018). Do mental health mobile apps work: Evidence and recommendations for designing high-efficacy mental health mobile apps. *MHealth*, *4*, 6. https://doi.org/10.21037/mhealth.2018.03.02.

Char, D. S., Shah, N. H., & Magnus, D. (2018). Implementing machine learning in health care—Addressing ethical challenges. *New England Journal of Medicine*, *378*(11), 981–983. https://doi.org/10.1056/NEJMp1714229.

Chassang, G. (2017). The impact of the EU general data protection regulation on scientific research. *Ecancermedicalscience*, *11*, 709. https://doi.org/10.3332/ecancer.2017.709.

Cheng, V. W. S., Davenport, T., Johnson, D., Vella, K., & Hickie, I. B. (2019). Gamification in apps and technologies for improving mental health and well-being: Systematic review. *JMIR Mental Health*, *6*(6), e13717. https://doi.org/10.2196/13717.

Christensen, H. (2010). Increasing access effectiveness: Using the internet to deliver low intensity CBT. In J. Bennett-Levy, D. Richards, P. Farrand, H. Christensen, K. Griffiths, & D. Kavanagh, et al. (Eds.), *Oxford guide to low intensity CBT interventions* (pp. 53–67). Oxford, UK: Oxford University Press.

Cohen, I. G., Amarasingham, R., Shah, A., Xie, B., & Lo, B. (2014). The legal and ethical concerns that arise from using complex predictive analytics in health care. *Health Affairs*, *33*(7), 1139–1147. https://doi.org/10.1377/hlthaff.2014.0048.

Connolly, S. L., Miller, C. J., Koenig, C. J., Zamora, K. A., Wright, P. B., Stanley, R. L., et al. (2018). Veterans' attitudes toward smartphone app use for mental health care: Qualitative study of rurality and age differences. *JMIR mHealth and uHealth*, *6*(8), e10748. https://doi.org/10.2196/10748.

Corcoran, C. M., & Cecchi, G. A. (2018). Computational approaches to behavior analysis in psychiatry. *Neuropsychopharmacology*, *43*(1), 225. https://doi.org/10.1038/npp.2017.188.

Cortez, N. G., Cohen, I. G., & Kesselheim, A. S. (2014). FDA regulation of mobile health technologies. *New England Journal of Medicine*, *371*(4), 372–379. https://doi.org/10.1056/NEJMhle1403384.

Dagum, P. (2018). Digital biomarkers of cognitive function. *NPJ Digital Medicine*, *1*(1), 10. https://doi.org/10.1038/s41746-018-0018-4.

Dankar, F. K., Ptitsyn, A., & Dankar, S. K. (2018). The development of large-scale de-identified biomedical databases in the age of genomics—Principles and challenges. *Human Genomics*, *12*, 19. https://doi.org/10.1186/s40246-018-0147-5.

Danks, D., & London, A. J. (2017). Algorithmic bias in autonomous systems. In *Proceedings of the 26th international joint conference on artificial intelligence* (pp. 4691–4697). Retrieved from http://dl.acm.org/citation.cfm?id=3171837.3171944.

Dickson, B. (2017, April 23). *Artificial intelligence has to deal with its transparency problems*. The Next Web. Retrieved from https://thenextweb.com/artificial-intelligence/2017/04/23/artificial-intelligence-has-to-deal-with-its-transparency-problems/.

Dobrygowski, D., & Hoffman, W. (2019, May 28). *We need to build up 'digital trust' in tech. Wired.* Retrieved from https://www.wired.com/story/we-need-to-build-up-digital-trust-in-tech/.

Duggal, R., Brindle, I., & Bagenal, J. (2018). Digital healthcare: Regulating the revolution. *BMJ, 360,* k6. https://doi.org/10.1136/bmj.k6.

East, M. L., & Havard, B. C. (2015). Mental health mobile apps: From infusion to diffusion in the mental health social system. *JMIR Mental Health, 2*(1). e10, https://doi.org/10.2196/mental.3954.

Enriquez, J. (2016, August 1). *FDA won't regulate general wellness products. Med Device Online.* Retrieved from https://www.meddeviceonline.com/doc/fda-won-t-regulate-general-wellness-products-0001.

Federal Drug Administration. (2019, September 26). *General wellness: Policy for low risk devices.* Retrieved December 1, 2019, from U.S. Food and Drug Administration website. http://www.fda.gov/regulatory-information/search-fda-guidance-documents/general-wellness-policy-low-risk-devices.

Fisher, C. E., & Appelbaum, P. S. (2017). Beyond googling: The ethics of using patients' electronic footprints in psychiatric practice. *Harvard Review of Psychiatry, 25*(4), 170–179. https://doi.org/10.1097/HRP.0000000000000145.

Fitzpatrick, K. K., Darcy, A., & Vierhile, M. (2017). Delivering cognitive behavior therapy to young adults with symptoms of depression and anxiety using a fully automated conversational agent (woebot): A randomized controlled trial. *JMIR Mental Health, 4*(2), e19.

Friedland, S. I. (2015). I spy: The new self-cybersurveillance and the "internet of things" *Washington and Lee Law Review, 72,* 1459.

Gaebel, W., Muijen, M., Baumann, A. E., et al. (2014). EPA guidance on building trust in mental health services. *European Psychiatry, 29*(2), 83–100. https://doi.org/10.1016/j.eurpsy.2014.01.001.

Gavish, Y. (2017, July 25). *#2: What you need to know about ML algorithms and why you should care. Medium.* Retrieved from https://medium.com/@yaelg/product-manager-guide-part-2-what-you-need-know-machine-learning-algorithms-models-data-performance-cff5a837cec2.

General Data Protection Regulation (GDPR). (2018). *General data protection regulation (GDPR).* Available at: https://gdpr-info.eu/.

Gianfrancesco, M. A., Tamang, S., Yazdany, J., & Schmajuk, G. (2018). Potential biases in machine learning algorithms using electronic health record data. *JAMA Internal Medicine, 178*(11), 1544–1547. https://doi.org/10.1001/jamainternmed.2018.3763.

Gilson, L. (2003). Trust and the development of health care as a social institution. *Social Science & Medicine, 56*(7), 1453–1468.

Glenn, T., & Monteith, S. (2014). Privacy in the digital world: Medical and health data outside of HIPAA protections. *Current Psychiatry Reports, 16*(11), 494. https://doi.org/10.1007/s11920-014-0494-4. 25218603.

Glymour, B., & Herington, J. (2019). Measuring the biases that matter: The ethical and casual foundations for measures of fairness in algorithms. In *Proceedings of the conference on fairness, accountability, and transparency* (pp. 269–278).

Gonfalonieri, A. (2019, April 22). *What is an AI algorithm? Medium.* Retrieved from https://medium.com/predict/what-is-an-ai-algorithm-aceeab80e7e3.

Goudge, J., & Gilson, L. (2005). How can trust be investigated? Drawing lessons from past experience. *Social Science & Medicine, 61*(7), 1439–1451. https://doi.org/10.1016/j.socscimed.2004.11.071.

Gravenhorst, F., Muaremi, A., Bardram, J., Grünerbl, A., Mayora, O., Wurzer, G., et al. (2015). Mobile phones as medical devices in mental disorder treatment: An overview. *Personal and Ubiquitous Computing, 19*(2), 335–353. https://doi.org/10.1007/s00779-014-0829-5.

Hall, M. A., Dugan, E., Zheng, B., & Mishra, A. K. (2001). Trust in physicians and medical institutions: What is it, can it be measured, and does it matter? *The Milbank Quarterly*, *79*(4), 613–639.

Harari, G. M., Lane, N. D., Wang, R., Crosier, B. S., Campbell, A. T., & Gosling, S. D. (2016). Using smartphones to collect behavioral data in psychological science: Opportunities, practical considerations, and challenges. *Perspectives on Psychological Science: A Journal of the Association for Psychological Science*, *11*(6), 838–854. https://doi.org/10.1177/1745691616650285.

Hedden, S., Kennet, J., Lipari, R., Medley, G., & Tice, P. (2015). *Behavioral health trends in the United States: Results from the 2014 National Survey on Drug Use and Health*. Retrieved from https://www.samhsa.gov/data/sites/default/files/NSDUH-DetTabs-2015/NSDUH-DetTabs-2015/NSDUH-DetTabs-2015.htm.

HIPAA. (1996). *Health insurance portability and accountability act of 1996*. Pub. L. No. 104-191.

Holzinger, A., Biemann, C., Pattichis, C. S., & Kell, D. B. (2017). *What do we need to build explainable AI systems for the medical domain?* ArXiv:1712.09923 [Cs, Stat]. Retrieved from http://arxiv.org/abs/1712.09923.

Horn, N. (2019, October 6). *What you need to know about monetizing healthcare data—MedCity News*. Medcity News. Retrieved from https://medcitynews.com/2019/10/what-you-need-to-know-about-monetizing-healthcare-data/.

Hsin, H., & Torous, J. (2018). Creating boundaries to empower digital health technology. *BJPsych Open*, *4*(4), 235–237. https://doi.org/10.1192/bjo.2018.37.

Hsin, H., Torous, J., & Roberts, L. (2016). An adjuvant role for mobile health in psychiatry. *JAMA Psychiatry*, *73*(2), 103. https://doi.org/10.1001/jamapsychiatry.2015.2839.

Huang, H. Y., & Bashir, M. (2017). Users' adoption of mental health apps: Examining the impact of information cues. *JMIR mHealth and uHealth*, *5*(6), e83. https://doi.org/10.2196/mhealth.6827.

Huckvale, K., Torous, J., & Larsen, M. E. (2019). Assessment of the data sharing and privacy practices of smartphone apps for depression and smoking cessation. *JAMA Network Open*, *2*(4), e192542. https://doi.org/10.1001/jamanetworkopen.2019.2542.

Insel, T. R. (2017). Digital phenotyping: Technology for a new science of behavior. *JAMA*, *318*(13), 1215–1216. https://doi.org/10.1001/jama.2017.11295.

Jacobsen, R. (2014, November 1). *10 Mobile apps that deliver advice and therapy—Scientific American*. Scientific American. Retrieved from https://www.scientificamerican.com/article/10-mobile-apps-that-deliver-advice-and-therapy/.

Kao, C. K., & Liebovitz, D. M. (2017). Consumer mobile health apps: Current state, barriers, and future directions. *PM & R: The Journal of Injury, Function, and Rehabilitation*, *9*(Suppl. 5), S106–S115. https://doi.org/10.1016/j.pmrj.2017.02.018.

Kasperbauer, T. J., & Wright, D. E. (2019). Expanded FDA regulation of health and wellness apps. *Bioethics*, *34*, 235. https://doi.org/10.1111/bioe.12674.

Kleiman, E. M., Turner, B. J., Fedor, S., Beale, E. E., Picard, R. W., Huffman, J. C., et al. (2018). Digital phenotyping of suicidal thoughts. *Depression and Anxiety*, *35*(7), 601–608. https://doi.org/10.1002/da.22730.

Knight, W. (2017). *There's a big problem with AI: Even its creators can't explain how it works*. MIT Technology Review. https://www.technologyreview.com/s/604087/the-dark-secret-at-the-heart-of-ai/. Published April 11, 2017. Accessed December 14, 2018.

Kobayashi, L. (2015). *The ethics of mobile health technology*. Retrieved September 22, 2015, from Public Health website: http://blogs.plos.org/publichealth/2014/10/16/health-apps/.

Kraft, S. A., Cho, M. K., Gillespie, K., Halley, M., Varsava, N., Ormond, K. E., et al. (2018). Beyond consent: Building trusting relationships with diverse populations in precision medicine research. *The American Journal of Bioethics*, *18*(4), 3–20. https://doi.org/10.1080/15265161.2018.1431322.

Kretzschmar, K., Tyroll, H., Pavarini, G., Manzini, A., & Singh, I. (2019). Can your phone be your therapist? Young people's ethical perspectives on the use of fully automated conversational agents (chatbots) in mental health support. *Biomedical Informatics Insights*, *11*, 1178222619829083. https://doi.org/10.1177/1178222619829083.

Lakshmanan, R. (2019, October 29). *Mozilla and element AI want to build "data trusts" in the artificial intelligence age*. The Next Web. Retrieved from https://thenextweb.com/artificial-intelligence/2019/10/29/mozilla-and-element-ai-want-to-build-data-trusts-in-the-artificial-intelligence-age/.

Lambert, M. J., & Barley, D. E. (2001). Research summary on the therapeutic relationship and psychotherapy outcome. *Psychotherapy: Theory, Research, Practice, Training*, *38*(4), 357–361. https://doi.org/10.1037/0033-3204.38.4.357.

Langley, G. C., & Klopper, H. (2005). Trust as a foundation for the therapeutic intervention for patients with borderline personality disorder. *Journal of Psychiatric and Mental Health Nursing*, *12*(1), 23–32. https://doi.org/10.1111/j.1365-2850.2004.00774.x.

Larsen, M. E., Huckvale, K., Nicholas, J., et al. (2019). Using science to sell apps: Evaluation of mental health app store quality claims. *NPJ Digital Medicine*, *2*(1), 18. https://doi.org/10.1038/s41746-019-0093-1.

Larsen, M. E., Nicholas, J., & Christensen, H. (2016). A systematic assessment of smartphone tools for suicide prevention. *PLoS One*, *11*(4), e0152285. https://doi.org/10.1371/journal.pone.0152285.

Lee, L. M. (2017). Ethics and subsequent use of electronic health record data. *Journal of Biomedical Informatics*, *71*, 143–146https://doi.org/10.1016/j.jbi.2017.05.022.

Lee, Y., & Lin, J. (2008). Linking patients' trust in physicians to health outcomes. *British Journal of Hospital Medicine*, *69*(1), 42–46. https://doi.org/10.12968/hmed.2008.69.1.28040.

Leigh, S., & Flatt, S. (2015). App-based psychological interventions: Friend or foe? *Evidence-Based Mental Health*, *18*(4), 97–99. https://doi.org/10.1136/eb-2015-102203.

Lipschitz, J., Miller, C. J., Hogan, T. P., Burdick, K. E., Lippin-Foster, R., Simon, S. R., et al. (2019). Adoption of mobile apps for depression and anxiety: Cross-sectional survey study on patient interest and barriers to engagement. *JMIR Mental Health*, *6*(1), e11334. https://doi.org/10.2196/11334.

Lipton, Z. C. L., & Steinhardt, J. (2018). Troubling trends in machine learning scholarship. In *ICML 2018 debate papers*.

LoCurto, J., & Berg, G. M. (2016). Trust in healthcare settings: Scale development, methods, and preliminary determinants. *SAGE Open Medicine*, *4*, 4. https://doi.org/10.1177/2050312116664224.

Luxton, D. D. (2016). An introduction to artificial intelligence in behavioral and mental health care. In D. D. Luxton (Ed.), *Artificial intelligence in behavioral and mental health care* (pp. 1–26). (chapter 1). https://doi.org/10.1016/B978-0-12-420248-1.00001-5.

Magee, J. C., Adut, S., Brazill, K., & Warnick, S. (2018). Mobile app tools for identifying and managing mental health disorders in primary care. *Current Treatment Options in Psychiatry*, *5*(3), 345–362. https://doi.org/10.1007/s40501-018-0154-0.

Malhotra, S., Chakrabarti, S., & Shah, R. (2019). A model for digital mental healthcare: Its usefulness and potential for service delivery in low- and middle-income countries. *Indian Journal of Psychiatry*, *61*(1), 27–36.

Marcus, G. (2018). *Deep learning: A critical appraisal*. ArXiv:1801.00631 [Cs, Stat]. Retrieved from http://arxiv.org/abs/1801.00631.

Martinez-Martin, N. (2019). What are important ethical implications of using facial recognition technology in health care? *AMA Journal of Ethics*, *21*(2), 180–187. https://doi.org/10.1001/amajethics.2019.180.

Martinez-Martin, N., & Char, D. (2018). Surveillance and digital health. *The American Journal of Bioethics: AJOB*, *18*(9), 67–68. https://doi.org/10.1080/15265161.2018.1498954.

Martinez-Martin, N., Dunn, L. B., & Roberts, L. W. (2018). Is it ethical to use prognostic estimates from machine learning to treat psychosis? *AMA Journal of Ethics*, *20*(9), E804–E811. https://doi.org/10.1001/amajethics.2018.804.

Martinez-Martin, N., Insel, T. R., Dagum, P., Greely, H. T., & Cho, M. K. (2018). Data mining for health: Staking out the ethical territory of digital phenotyping. *NPJ Digital Medicine*, *1*(1), 68. https://doi.org/10.1038/s41746-018-0075-8.

Martinez-Martin, N., & Kreitmair, K. (2018). Ethical issues for direct-to-consumer digital psychotherapy apps: Addressing accountability, data protection, and consent. *JMIR Mental Health*, *5*(2), e32. https://doi.org/10.2196/mental.9423.

Meng, J., Hussain, S. A., Mohr, D. C., Czerwinski, M., & Zhang, M. (2018). Exploring user needs for a mobile behavioral-sensing technology for depression management: Qualitative study. *Journal of Medical Internet Research*, *20*(7), e10139. https://doi.org/10.2196/10139.

Mey, A., Knox, K., Kelly, F., Davey, A. K., Fowler, J., Hattingh, L., et al. (2013). Trust and safe spaces: Mental health consumers' and carers' relationships with community pharmacy staff. *The Patient*, *6*(4), 281–289. https://doi.org/10.1007/s40271-013-0032-1.

mHealthIntelligence. (2017). *Health systems target telehealth to fill a mental healthcare gap*. mHealthIntelligence. https://mhealthintelligence.com/news/health-systems-target-telehealth-to-fill-a-mental-healthcare-gap. Published December 18, 2017. Accessed January 28, 2018.

Miller, D. J., & Thelen, M. H. (1986). Knowledge and beliefs about confidentiality in psychotherapy. *Professional Psychology: Research and Practice*, *17*(1), 15–19. https://doi.org/10.1037/0735-7028.17.1.15.

Mittelstadt, B. D., Allo, P., Taddeo, M., Wachter, S., & Floridi, L. (2016). The ethics of algorithms: Mapping the debate. *Big Data & Society*, *3*(2), 2053951716679679. https://doi.org/10.1177/2053951716679679.

Mohr, D. C., Riper, H., & Schueller, S. M. (2017). A solution-focused research approach to achieve an implementable revolution in digital mental health. *JAMA Psychiatry*, *75*, 113–114. https://doi.org/10.1001/jamapsychiatry.2017.3838.

Mols, B. (2019). *In black box algorithms we trust (or do we?)*. Retrieved August 24, 2019, from https://cacm.acm.org/news/214618-in-black-box-algorithms-we-trust-or-do-we/fulltext.

Morganstein, J. (2016). Mobile applications for mental health providers. *Psychiatry*, *79*(4), 358–363. https://doi.org/10.1080/00332747.2016.1237753.

Neary, M., & Schueller, S. M. (2018). State of the field of mental health apps. *Cognitive and Behavioral Practice*, *25*(4), 531–537. https://doi.org/10.1016/j.cbpra.2018.01.002.

Nebeker, C., Harlow, J., Espinoza Giacinto, R., Orozco-Linares, R., Bloss, C. S., & Weibel, N. (2017). Ethical and regulatory challenges of research using pervasive sensing and other emerging technologies: IRB perspectives. *AJOB Empirical Bioethics*, *8*(4), 266–276. https://doi.org/10.1080/23294515.2017.1403980.

Nebeker, C., Murray, K., Holub, C., Haughton, J., & Arredondo, E. M. (2017). Acceptance of mobile health in communities underrepresented in biomedical research: Barriers and ethical considerations for scientists. *JMIR mHealth and uHealth*, *5*(6), e87. https://doi.org/10.2196/mhealth.6494.

Nebeker, C., Torous, J., & Bartlett Ellis, R. J. (2019). Building the case for actionable ethics in digital health research supported by artificial intelligence. *BMC Medicine*, *17*(1), 137. https://doi.org/10.1186/s12916-019-1377-7.

Ng, M. M., Firth, J., Minen, M., & Torous, J. (2019). User engagement in mental health apps: A review of measurement, reporting, and validity. *Psychiatric Services*, *70*(7), 538–544. https://doi.org/10.1176/appi.ps.201800519.

Ngiam, K. Y., & Khor, I. W. (2019). Big data and machine learning algorithms for health-care delivery. *The Lancet Oncology*, *20*(5), e262–e273. https://doi.org/10.1016/S1470-2045(19)30149-4.

NHS. (2014). *Psychological therapies, annual report on the use of IAPT services—England, 2013-14.* Retrieved December 3, 2019, from NHS Digital website: https://digital.nhs.uk/data-and-information/publications/statistical/psychological-therapies-annual-reports-on-the-use-of-iapt-services/annual-report-2013-14.

Nicholas, J., Larsen, M. E., Proudfoot, J., & Christensen, H. (2015). Mobile apps for bipolar disorder: A systematic review of features and content quality. *Journal of Medical Internet Research, 17*(8), e198. https://doi.org/10.2196/jmir.4581.

Nixon, B. (2018, February 12). *Can wearable technology and predictive analytics prevent workplace violence?* Retrieved December 5, 2019, from HR Daily Advisor website: https://hrdailyadvisor.blr.com/2018/02/12/can-wearable-technology-predictive-analytics-prevent-workplace-violence/.

NPR. (2018). *Facebook increasingly reliant on A.I. to predict suicide risk.* Retrieved August 28, 2019, from NPR.org website: https://www.npr.org/2018/11/17/668408122/facebook-increasingly-reliant-on-a-i-to-predict-suicide-risk.

Nutt, A. E. (2017, December 3). *'The Woebot will see you now'—The rise of chatbot therapy.* Washington Post. Retrieved from https://www.washingtonpost.com/news/to-your-health/wp/2017/12/03/the-woebot-will-see-you-now-the-rise-of-chatbot-therapy/.

O'Loughlin, K., Neary, M., Adkins, E. C., & Schueller, S. M. (2019). Reviewing the data security and privacy policies of mobile apps for depression. *Internet Interventions, 15,* 110–115. https://doi.org/10.1016/j.invent.2018.12.001.

Onnela, J. P., & Rauch, S. L. (2016). Harnessing smartphone-based digital phenotyping to enhance behavioral and mental health. *Neuropsychopharmacology, 41*(7), 1691–1696. https://doi.org/10.1038/npp.2016.7.

Owen, J. E., Kuhn, E., Jaworski, B. K., McGee-Vincent, P., Juhasz, K., Hoffman, J. E., et al. (2018). VA mobile apps for PTSD and related problems: Public health resources for veterans and those who care for them. *MHealth, 4,* 28. https://doi.org/10.21037/mhealth.2018.05.07.

Ozawa, S., & Sripad, P. (2013). How do you measure trust in the health system? A systematic review of the literature. *Social Science & Medicine, 91,* 10–14. https://doi.org/10.1016/j.socscimed.2013.05.005.

Parikh, R. B., Obermeyer, Z., & Navathe, A. S. (2019). Regulation of predictive analytics in medicine. *Science (New York, N.Y.), 363*(6429), 810–812. https://doi.org/10.1126/science.aaw0029.

Pasquale, F. (2016). *The black box society: The secret algorithms that control money and information.* Cambridge, MA: Harvard University Press.

Pavlou, P. A. (2003). *Consumer acceptance of electronic commerce: Integrating trust and risk with the technology acceptance model (SSRN Scholarly Paper No. ID 2742286).* Retrieved from Social Science Research Network website: https://papers.ssrn.com/abstract=2742286.

Payne, H. E., Lister, C., West, J. H., & Bernhardt, J. M. (2015). Behavioral functionality of mobile apps in health interventions: A systematic review of the literature. *JMIR mHealth and uHealth, 3*(1), e20. https://doi.org/10.2196/mhealth.3335.

Possemato, K., Kuhn, E., Johnson, E., et al. (2016). Using PTSD Coach in primary care with and without clinician support: A pilot randomized controlled trial. *General Hospital Psychiatry, 38,* 94–98. https://doi.org/10.1016/j.genhosppsych.2015.09.005. 26589765.

Powell, A., Yue, Z., Shan, C., & Torous, J. B. (2019). The monetization strategies of apps for anxiety management: An international comparison. *Journal of Technology in Behavioral Science, 4,* 67–72. https://doi.org/10.1007/s41347-019-00093-y.

Price, W. N. (2017). *Regulating black-box medicine (SSRN Scholarly Paper No. ID 2938391).* Retrieved from Social Science Research Network website: https://papers.ssrn.com/abstract=2938391.

Price, M., Yuen, E. K., Goetter, E. M., Herbert, J. D., Forman, E. M., Acierno, R., et al. (2014). mHealth: A mechanism to deliver more accessible, more effective mental health care. *Clinical Psychology & Psychotherapy*, *21*(5), 427–436. https://doi.org/10.1002/cpp.1855.

PsyberGuide. (2019). *The mental health app guide designed with you in mind*. Retrieved October 25, 2019, from PsyberGuide website: https://psyberguide.org/.

Rahim, Z. (2019, September 10). *Trump administration considers monitoring smartphones of people with mental health problems*. The Independent. Retrieved from https://www.independent.co.uk/news/world/americas/us-politics/trump-mental-health-guns-shootings-tracking-phones-a9098946.html.

Reilly, M. (2017). *Is facebook targeting advertising at depressed teens?* MIT Technology Review. https://www.technologyreview.com/s/604307/is-facebook-targeting-ads-at-sad-teens/ Accessed October 26, 2017. Cached at: http://www.webcitation.org/6x2LOfMBM.

Rendina, H. J., & Mustanski, B. (2018). Privacy, trust, and data sharing in web-based and mobile research: Participant perspectives in a large nationwide sample of men who have sex with men in the United States. *Journal of Medical Internet Research*, *20*(7). e233, https://doi.org/10.2196/jmir.9019.

Research2guidance. (2017, November 7). *325000 mobile health apps available in 2017*. Retrieved November 1, 2019, from Research2guidance website: https://research2guidance.com/325000-mobile-health-apps-available-in-2017/.

Reynolds, E. (2017). *How effective are mental health apps?* WIRED UK. http://www.wired.co.uk/article/mental-health-apps. Accessed 24 August 2017.

Rocher, L., Hendrickx, J. M., & de Montjoye, Y. A. (2019). Estimating the success of re-identifications in incomplete datasets using generative models. *Nature Communications*, *10*(1), 1–9. https://doi.org/10.1038/s41467-019-10933-3.

Rothstein, M. A. (2007). Health privacy in the electronic age. *The Journal of Legal Medicine*, *28*(4), 487–501. https://doi.org/10.1080/01947640701732148.

Saeb, S., Lattie, E. G., Schueller, S. M., Kording, K. P., & Mohr, D. C. (2016). The relationship between mobile phone location sensor data and depressive symptom severity. *PeerJ*, *4*, e2537. https://doi.org/10.7717/peerj.2537.

Schairer, C. E., Rubanovich, C. K., & Bloss, C. S. (2018). How could commercial terms of use and privacy policies undermine informed consent in the age of mobile health? *AMA Journal of Ethics*, *20*(9), 864–872. https://doi.org/10.1001/amajethics.2018.864.

Schueller, S. M., Neary, M., O'Loughlin, K., & Adkins, E. C. (2018). Discovery of and interest in health apps among those with mental health needs: Survey and focus group study. *Journal of Medical Internet Research*, *20*(6), e10141. https://doi.org/10.2196/10141.

Shoenbill, K., Fost, N., Tachinardi, U., & Mendonca, E. A. (2014). Genetic data and electronic health records: A discussion of ethical, logistical and technological considerations. *Journal of the American Medical Informatics Association: JAMIA*, *21*(1), 171–180. https://doi.org/10.1136/amiajnl-2013-001694.

Shuren, J., Patel, B., & Gottlieb, S. (2018). FDA regulation of mobile medical apps. *JAMA*, *320*, 337–338.

Siau, K., & Wang, W. (2018). Building trust in artificial intelligence, machine learning, and robotics. *Cutter Business Technology Journal*, *31*(2), 47–53.

Singer, N. (2014, June 28). *When a health plan knows how you shop*. The New York Times. Retrieved from https://www.nytimes.com/2014/06/29/technology/when-a-health-plan-knows-how-you-shop.html.

Singer, N. (2018, February 25). *How companies scour our digital lives for clues to our health*. The New York Times. Retrieved from https://www.nytimes.com/2018/02/25/technology/smartphones-mental-health.html.

Soni, H., Grando, A., Murcko, A., Bayuk, M., Chandrashekar, P., Mukundan, M., et al. (2018). Current state of electronic consent processes in behavioral health: Outcomes from an observational study. *AMIA Annual Symposium Proceedings, 2017*, 1607–1616.

Spector-Bagdady, K. (2015). Reconceptualizing consent for direct-to-consumer health services. *American Journal of Law & Medicine, 41*(4), 568–616. https://doi.org/10.1177/0098858815622191.

Torous, J., Andersson, G., Bertagnoli, A., Christensen, H., Cuijpers, P., Firth, J., et al. (2019). Towards a consensus around standards for smartphone apps and digital mental health. *World Psychiatry, 18*(1), 97–98. https://doi.org/10.1002/wps.20592.

Torous, J. B., Chan, S. R., Gipson, S. Y.-M. T., Kim, J. W., Nguyen, T.-Q., Luo, J., et al. (2018). A hierarchical framework for evaluation and informed decision making regarding smartphone apps for clinical care. *Psychiatric Services, 69*(5), 498–500. https://doi.org/10.1176/appi.ps.201700423.

Torous, J., & Nebeker, C. (2017). Navigating ethics in the digital age: Introducing connected and open research ethics (CORE), a Tool for researchers and institutional review boards. *Journal of Medical Internet Research, 19*(2), e38. https://doi.org/10.2196/jmir.6793.

Torous, J., Nicholas, J., Larsen, M. E., Firth, J., & Christensen, H. (2018). Clinical review of user engagement with mental health smartphone apps: Evidence, theory and improvements. *Evidence-Based Mental Health, 21*(3), 116–119. https://doi.org/10.1136/eb-2018-102891.

Torous, J., Onnela, J.-P., & Keshavan, M. (2017). New dimensions and new tools to realize the potential of RDoC: Digital phenotyping via smartphones and connected devices. *Translational Psychiatry*, 7(3), e1053. https://doi.org/10.1038/tp.2017.25.

Torous, J., & Roberts, L. W. (2017a). The ethical use of mobile health technology in clinical psychiatry. *The Journal of Nervous and Mental Disease, 205*(1), 4–8. https://doi.org/10.1097/NMD.0000000000000596.

Torous, J., & Roberts, L. W. (2017b). Needed innovation in digital health and smartphone applications for mental health: Transparency and trust. *JAMA Psychiatry, 74*(5), 437–438. https://doi.org/10.1001/jamapsychiatry.2017.0262.

Towards trustable machine learning. (2018). *Nature Biomedical Engineering, 2*(10), 709. https://doi.org/10.1038/s41551-018-0315-x.

Tozzi, J. (2019, May 29). *Mental health app raises $50 million for expansion*. Bloomberg.Com. Retrieved from https://www.bloomberg.com/news/articles/2019-05-29/mental-health-app-raises-50-million-as-digital-therapy-grows.

Tramèr, F., Atlidakis, V., Geambasu, R., Hsu, D., Hubaux, J. P., Humbert, M., et al. (2015). *FairTest: Discovering unwarranted associations in data-driven applications*. ArXiv:1510.02377 [Cs]. Retrieved from http://arxiv.org/abs/1510.02377.

Udesky, L. (2010). The ethics of direct-to-consumer genetic testing. *The Lancet, 376*(9750), 1377–1378. https://doi.org/10.1016/S0140-6736(10)61939-3.

Vaidyam, A. N., Wisniewski, H., Halamka, J. D., Kashavan, M. S., & Torous, J. B. (2019). Chatbots and conversational agents in mental health: A review of the psychiatric landscape. *The Canadian Journal of Psychiatry, 64*(7), 456–464. https://doi.org/10.1177/0706743719828977.

Vellido, A., Martín-Guerrero, J. D., & Lisboa, P. J. G. (2012). Making machine learning models interpretable. In *Proceedings of the European symposium on artificial neural networks, computational intelligence and machine learning*.

Vincent, J. (2006). Emotional attachment and mobile phones. *Knowledge, Technology, and Policy, 19*, 39–44. https://doi.org/10.1007/s12130-006-1013-7.

Vis, C., Mol, M., Kleiboer, A., Bührmann, L., Finch, T., Smit, J., et al. (2018). Improving implementation of eMental health for mood disorders in routine practice: Systematic review of barriers and facilitating factors. *JMIR Mental Health, 5*(1), e20. https://doi.org/10.2196/mental.9769.

Wachter, S., & Mittelstadt, B. (2019). *A right to reasonable inferences: Re-thinking data protection law in the age of big data and AI (SSRN Scholarly Paper No. ID 3248829)*. Retrieved from Social Science Research Network website: https://papers.ssrn.com/abstract=3248829.
Weisel, K. K., Fuhrmann, L. M., Berking, M., Baumeister, H., Cuijpers, P., & Ebert, D. D. (2019). Standalone smartphone apps for mental health—A systematic review and meta-analysis. *NPJ Digital Medicine*, *2*(1), 1–10. https://doi.org/10.1038/s41746-019-0188-8.
Wicken, C., Nevin, R., & Ritchie, E. C. (2015). U.S. Military surveillance of mental disorders, 1998–2013. *Psychiatric Services*, *67*(2), 248–251. https://doi.org/10.1176/appi.ps.201500001.
Yang, Y. T., & Silverman, R. D. (2014). Mobile health applications: The patchwork of legal and liability issues suggests strategies to improve oversight. *Health Affairs*, *33*(2), 222–227. https://doi.org/10.1377/hlthaff.2013.0958.
Yu, K. H., Beam, A. L., & Kohane, I. S. (2018). Artificial intelligence in healthcare. *Nature Biomedical Engineering*, *2*(10), 719. https://doi.org/10.1038/s41551-018-0305-z.
Zur Institute. (2018, December 11). *TeleMental health services across state lines*. Retrieved October 7, 2019, from Zur Institute website: https://www.zurinstitute.com/telehealth-across-state-lines/.

Further reading

Blunt, D. R. (2006). *Confidentiality, informed consent, and ethical considerations in reviewing the client's psychotherapy records. (ERIC Document Reproduction Service No. ED490794)*.
Mittelstadt, B. D., & Floridi, L. (2016). The ethics of big data: Current and foreseeable issues in biomedical contexts. *Science and Engineering Ethics*, *22*(2), 303–341. https://doi.org/10.1007/s11948-015-9652-2.
United States Code, 2006 Edition, supplement 4, title 21—Food and drugs. 21 U.S. Code § 360j.
Wahle, F., Kowatsch, T., Fleisch, E., Rufer, M., & Weidt, S. (2016). Mobile sensing and support for people with depression: A pilot trial in the wild. *JMIR mHealth and uHealth*, *4*(3), e111. https://doi.org/10.2196/mhealth.5960.

CHAPTER FOUR

Tailoring reality—The ethics of DIY and consumer sensory enhancement

Imre Bárd*
Methodology Department, London School of Economics and Political Science, London, United Kingdom
*Corresponding author: e-mail address: i.bard@lse.ac.uk

Contents

1. Introduction

The senses are our windows onto the world, and technologies that modulate perception directly influence our cognition, emotions and behavior. Although there are a number of different interpretations with regard to the types of interventions and devices that should fall under the category of neurotechnologies, sensory enhancement methods are usually excluded (Eaton & Illes, 2007; Kreitmair, 2019; Müller & Rotter, 2017). I believe this

Developments in Neuroethics and Bioethics, Volume 3
ISSN 2589-2959
https://doi.org/10.1016/bs.dnb.2020.03.004

is an omission that should be addressed. In this chapter I will argue that there are a variety of technologies for sensory modification, which warrant consideration by neuroethicists. Although the range of tools and technologies that one might consider sensory is vast, from telescopes to stethoscopes and many more, my focus will be on devices, which are intended to be used in an ongoing or even permanent manner to mediate our experience of the world. In addition, I will consider only non-therapeutic applications, that is, sensory modification technologies that confer capacities normally outside the scope of human perception. Admittedly, the boundaries may be blurry but restorative technologies will be considered only to the extent that they enable non-therapeutic outcomes by opening up new perceptual capacities.

I will start with a brief overview of why perception and the senses, as well as technologies for their modulation are important to consider. Then, I will introduce a conceptual framework for the categorization of sensory technologies, and describe the main methods for achieving sensory modification. Subsequently, I will discuss some of the ethical issues that emerge, structured around two spheres of use, the Do-It-Yourself (DIY), body hacker, or grinder community, whose members experiment on themselves to extend their perception, and the rapidly growing field of consumer augmented reality, developed by some of the world's largest technology companies. I believe these two domains embody competing visions with regard to the future of sensory enhancement and they present a related set of challenges with regard to positive and negative freedoms. In the case of DIY body modification, we confront the question pertaining to the scope of individual choice in relation to self-enhancement, that is, the *freedom to* pursue such practices. On the other hand, consumer augmented reality challenges us to consider the potentially autonomy-constraining effects of new consumer technologies, calling on us to find ways of ensuring *freedom from* such adverse influences.

2. A neglected topic in neuroethics

Studying perception and its alterations through pharmacological or technological means yields important insights into the organization of the brain (Hillenbrand, Raveh, & Amedi, 2018), the workings of the mind (Carhart-Harris et al., 2014; Welch, 1978) and provides an empirical basis for philosophical discussions about the nature of sensory experience (Fish, 2010), the individuation of the senses (Biggs, Matthen, & Stokes, 2014), synesthesia (Macpherson, 2007), and more. While classical theories considered cognition to consist of computational processes executed on amodal mental representations, this approach has come under heavy criticism for its failure to adequately account for the body as a key factor in shaping an organism's cognitive architecture and capacities (Clark, 1999; Engel, Maye, Kurthen, & König, 2013). Research has demonstrated the fundamental importance of embodied, sensorimotor activity for the development of cognitive functions including the formation of high-level concepts (Beauchamp & Martin, 2007), language (Johnson, 1987), and mathematics (Lakoff & Núñez, 2000). According to these accounts, perception and action are intertwined and cognition is best understood as an embodied activity, which arises from an organism's dynamic, multisensory interactions with its environment (Noë & Noë, 2004; O'Regan & Noë, 2001). Theories of embodied cognition come in several varieties but a key insight is that the kinds of thoughts we are capable of are inseparably intertwined with the morphology of our bodies and the perceptual system we are equipped with (Pfeifer & Bongard, 2006). A highly influential strand of this view is the enactive approach to cognition, according to which every organism "brings forth," or enacts its world through interactions with the environment, which are constrained by the organism's perceptual and behavioral repertoire (Varela, Thompson, & Rosch, 1991). Arguably, in the case of humans, technology plays a pivotal role in this process as there is an additional feedback loop between our biological make-up and the technologies we create. Therefore, the effects of our most ancient techniques as well as those of our most advanced technologies are inscribed deep in the body. Research shows that learning to read causes wide-ranging cortical reorganization (Skeide et al., 2017), while emerging evidence suggests that smartphone use may have detrimental effects on memory (Wilmer, Sherman, & Chein, 2017) and that the mere presence of one's smartphone can reduce cognitive capacity (Ward, Duke, Gneezy, & Bos, 2017). In Clark's famous formulation, we are natural-born cyborgs, for our cognition unavoidably intermingles with the world around us and with the technologies we create, for better and for worse (Clark, 2004).

Despite the importance of perception and the senses within neuroscience, philosophy, and our everyday lived experience, scholars in neuroethics have largely ignored the subject. To some extent, this lack of attention may be explained by the fact that clinically relevant technological breakthroughs have been more pronounced—or at least more readily anticipated—in other domains, such as psychopharmaceuticals, neurosurgery, and brain imaging, which have largely preoccupied the field. The most notable exceptions are cochlear implants and, to a lesser extent, visual prosthetics, which have attracted some attention from neuroethicists (Hansson, 2015; Komesaroff, Komesaroff, & Hyde, 2015; Sparrow, 2005; Tucker, 1998), while the topic of sensory enhancement is discussed even more rarely (Jebari, 2015; Wolbring, 2013).

Yet, there are multiple avenues through which perception modification technologies are likely to become widespread over the coming years. First, with the ageing of the population in developed countries, the rate of sensory deficiencies, particularly those related to vision and hearing, is expected to increase significantly. For example, currently incurable conditions, such as age-related macular degeneration is predicted to affect 288 million people worldwide by 2040 (Wong et al., 2014), and the World Health Organization (WHO) predicts that by 2050 over 900 million people will have disabling hearing loss (World Health Organization, 2019). Meeting this medical need presents a strong incentive for continued medical research as well as corporate innovation. At the same time, the military is likely to be another key driver of advances in sensory technologies. A report from late 2019 by the US Department of Defence's Biotechnologies for Health and Human Performance Council identified four potential military use-cases emerging from scientific advances over the next 30 years. Two of these use-cases pertain to sensory enhancement, namely ocular and auditory augmentations (Emanuel et al., 2019).

Admittedly, a number of these technologies may ultimately be implanted in the body (Royal Society, 2019). However, there is currently a significant interest in the technology industry to find the successor to the handheld screen-based computational paradigm. We can already observe the shortening of the chain of command between humans and machines, moving toward ever more intuitive and natural interfaces; from clicking and typing, through touch and speech, to natural gestures, gaze control, and prediction-based interactions where the system anticipates the user's intentions on the basis of biometric and other measurements (Deloitte, 2016). Further trends in technology development, such as the growing miniaturization of sensors and other components, ever faster data connectivity and increasing processing power act as enabling forces for powerful wearable technologies ranging from smart glasses to AI-enabled hearing aids (Morvan, Ovanesoff, Carrell-Billiard, & Hintermann, 2019). The market size for smart wearables is already expanding rapidly and is expected to grow by 19% annually from 2020 to 2025 (Mordor Intelligence, 2019).

The rapid reduction of the cost of sensors and other hardware have made it possible for Do-It-Yourself (DIY) enthusiasts to experiment and construct their own devices for sensory exploration as well. Taken together, these developments make the blending of physical and virtual environments possible and allow for new ways of channeling unprecedented amounts of information to the human sensory system.

3. What are perception modification technologies?

Wright and Ward proposed a framework for conceptualizing what they called "sensory tools." On their account, these tools mediate the ways in which one perceives their environment, allowing for the execution of sensory operations (Wright & Ward, 2018). The authors also devised a useful schematic to illustrate how sensory tools function, distinguishing between simple devices that rely on a natural coupling system, and advanced devices, which rely on an algorithmic coupling system (see Fig. 1). A simple sensory tool "is one that works primarily because of a natural physical or mechanical relationship between the source and target modalities" (Wright & Ward, 2018:194). For a simple sensory tool the most straightforward example would be a pair of glasses, where the lenses directly bend or spread light to change the focal point on the wearer's retina. In contrast, advanced sensory

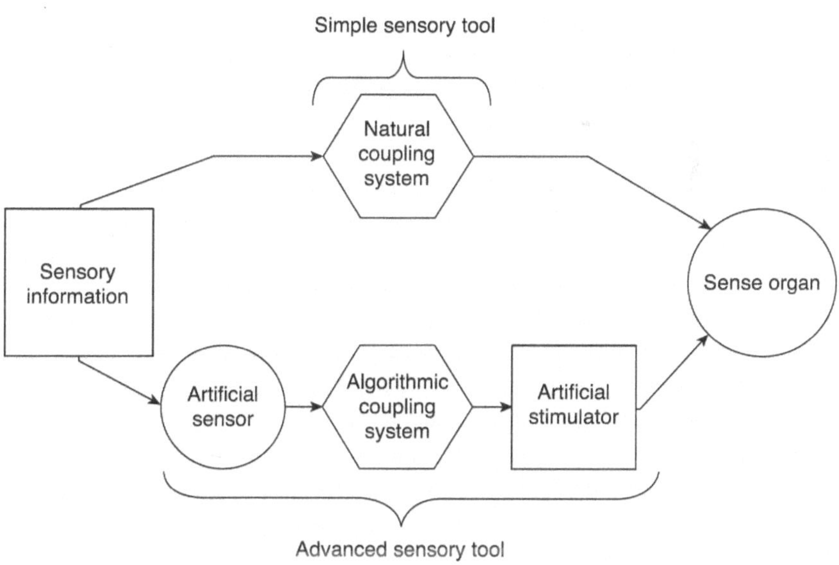

Fig. 1 Comparison of simple and advanced sensory tools. © The British Academy 2018. *Reproduced by permission from Wright, T.D., & Ward, J., Sensory substitution devices as advanced sensory tools. In F. Macpherson (Ed.),* Sensory substitution and augmentation, *p. 194, Oxford: Oxford University Press.*

tools are characterized by some degree of algorithmic processing of the information that is picked up by sensors and then conveyed to the sense organ via an artificial stimulator. For example, a cochlear implant would qualify as an advanced sensory tool since it comprises an artificial sensor, in this case a microphone that picks up sounds from the environment, which are algorithmically processed and coded into patterns of electrical stimulations that are applied to the auditory nerve by tiny electrodes in the cochlea.

Although this framework is broad enough to capture sensory technologies of all kinds, including ones that are only used intermittently and for specific purposes, such as a periscope or a magnifying glass, I intend to focus on examples where the technology is intended to be used over extended periods in such a way that they may no longer be described as mere tools, but as extensions of the individual. While I mostly agree with the framework developed by Wright and Ward, there is one important point where I wish to diverge, which has to do with their restriction of the framework to *sensory* information. Wright and Ward embrace a rather broad interpretation of sensory, which includes the classical modalities of vision, hearing, taste, smell and touch, but also those forms of sensing that are not normally available to humans. However, they restrict their framework to apply only to "raw sensory information rather than information that is categorized or semanticized" (Wright & Ward, 2018:192). For example, an assistive device equipped with computer vision that was capable of recognizing objects and reading out the words denoting what the person was looking at, would not count as a sensory tool because the algorithmic conversion occurs not at the sensory, but at the symbolic level. This distinction is meaningful for the proper definition of what counts as sensory information; nevertheless, I believe the general schema offered by Wright and Ward can be used to consider a broader range of technologies that mediate how we come to experience our environment in an ongoing manner. From the perspective of an ethical assessment it is of limited importance whether this mediation is restricted to raw sensory information or whether it includes categorized, symbolic information as well. For example, we may imagine a wearable device comprised of a camera embedded in a pair of smart glasses, which captured visual information in the context of a conversation and with the help of facial recognition, voice analysis and emotion detection algorithms could categorize an interlocutor's momentary stress level and affective state. This kind of technology would rightly be considered a device which mediates how we experience the world around us and certainly raises a host of ethical issues, but would not qualify as a sensory tool on Wright's and Ward's account.

To resolve this issue we may rely on the distinction between sensation and perception. Whereas sensation is usually understood to refer to the experience of sensory stimulation itself, perception refers to the process whereby the brain interprets sensory information to arrive at something meaningful (Goldstein, 2008). In the case of our hypothetical device then, we are outsourcing a part of the perceptual interpretation process to an algorithm. Thus, while it is true that our hypothetical emotion detection device does not operate at the sensory level, it is also true that it allows us to *perceive* an aspect of the world that is otherwise not available, namely, a person's affective state and stress level. Therefore, it appears appropriate to use the schema depicted in Fig. 1 to describe both sensory and perception modification/augmentation technologies.

With these caveats, I will use Wright's and Ward's overall schema and their distinction between natural and algorithmic coupling to describe different technologies, before proceeding to discuss some of the ethical issues they present.

3.1 Natural coupling

3.1.1 Magnetic implants

The most widely discussed, and to date perhaps the only perceptual enhancement intervention in this category is the implantation of rare earth magnets. The procedure involves making an incision on a person's finger and inserting a small, approximately 3 × 1 mm, Neodymium or Alnico magnet wrapped in a layer of biocompatible coating, such as medical grade silicone (Hameed, Harrison, Gasson, & Warwick, 2010). The purpose of getting a magnetic implant is to enable a person to sense electromagnetic fields. This occurs due to the fact that strong electromagnetic fields cause the implanted magnet to move or oscillate slightly, in synchrony with the electromagnetic field. This stimulates mechanoreceptors in the skin and causes a vibro-tactile sensation (Harrison, 2015). The intensity of the vibration is proportional to the frequency of the electromagnetic field, allowing implantees to feel and differentiate between various sources and to sense when wires or electrical appliances are live. The practice originates in the body modification community and started to become more widespread in the late 2000s and early 2010s (Norton, 2006). The implantation itself is usually performed in piercing or body modification studios.

Another application of the magnetic implant is to embed it in the tragus, which is the cartilage next to the ear canal's opening. Using an induction coil to transmit audio information, the magnet can function as an implanted earphone (Doerksen, 2017).

However, the magnetic implant also comes with a variety of risks, such as corrosion and infection at the implantation site if the protective sheath is breached or degrades. In addition, first-person accounts report that over a few years' time the magnet's power weakens and the acquired perceptual capabilities of the user fade, before disappearing altogether (Robertson, 2017).

3.2 Algorithmic coupling systems

This section will consider two broad types of technologies in the category of advanced devices, which rely on an algorithmic coupling system: sensory substitution and augmented reality.

3.2.1 Perceptual enhancement via sensory substitution

Sensory substitution devices translate information from one sensory modality into another (Renier & De Volder, 2013). In the 1960s, neuroscientist Paul Bach-y-Rita was investigating brain plasticity by studying whether it is possible for congenitally blind individuals to acquire the sense of vision. Together with colleagues, he conducted a series of ground-breaking experiments, which demonstrated the phenomenon of sensory substitution and established a new field of research. Bach-y-Rita's and colleagues' novel experimental setup consisted of a TV camera, a monitor, an electronic commutator device, and an array of 400 vibrating metal pins arranged in a 20×20 grid, which was attached to the back of a modified dentist's chair. The camera captured images of objects placed in front of it and the commutator device converted the visual signals into patterns of vibrations that a person sitting in the chair could feel in their lower back. The location of the vibrating pins was mapped unto the location of pixels on the monitor, and the strength of the vibration was proportional to the luminance of pixels on the screen, such that brighter pixels produced a stronger vibration (Bach-y-Rita, Collins, Saunders, White, & Scadden, 1969). This was the first sensory substitution device (SSD), which came to be known as the Tactile-Vision Substitution System (TVSS). After a few hours of training, users seated in the chair could recognize a range of objects, and judge their absolute size and distance based on the patterns of vibrations in their lower back (Bach-y-Rita et al., 1969). Importantly, subjects experienced the location of objects to be in front of them, despite the stimulus being presented to their backs (Guarniero, 1974). Bach-y-Rita also reported that upon zooming of the camera's lens subjects instinctively moved their head and body backwards, as if the object on the screen was moving toward them (Bach-y-Rita, 2002). Although there had already been a number of devices as early as the 1910s that relied on the concept of representing visual information as sound or touch, these were developed only in the context of allowing the blind to read and were considered reading machines (Koestler, 2004). In contrast, Bach-y-Rita's experiments opened up a much broader horizon by asking whether humans could learn new sensory systems.

Since then, a variety of sensory substitution devices have been developed. The Tongue Display Unit (TDU) converts visual information into patterns of small electrical stimulations delivered via a grid placed on the tongue (Sampaio, Maris, & Bach-y-Rita, 2001), capitalizing on its extremely high spatial acuity (Van Boven & Johnson, 1994). Another device, called the vOICe (capitalized letters representing "Oh, I see"), attempts to translate visual information into sound by using a head-mounted camera and a small computer that maps the visual field at regular intervals and creates a corresponding soundscape (Meijer, 1992). Long-term users of the vOICe learn to recognize the patterns in the soundscape and can develop great proficiency in navigating their environment using the device, even acquiring vision-like experiences (Ward & Meijer, 2010). The vOICe is available as a smartphone application as well.[1] While existing vision-to-sound SSDs are only capable of mapping the spatial features of a visual scene, such as an object's shape, distance and movement, current research is building on insights from synesthesia in order to incorporate color information as well (Hamilton-Fletcher, Mengucci, & Medeiros, 2016).

In 2015, Scott Novich and David Eagleman presented a sound-to-touch sensory substitution device called the Versatile Extra-Sensory Transducer (VEST) (Eagleman, 2015). The device is intended to assist individuals with impaired hearing and comes in the form of a vest, which can be worn underneath one's clothing. It translates sounds picked up via a microphone into patterns of tactile stimulation along the torso (Novich, 2015). With some training, deaf individuals can use the device to recognize spoken language. More recently, the company co-founded by Eagleman and Novich released the Neosensory Buzz, a wristband that also translates environmental sounds into vibrational patterns. According to the company's website, with practice the associations between sounds and vibrations become automatic and "a new sense is born."[2]

So far, the chemical senses have not proved amenable to this kind of technology. For the sense of smell, it has been suggested that the reason

1 https://www.seeingwithsound.com/.
2 https://neosensory.com/.

might be the largely synthetic nature of olfaction. In other words, there do not appear to be primary building blocks or basic components to which olfactory experience could be reduced (Wilson & Stevenson, 2006). In the case of gustation, although we know that multisensory integration plays an important role in our perception of flavors (Spence, 2015), it is unclear how a mapping of the five basic tastes to sounds or tactile stimuli could produce an experience comparable to taste qualities in a person who never experienced flavors (Macpherson, 2018). Similarly, while there is great interest in haptic technologies at the moment (Jewitt, Price, Leder Mackley, Yiannoutsou, & Atkinson, 2020), my research has not uncovered any SSDs that would translate touch into a different sensory modality.

However, despite bold visions with regard to the potential of sensory substitution-based technologies, their uptake as assistive technologies to date has been extremely limited. Critics have argued that fundamental imbalances with regard to the cortical area available to tactile and auditory processing as compared to vision make it unlikely for these modalities to truly compensate for the lack of sight. Moreover, sensory substitution devices are incapable of capturing the hedonic aspects of the sense modality being substituted for, which might further explain their limited use (Spence, 2014, 2018).

Although SSDs emerged in the context of restorative and assistive technologies, since the early days of this research there has been interest in applying them for augmentation purposes (Bach-y-Rim, Webster, Tompkins, & Crabb, 1987). It has been proposed that the principle of sensory substitution may be extended to feed new kinds of information to the nervous system, thereby allowing for the acquisition of new senses. To illustrate the concept, a German research group developed a wearable device to deliver tactile stimulation around the waist to signal the direction of north, giving wearers constant information about their orientation in space. Through a series of experiments they demonstrated that "newly acquired sensory information can have profound effects on performance and perceptual experience" (Nagel, Carl, Kringe, Märtin, & König, 2005:R15). Similarly, the VEST developed by Novich and Eagleman has been proposed as a model for creating non-invasive devices to extend human perception and extract features from a variety of high-dimension data by leveraging neuroplasticity and the brain's remarkable pattern recognition ability (Novich, 2015).

3.2.2 Augmented reality

The concept of augmented reality (AR) seems like a suitable synonym for sensory augmentation or perceptual enhancement, nevertheless it denotes a certain set of technologies that are quite distinct from the sensory substitution-based methods discussed in the previous section. The term itself was coined in 1992 by Tom Caudell and David Mizell to describe the technology of overlaying computer-generated graphics on the real world (Caudell & Mizell, 1992). According to an early, operational definition by Milgram and Kishino, augmented reality refers to "any case in which an otherwise real environment is "augmented" by means of virtual (computer graphic) objects" (1994). Thus, along the reality-virtuality continuum—also proposed by Milgram and Kishino (1994)—augmented reality sits between the real physical environment and fully immersive virtual reality. Although this suggests that AR can only be visual, there are auditory and haptic AR applications as well (Romanus et al., 2020; Stone & Tata, 2017). Finally, Azuma (1997) offered three criteria to delineate AR, which state that AR combines the real and the virtual, it is real-time and interactive, and it is registered in 3D.

The most prominent form of augmented reality is visual, and it may be displayed in various ways, such as projected onto surfaces or viewed via screens, glasses, head-mounted displays, mobile phones, contact lenses or projected straight into the eye via virtual retinal displays or light field displays. Generating AR content involves Simultaneous Localization and Tracking (SLAM) methods as well as depth tracking technology. These are used to create a live digital representation of the user's environment, while also keeping track of the precise location and movement of the user in that mapped space in order to place virtual objects accurately. At present, we may distinguish between four main types of AR (Edwards-Stewart, Hoyt, & Reger, 2016; Thinkmobiles, n.d.):

- *marker-based AR*, where an image, such as a QR code or a 3D-mapped object serves as the trigger for activating the AR content. For example, a picture printed on a breakfast cereal box, or indeed the whole box itself, when viewed via an AR device, might act as a trigger to activate the AR content;
- *markerless AR*, which relies on GPS coordinates and the device's compass and gyroscope to position the digital content, such as a navigation support app that overlays arrows and other directional cues and information on a live view of the user's environment;
- *projection-based AR*, where light is projected onto a surface to make it responsive to interaction, for example, augmenting a physical object such that touching certain parts can trigger digital events (i.e., turning an ordinary desk surface into a virtual keyboard);
- *superimposition-based AR*, where a digital image is partly or completely superimposed on an object, for example, superimposing an internal view, such as an X-ray or other medical scan on a live view of a patient.

An important figure in the history of AR is Steve Mann, who has been described as the father of wearable computing. Mann holds several patents related to image processing, wearable computing, brain-computer interfaces (BCIs) and other technologies, and he is among the founders of the BCI company, InteraXon, and AR headset manufacturer, Meta. While Mann is an unusually prolific and influential inventor, his work illustrates how AR intersects with a number of other technologies, such as BCIs and the Internet of Things, and is part of a larger ecosystem of digital technologies. As a pioneer of the field, Mann has been wearing some version of computerized eyeware during most waking hours over the past 40 years (Mann, 2013). The Eyetap device allows Mann to enhance his vision giving him an ultra-high dynamic range view of the world (Mann et al., 2012) and allows him to filter out certain parts of his visual scene, which Mann has been using to remove ads for cigarettes in real-time (Bilton, 2012). Mann is critical of conventional uses of AR as digital pop-up messages thrown on top of one's field of view, which merely results in information overload. Instead, he prefers to use the term mediated, or aug-mediated reality to describe the ways in which his eyewear enables him to experience reality more fully (Mann, 2013). Mann's vision for human-computer interaction, which he has been practicing since the 1970s is rooted in the conviction that natural computer interfaces should twist around and adapt to the human, instead of forcing humans to bend over uncomfortable and clunky input methods like keyboards and mice (Computer History Museum, 2019).

Today, consumers mostly experience mobile AR, which uses the camera of a smartphone to overlay digital content on the live view of the physical environment on screen. AR-based applications are among the fastest growing segments of the smartphone app industry, with blockbusters like Niantic's Pokémon Go—an AR game in which users collect and interact with digital characters appearing in various real-world locations—grossing over $3 billion between 2016 and 2019 (Statt, 2020).

Mobile AR may be seen as an interim step before the arrival of everyday smart glasses, similar to those pioneered by Mann. These allow for the fluid and continuous blending of digital content with one's physical environment. Google's prototype smart glasses, released in 2013, ran up against considerable public resistance as people felt their privacy was being intruded upon by the device's onboard camera that could be used to overtly take photos and videos (Sarpu, 2014). However, the product was subsequently rebranded and is currently widely used in the context of specific industrial applications (Naughton, 2017). Similarly, Microsoft's HoloLens is primarily used for guided instruction programs in various industries from medical training to factory assembly, where rich, animated content is overlaid directly on a trainee's field of vision (Gavish et al., 2015). Currently available headsets are often bulky and not yet suitable for everyday use. However, the technology's development points toward continuously shrinking the device to a comfortable form factor. Over the past years, the size of augmented reality glasses and headsets has reduced significantly, while their computational capacity has continued to increase. Apple and Facebook are expected to release consumer smart glasses by 2023 and both companies hold a number of patents in this area of innovation, with significant recent hires and acquisitions (Robertson, 2019; Vincent, 2019). In early 2020 Mojo Vision showcased a prototype AR contact lens that seeks to bring the technology even closer to the body by realizing what the company calls "invisible computing." The product received the FDA's Breakthrough Device Designation and Mojo Vision intends to first release it as a medical device before rolling it out to consumers (Heater, 2020).

Augmented reality is expected to become drastically more realistic and seamless with the arrival of 5G connectivity and edge computing. This will allow for extremely fast data connections between network servers—which perform the calculations and render photorealistic visuals at high speed—and the smart glasses/contact lenses which display the content itself. Building on such advances, several companies, including Google and Facebook, are working on AR cloud technology, which will act as an underlying digital content layer mapped unto objects and physical locations. The AR cloud is regarded as the infrastructure for the dawn of a novel, spatial computing paradigm, which will open entirely new forms of interaction, new services and vast new monetization opportunities (Koetsier, 2019). Importantly, augmented reality technologies will be integrated with the Internet of Things and be powered by the rapidly advancing capabilities of AI and machine learning, resulting in a new ecosystem of technologies that enable our physical environment to be constantly enriched by an additional layer of digital information. Besides major technology companies working to advance this, there is also an open source effort driven by the Open AR Cloud Association, a US-based non-profit,[3] that seeks to ensure that this new infrastructure of experiencing reality is built on the values of freedom, transparency, privacy and dignity.

3 https://www.openarcloud.org.

If and when large tech companies release AR glasses or contact lenses that are fit for everyday use, then this type of device can rapidly become the most widely used neurotechnology by a considerable margin. In 2019 alone, Apple sold over 185 million iPhones (Graham, 2020) and its sales of over 31 million Apple Watch units exceeded the performance of the entire Swiss watch industry (Reichert, 2020). Thus, the size of the market for these technologies, and hence the potential scope of their impact is enormous.

4. Ethical issues

There are multiple ways in which one could approach the discussion of ethical issues around the perception modification technologies described in the previous section. There are questions related to the physical properties of the system, denoting the risks and harms around the way it is attached to the body and the way in which it delivers stimulation to the nervous system, including its short- and long-term effects. Furthermore, there are potential ethical issues related to the phenomenological characteristics of the perception modification itself, that is, the contents of the experience. Finally, in the case of advanced devices, we might interrogate the workings of the algorithms that perform the mediation, looking at issues such as transparency, privacy, and autonomy.

While these questions could be addressed with reference solely to the technologies themselves, I believe they take on different meanings depending on whether we are discussing them in the context of a DIY practice or a consumer product. Therefore, in the subsequent section I will look at these two broad spheres of use, drawing out what I believe to be the distinctive issues in each domain. It is important to emphasize that while my discussion of the DIY space describes practices we are already confronting, my consideration of the issues around the consumer context will engage in a modest form of speculation as I try to anticipate the kinds of issues that might arise when potent AR devices are available for everyday use.

5. Hacking the senses—DIY sensory enhancement

The practice of technologically extending the body and pushing the limits of perception has been at the forefront of certain artists' work since at least the 1960s (Schwartzman, 2011). However, with advances in technology, the proliferation of cheap and accessible electronics components, and the growing cultural ubiquity of transhumanist ideas, the biohacker movement was born. Its members largely rely on the hacker ethic of hands-on experimentation, knowledge sharing and collaboration (Levy, 1984) and apply these principles to their own bodies. Biohacking methods range from relatively simple interventions like dietary supplementation, self-tracking, sleep optimization with melatonin or various devices (Platoni, 2015), all the way to more radical methods involving the implantation of devices into the body. Members in the latter category are also referred to as grinders. They take a pragmatic approach to transhumanism (Warwick, 2016) and their proactionary ethos (Fuller & Lipinska, 2014) is expressed succinctly on the biohack.me website, an important online resource of the community:

> "Grinders practice functional (sometimes extreme) body modification in an effort to improve the human condition. We hack ourselves with electronic hardware to extend and improve human capacities. Grinders believe in action, our bodies the experiment."[4]

The practice also takes inspiration from the broader body modification community and shares affinity with citizen science, as well as aspects of the quantified-self movement (Yetisen, 2018). In many respects, DIY sensory enhancement practices, as well as the grinder movement can be situated within the broader socio-cultural context of citizen science (Wexler, 2017). However, in comparison to DIY practices with other neurotechnologies like electrical brain stimulation (Jwa, 2015; Wexler, 2016), there is very little empirical research about body hackers/grinders. Based on available studies, the practice of bodyhacking plays a central role in the identity and self-perception of those engaged in such activities and this identity construction has a strong communal aspect to it. While practitioners of DIY neurostimulation congregate mostly online and pursue brain stimulation alone as a personal practice, grinders regularly meet at large festivals called grindfests or body hacking conferences, where they share knowledge and experiences, discuss new technologies and get magnets or RFID chips implanted. In addition, body hacking is associated with a certain disregard for societal conventions and a desire to step outside prevailing frames of reference and normative conceptions about the body and its proper functioning (Doerksen, 2018). Grinders are often associated with the image of makeshift basement laboratories and a certain recklessness (Popper, 2012) but members of the community have spoken out against this image, emphasizing that grinders are often highly trained and knowledgeable, who apply similar

[4] https://wiki.biohack.me/Who_We_Are.

safety measures as those used in hospitals, and even strive to innovate on existing practices and technologies (DEFCON Conference, 2017).

Arguably the most well-known individual in this space is Neil Harbisson, whose cyborg artist celebrity status sets him somewhat apart from the usual association of bodyhacking with the underground. Nevertheless, he serves as a key reference and source of inspiration for those engaged in sensory enhancement. Harbisson was born with a rare medical condition called achromatopsia which results in total color blindness. Since the early 2000s, Harbisson has been using a device co-developed with Adam Montandon, called the eyeborg (Wade, 2005). It consists of a camera, a microcomputer and a pair of headphones and it converts electromagnetic light waves into sound frequencies, thereby making colors audible. Due to gradual improvements, the device was developed into an antenna that has been surgically attached to Harbisson's head in such a way that he perceives color information through bone conduction. The antenna's capability has been extended so Harbisson is able to hear ultraviolet and infrared frequencies as well, and he can receive color information sent directly to his implant via the internet. The implantation procedure was carried out by an anonymous surgeon because Harbisson had difficulty convincing hospital bioethics committees to permit the operation (Harbisson, 2018). As a self-proclaimed cyborg, Harbisson has become very active in promoting the creation and use of technology to extend human perception and in 2010 he and fellow cyborg artist Moon Ribas launched the Cyborg Foundation to advance this aim.[5] In contrast to the underground quality associated with grinders, Harbisson has acquired considerable fame and has evangelized the notion of merging technologies with the human body in university lectures, conferences documentaries and mainstream media.

The body hacker community functions very much like a community in the sense of being bound together by a strong mutual interest in enhancing the body and expanding the senses, as well as a shared commitment to openness and information exchange. A small number of companies have also emerged that offer sensory extension products, which makes body hacking more widely accessible. For example, Dangerous Things, a US-based company sells chips and magnetic implants, as well as equipment for self-implantation, such as sterile scalpels and a "pain management kit" comprised of a sterile syringe and needle as well as lidocaine. In 2017, Cyborg Nest, a British company co-founded by Harbisson, released the North Sense, a small

[5] https://www.cyborgfoundation.com/.

device based on the idea of sensory substitution, which can be attached to the body with the help of transdermal titanium bars, which act as an anchoring system. The device gives off a gentle vibration whenever the wearer is facing north. The device was intended to be attached to the body during all waking hours in order to "experience the world through an extended frame."[6] The North Sense enjoyed very wide media coverage and brought the idea of sensory enhancement to a far broader audience than before (Holmes, 2016; Nye, 2018; Thaddeus-Johns, 2017). It was Cyborg Nest's first attempt at creating a commercial sensory enhancement product, but it was released as an experimental prototype, with the majority of customers representing early adopters with a strong interest in human enhancement (Emslie, 2017). The first batch of the North Sense sold out quickly and there is no information about a re-release but according to Cyborg Nest's website the company is working on a new range of sensory enhancement products.

5.1 Sensory enhancement as an end in itself?

Although more empirical work is needed to understand the motivations and aspirations of those who pursue DIY sensory enhancement, existing research seems to suggest that grinders and body hackers view it as an intrinsically valuable pursuit on the path to constructing an authentic self, irrespective of whether they acquire other advantages through the augmentation (Doerksen, 2018). This view sometimes connects to larger narratives about the desire and need to overcome the limitations of the human body (O'Connell, 2018). I believe there are at least two reasons why the esthetic component motivating sensory enhancement is an important factor that sets it apart from the majority of the enhancement debate. First, one of the most important ethical concerns with regard to enhancement is that it creates externalities for others by introducing competition and coercing those who are reluctant to enhance themselves to keep up with those who chose to do so. In this context, Cohen distinguishes between positional and absolute goods, where the former are goods worth having inasmuch as others lack them because they confer an advantage, while absolute goods are desirable irrespective of whether others possess them (Cohen, 2013). In line with this distinction, I would argue that the capacity to sense magnetic fields or to perceive north is not understood by biohackers as a positional good that is desirable for its utility. Rather, it is conceived of as an opportunity to experience the world more completely, which is evidenced by the

[6] https://web.archive.org/web/20180125105246/https://cyborgnest.net/.

fact that first-person accounts often emphasize the qualitative aspects of sensory enhancement over their usefulness.[7] One of the first recipients of the magnetic implant described the experience as follows:

> *"The implant has changed my perception of the world around me in a small but significant way. Information is constantly flowing around us, and we remain blissfully unaware of most of it. Having a tiny bit of that data stream pulled into your conscious awareness is a shocking experience. Functionally I have changed very little, but I am now more aware of what it is I don't feel. There is an untold amount of information flowing around us that we don't experience; my implant makes me think about this more" (Larratt, 2004).*

In addition, DIY sensory enhancement relies on technologies and interventions that are completely outside the scope of medicine. On the one hand, this means that their risks are lesser known and we mostly lack the moral frameworks to assess the practice. However, it also means that these pursuits are free of the central moral dichotomy between treatment and enhancement, which has defined much of the debate about enhancement (Cabrera, Fitz, & Reiner, 2015; Schwartz, 2005). Individuals will have different positions with regard to whether the experience ofbeing able to sense electromagnetic fields or to perceive the direction of north is valuable enough for them to undergo a painful procedure and to expose themselves to risks, such as infections. Currently, for most people, that calculus seems to tilt toward the negative and toward foregoing the novel experience. However, for some, this is not the case. These factors make DIY sensory enhancement a highly worthwhile future research topic for neuroethics.

5.2 Who is responsible when something goes wrong?

If we accept that DIY sensory augmentation amounts to a non-competitive form of enhancement that plays an important role in the identity of those who engage in this practice, then a case can be made for why individuals should have the freedom to pursue it, even though it involves risks and harms. However, sensory enhancement is not an entirely do-it-yourself or individual practice, as only a very small number of biohackers make their own sensory enhancements, while the majority purchase them from companies started by other biohackers. Similarly, only a very small minority would implant magnets or devices like the North Sense in themselves, and instead, most rely on body modification professionals to do so. Hence, there are two additional actors whose responsibility has to be considered:

[7] http://feelingwaves.blogspot.com/.

the companies manufacturing and selling sensory enhancement products, and the body modification professionals who install them on customers.

At present, both spheres are largely unregulated. The Dangerous Things company website includes a disclaimer that their products are not certified by any regulatory agency and that individuals use them at their own risk. Importantly, the website does not even claim that the products are safe. Customers either perform the implantation themselves, or seek out a body modification professional. Cyborg Nest emphasized on their website that they use high quality components in their product, such as materials approved for implantation in the human body like medical grade titanium and silicone but the device didn't undergo any form of licensing or authorization before being released to the public. Although it is marketed as a "new sense" the meaning of this is unclear from a legal perspective. What customers actually purchase are a set of piercings and a small, custom designed and manufactured microcomputer that is a simple consumer electronics product. The insertion of the titanium bar piercings has to be done in a suitable body modification, which is the responsibility of the customer.

Piercing and tattooing salons are subject to some degree of licensing in most countries and they are expected to comply with certain hygiene standards. In the United Kingdom, piercing studios have to register with their local council and their premises are likely to be regularly inspected for hygiene, but there are no mandatory education and training programs for piercers and quality insurance is essentially left to the industry (UK Government, n.d.). Groups such as the Association of Professional Piercers in the US and the UK act as voluntary organizations spreading safe piercing practices.[8] More invasive forms of body modification, such as subdermal and transdermal implants are not covered by UK law and in the US regulations can vary from state to state (Pegg, 2017). As a consequence, the practice of grinding exists in a legal limbo, where members of the community are simultaneously committed to sharing knowledge, commercializing their products and operating in a way that avoids regulation (Doerksen, 2018). A recent case in the United Kingdom made the practice of body modification even more uncertain. A practitioner was convicted of causing grievous bodily harm for performing certain procedures such as tongue-splitting and the removal of a man's nipple. Although the clients had consented to the procedures, the court ruled that the interventions were surgical in nature, carried out by an untrained person and posed serious risk to the

[8] See https://www.ukapp.org.uk/ and https://www.safepiercing.org/.

client and were performed for no good medical reason, irrespective of whether the act was consensual (Pegg, 2019). While it is expected that future regulation will clarify the status of body modification, at present—in the United Kingdom at least—there is no clarity with regard to what procedures might result in similar convictions. The body modification techniques that were implicated in the case are unrelated to current DIY sensory enhancement practices, the more invasive their interventions become, the more they will be exposed to similar consequences. Thus, at present, there is a growing community of individuals interested in pursuing sensory and other enhancements, as well as a handful of companies emerging from this community, working on developing a variety of products but the occupation with the necessary training and regulation to perform the installation procedure on the body does not yet exist. Given that sensory enhancement products affect the structure and function of the human body, they satisfy at least one of the US Food and Drug Administration's criteria for medical devices (FDA, 2019) and could therefore fall under the agency's regulatory purview. However, so far health regulatory agencies have not taken any enforcement action against grinding or sensory augmentation practices, similar to how direct-to-consumer neurotechnology companies operate (Wexler & Reiner, 2019). If sensory enhancement devices were to be classified as medical devices that would spell the end of the DIY movement as body modification experts would no longer be allowed to implant them at all. Such a development might move the practice of sensory enhancement toward the cosmetic surgery industry, while driving up costs and reducing access. At the same time, the practice of self-experimentation is unlikely to halt, suggesting that overly stringent regulation might push the practice further underground and toward riskier conditions. Although this is arguably an undesirable outlook, the current legal gray area in which DIY sensory enhancement unfolds, is equally suboptimal.

5.3 Going ahead

DIY practices involving various branches of science have encountered mixed reactions from the scientific community. DIY biology, the citizen science practice involving biological experimentation in amateur laboratories has attracted both praise as a creative and responsible democratization of science, and criticism as a potentially risky endeavor (Kuiken, 2016; Wolinsky, 2009). The phenomenon of DIY neurostimulation was greeted with concern as well but researchers also perceived an ethical obligation to engage in discussion with the community and offered practical guidance (Fitz & Reiner, 2015; Wurzman, Hamilton, Pascual-Leone, & Fox, 2016). When it comes to bodyhacking and the DIY sensory enhancement or grinder community, this kind of engagement is still in its very early stages and mostly takes place through highly informal channels, such as panel discussions organized at body hacking conferences (BDYHAX, 2017). So far there are no known cases of serious injury or adverse events related to DIY sensory enhancement practices. However, the inherently risky nature of these activities makes it almost unavoidable that some individuals will be harmed. Given the largely reactive mode of operation characteristic of regulatory agencies, such an incident is likely to result in draconian rules with regard to biohacking, which could make the practice more unsafe, by pushing it underground. Therefore, it is important to engage in early conversations about how to create the conditions for the creative exploration that characterizes DIY sensory enhancement to proceed safely and with adequate oversight.

In the meantime, it is important for individuals interested in experimenting with sensory enhancement products to be aware of the uncertainties surrounding the practice, as there are no established measures and assessments of the possible risks. Companies operating in this space must embrace responsible communication practices and avoid overstating the effects of their products. Moreover, media reporting about biohacking should refrain from sensationalism.

The citizen science movement offers a fruitful framework for interpreting the practices of DIY sensory enhancement and body hacking (Wexler, 2017). However, more research is needed to understand the unique features of this practice, most notably the elevated risks presented by experimentation with invasive methods, the ways in which scientific knowledge is both mobilized and questioned by the community, the legal lacuna concerning the liability for bodyhacking as well as the tensions that emerging companies must navigate as they seek to commercialize their products, while attempting to maintain their commitment to the open source mentality of the community (Furness, 2017).

6. Mass customized realities

The DIY sensory enhancement practices discussed in the previous section currently involve a small subculture of grinders and body hackers. In contrast, the technologies discussed below have the potential to impact millions of people around the world, if augmented reality glasses or contact lenses become as ubiquitous as mobile phones are today. Given that augmented reality has been described as the "operating system of the future" (Koetsier, 2019) and heralded as the next computational paradigm, this prospect is not entirely far-fetched.

There are a number of potential issues that arise in relation to the widespread use of augmented reality technologies. First, there are the potential health effects of continued exposure to AR delivered via glasses, contact lenses or retinal displays. Second, there are concerns related to the ways in which the algorithms in the device operate. In recent years a rich literature has developed on the ethics of algorithms, which has highlighted questions related to transparency, privacy and security, as well as autonomy and manipulation (Mittelstadt, Allo, Taddeo, Wachter, & Floridi, 2016). Finally, there are justice-related concerns around AR's potential discriminatory effects. It is important to emphasize that I am focusing not on the currently widespread mobile-based AR applications but on the kinds of AR implementations that allow ongoing use via glasses or contact lenses, which are beginning to enter the market and will likely proliferate over the coming years.

6.1 Health effects

The most straightforward question with regard to the health effects of AR glasses or contact lenses is related to adverse effects on the eyes and vision. Although optical see-through systems—like the ones currently on the market and under development—avoid earlier problems related to eye strain, there are some unanswered questions with regard to long-term use. For example, systems that display content to one eye only may produce a condition known as phoria, which is a misalignment of the eyes. In addition, having to process two different streams of information coming from each eye can lead to binocular rivalry and cause discomfort (Due, 2014). Furthermore, conventional AR glasses that rely on stereoscopy, can lead to an effect known as vergence-accommodation conflict. This occurs because the illusion of 3D in augmented reality is created via binocular disparities between the images presented to each eye. The brain uses such disparities to create our perception of depth. However, the lenses displaying the content are located at a fixed distance from the eyes. This creates a conflict between the perceived distance of the digital object and the actual focusing distance required for the eyes to keep the image blur-free. This conflict can lead to focusing problems, visual fatigue and eyestrain (Zabels et al., 2019). In addition, concerns have been raised about the level of blue light emitted by AR devices as continued exposure may have adverse effects on circadian rhythms (Legerton, Segre, & Marsh, 2018). Furthermore, the digital information displayed on AR devices requires users to shift their gaze which can cause distractions and increase the likelihood of accidents (Sabelman & Lam, 2015).

6.2 Privacy

There are several privacy-related concerns with augmented reality technology, which may be grouped into two broad categories: issues around the privacy of the user, and issues around the privacy of others.

The embedded camera on AR devices makes it possible to capture audio, photo or video material anywhere, including in public places without consent (Hofmann, Haustein, & Landeweerd, 2017). Using facial recognition technology, persons can also be identified even in situations where individuals might have a reasonable expectation of privacy (Acquisti, Gross, & Stutzman, 2014). Moreover, technologies such as emotion detection embedded in AR devices can reveal information others may not be willing to disclose (Mehta, Siddiqui, & Javaid, 2018; Sedenberg & Chuang, 2017; Sedenberg, Wong, & Chuang, 2018). AR is not only capable of collecting information about others but also to attach and overlay digital information on individuals, which might open avenues for new forms of harassment and bullying as individuals are tagged and marked with certain AR content against their desire or consent.

In addition, a new set of questions is starting to emerge around the ethics and legality of adding digital content to certain physical locations. In other words, who has the right to enrich a certain location with a digital layer of information, and what norms should guide this practice? The AR game Pokémon Go, mentioned in an earlier section, offers an illustration of the kinds of issues that need to be addressed. For example, homeowners across the USA and Canada have claimed that their property was negatively affected by the sudden surge of players who were led to their residential area by the AR app, and a number of lawsuits were filed with regard to this nuisance (Mitchell, 2016). The content added to a particular location raises a whole host of legal questions with regard to obscenity, defamation, and political and commercial advertisement (Judge & Brown, 2018). Moreover, cultural institutions, such as the Auschwitz-Birkenau

State Museum had to plead with the game's developer Niantic to not place virtual characters on the museum's premises because it was disrespectful (Associated Press, 2016).

With regard to potential violations of the AR user's privacy, it is important to recall that the effective functioning of AR devices makes it necessary for the system to continuously track the wearer's physical location and using the onboard camera and sensors, create a detailed map of the environment. This means that in addition to all the information that current smartphones routinely collect about users, AR devices would also include a live-feed of the person's point of view, as well as the sensor data that an AR device might be equipped with. For example, based on existing patents, future AR devices might be capable of inferring a user's level of attention and their emotional state using eye tracking and other sensor data (Asbun, Reznik, Zeira, Sternberg, & Neff, 2018).

Finally, there are also concerns related to the security of AR devices and their vulnerability to third parties gaining unauthorized access (Roesner, Kohno, & Molnar, 2014).

Currently, a number of legislative efforts seek to control the ways in which personal data may be collected and processed by technology companies. Europe's General Data Protection Regulation (GDPR), which came into effect in May 2018 (European Commission, 2016) is perhaps the most stringent and wide ranging of these efforts, which served as a model for other data protection regulations around the world. The GDPR is built on key principles, which are lawful, fair and transparent processing; purpose limitation; data minimization; accuracy; storage limitation; and integrity and confidentiality. The GDPR allows individuals to make decisions about how their data may be shared, and it does not permit users to opt-out of protections that limit a company's data collection to the purpose of the specific contract under which they provide their services. This is meant to ensure that fundamental protections are guaranteed to stay in place. However, scholars have argued that approaches to privacy and data protection that rely on the individual to manage their privacy choices are limited (Lehtiniemi & Kortesniemi, 2017). Moreover, recent research has shown that compliance with the GDPR's requirement to obtain consent for data processing is very low. Websites use dark patterns—certain features of user interface design that guide and direct user choice in certain directions—to defer users from opting for the most privacy-preserving options (Nouwens, Liccardi, Veale, Karger, & Kagal, 2020).

Admittedly, these points are drawn from our existing experience with the privacy challenges of the internet, however, several of the difficulties we encountered in the online space so far will get transposed to augmented reality, as it becomes a new and perhaps dominant interface.

6.3 Autonomy and manipulation

Although privacy and data protection concerns tend to dominate discussions about augmented reality (Hofmann et al., 2017), these only address the input side of the algorithms that are involved in the AR experience, that is, the information that systems may collect about the user and their environment. What is at least equally important is the inseparable question of the algorithm's output, that is, what users actually experience as augmented reality content. As discussed before, augmented reality systems allow for the modulation of a user's experience of their surroundings. This includes adding digital information on top of the physical environment, as well as the real-time modulation or replacement of elements in the visual scene. This raises the prospect of malicious third-party actors injecting or modifying AR content to cause harm, confusion, or distress to an unsuspecting user (Lebeck, Ruth, Kohno, & Roesner, 2017). Besides such interference, there are also questions with regard to the degree to which the normal and intended operation of AR services built on top of existing technology platforms might compromise autonomy. By normal operation, I mean business models predicated on the continuous collection of personal data in order to deliver mass customized digital services. Shoshanna Zuboff has described this post-industrialist form of production as surveillance capitalism. In this system data is the most valuable resource and technology companies are engaged in a relentless effort to find ways of transforming all aspects of human behavior into "raw material" that can be exploited for the generation of revenue (Zuboff, 2019). In this respect, augmented reality offers an unprecedented opportunity to gather new forms of behavioral data. However, it is important to keep in mind that the underlying reason and purpose of all this data collection and profiling is to enable customized service delivery. In other words, data collection is just the means by which users' digital experience can be tailored. In recent years, there has been a growing concern about the effects of personalized information filtering mechanisms on online platforms, which might contribute to the creation of echo chambers, where individuals are only exposed to content they already agree with, contributing to increased societal polarization and a distorted worldview (Pariser, 2011). The magnitude of this effect is still debated as rigorous empirical research has yielded inconclusive evidence (Flaxman, Goel, & Rao, 2016; Nechushtai & Lewis, 2019). However, in the wake of the Cambridge Analytica scandal related to electoral manipulation (Cadwalladr & Graham-Harrison, 2018), there has been increased attention to the potential of microtargeting—that is the ability of delivering personalized messages and content to users on the basis of digital profiling—to manipulate individuals, especially in the context of political decision-making (Gorton, 2016; Zuiderveen Borgesius et al., 2018). Scholars have argued that these tools enable hypernudging, a highly potent and pervasive form of algorithmically delivered behavior change (Yeung, 2017).

While strengthening data protection and privacy rights can curb the flow of data that feeds mass customization and hypernudging algorithms, a policy approach focused on the confidentiality aspects of privacy is unlikely to be sufficient. The well-intentioned efforts to create systems that embed privacy requirements at the level of technology development itself has contributed to the emergence of cryptographic and other methods that allow companies to continue to provide their highly targeted and customized services without having access to any personal data whatsoever (Veale, 2020). Thus, even in a world of ideal data protection laws, manipulative services could still be rendered. This is why scholars have called for a right to reasonable inferences (Wachter, 2019; Wachter & Mittelstadt, 2019) to ensure that users maintain control over not only what data informs algorithms, but also how the outputs and inferences of those algorithms come to affect them.

Considering the fact that it will be the same large technology companies developing the AR cloud infrastructure, which already hold unprecedented amounts of information about billions of users, there is a real risk that widespread adoption of augmented reality can accelerate us toward a fragmented future where we no longer perceive the same reality.

7. Conclusion

In this chapter I have argued that ethical issues related to technologies that mediate how humans perceive the world around them in an ongoing manner are worthy of more attention and consideration by neuroethicists. I have provided an overview of relevant technologies and discussed some of the ethical issues that they present. In particular, I have focused on two, in some ways diametrically opposed domains centered around the extension of our perceptual experience. On the one hand, biohackers engaged in DIY sensory enhancement and situated outside medical and corporate establishments are seeking to invent and share ways of broadening their sensory horizons. Their pursuits are risky and currently operate in an ethical and legal gray zone that needs to be resolved. On the other hand, some of the largest technology companies are in pursuit of the next computational paradigm that will integrate the digital and the physical worlds in a more seamless way than has been possible before. Although this technology opens up a vast range of opportunities, the currently dominant business models of extractive data accumulation and mass customization raise the prospect of manipulative uses of technology. To some extent, the two spheres represent competing visions with regard to our future relationship with sensory enhancement. We need robust and timely engagement in order to ensure that both develop in desirable directions.

References

Acquisti, A., Gross, R., & Stutzman, F. D. (2014). Face recognition and privacy in the age of augmented reality. *The Journal of Privacy and Confidentiality*, *6*(2), 1.

Asbun, E., Reznik, Y., Zeira, A., Sternberg, G. S., & Neff, R. (2018). *U.S. Patent No. 9,922,253*. Washington, DC: U.S. Patent and Trademark Office.

Associated Press. (2016, July 13). *Auschwitz memorial says playing 'Pokemon Go' not allowed*. https://apnews.com/de0fdd168b514cdf9f60ee9f083d83ba/auschwitz-memorial-says-playing-pokemon-go-not-allowed. Last accessed 31 January 2020.

Azuma, R. (1997). A survey of augmented reality. *Presence: Teleoperators and Virtual Environments*, *6*(4), 355–385.

Bach-y-Rim, P., Webster, J. G., Tompkins, W. J., & Crabb, T. (1987). Sensory substitution for space gloves and for space robots. In *Vol. 2. Proceedings of the workshop on space telerobotics* (pp. 51–57). Jet Propulsion Laboratory, California Institute of Technology. (SEE N89-26492 20-63).

Bach-y-Rita, P. (2002). Sensory substitution and qualia. In A. Noë & E. Thompson (Eds.), *Vision and mind* (pp. 497–514). Cambridge, MA: MIT Press.

Bach-y-Rita, P., Collins, C. C., Saunders, F. A., White, B., & Scadden, L. (1969). Vision substitution by tactile image projection. *Nature*, *221*(5184), 963–964.

BDYHAX. (2017, July 26). *BDYHAX 2017—Ethics in bodyhacking [video file]*. https://www.youtube.com/watch?v=sDbtv_yhLqM. Last accessed 31 January 2020.

Beauchamp, M. S., & Martin, A. (2007). Grounding object concepts in perception and action: Evidence from fMRI studies of tools. *Cortex*, *43*(3), 461–468.

Biggs, S., Matthen, M., & Stokes, D. (2014). Sorting the senses. In D. Stokes, M. Matthen, & S. Biggs (Eds.), *Perception and its modalities* (pp. 1–19). Oxford University Press.

Bilton, N. (2012, August 7). *One on one: Steve Mann, wearable computing pioneer*. The New York Times.

Cabrera, L. Y., Fitz, N. S., & Reiner, P. B. (2015). Empirical support for the moral salience of the therapy-enhancement distinction in the debate over cognitive, affective and social enhancement. *Neuroethics*, *8*(3), 243–256.

Cadwalladr, C., & Graham-Harrison, E. (2018). *Revealed: 50 million Facebook profiles harvested for Cambridge Analytica in major data breach. The Guardian.* https://www.theguardian.com/news/2018/mar/17/cambridge-analytica-facebook-influence-us-election. Accessed 31 January 2020.

Carhart-Harris, R. L., Leech, R., Hellyer, P. J., Shanahan, M., Feilding, A., Tagliazucchi, E., et al. (2014). The entropic brain: A theory of conscious states informed by neuroimaging research with psychedelic drugs. *Frontiers in Human Neuroscience, 8*, 20.

Caudell, T., & Mizell, D. (1992). Augmented reality: An application of heads-up display technology to manual manufacturing processes. In B. D. Shriver (Ed.), *Vol. 2. Proceedings of the twenty-fifth Hawaii international conference on systems sciences.*

Clark, A. (1999). An embodied cognitive science? *Trends in Cognitive Sciences, 3*(9), 345–351.

Clark, A. (2004). *Natural-born cyborgs—Minds, technologies, and the future of human intelligence.* Oxford University Press.

Cohen, I. G. (2013). What (if anything) is wrong with human enhancement; what (is anything) is right with it. *Tulsa Law Review, 49*, 645.

Computer History Museum. (2019, April 1). *Oral history of Steve Mann [video file].* Retrieved from https://www.youtube.com/watch?v=oPGEXxXTUHU. Last accessed 31 January 2020.

DEFCON Conference. (2017, October 30). *DEF CON 25 biohacking village—Jeffrey Tibbetts—Biohackers die [video file].* https://www.youtube.com/watch?v=nn0kglB46BU. Last accessed 31 January 2020.

Deloitte. (2016). *Tech trends 2016—Innovating in the digital era.* Deloitte University Press.

Doerksen, M. D. (2017). Electromagnetism and the Nth sense: Augmenting senses in the grinder subculture. *The Senses and Society, 12*(3), 344–349.

Doerksen, M. D. (2018). *How to make sense: Sensory modification in grinder subculture.* PhD thesis, Concordia University.

Due, B. L. (2014). The future of smart glasses: An essay about challenges and possibilities with smart glasses. *Working Papers on Interaction and Communication, 1*(2), 1–21.

Eagleman, D. (2015). *Can we create new senses for humans [video file].* https://www.ted.com/talks/david_eagleman_can_we_create_new_senses_for_humans. Accessed 31 January 2020.

Eaton, M. L., & Illes, J. (2007). Commercializing cognitive neurotechnology—The ethical terrain. *Nature Biotechnology, 25*(4), 393–397.

Edwards-Stewart, A., Hoyt, T., & Reger, G. (2016). Classifying different types of augmented reality technology. *Annual Review of Cybertherapy and Telemedicine, 14*, 199–201.

Emanuel, P., Walper, S., DiEuliis, D., Klein, N., Petro, J. B., & Giordano, J. (2019). *Cyborg soldier 2050: Human/machine fusion and the implications for the future of the DOD.* U.S. Army Combat Capabilities Development Command Chemical Biological Center.

Emslie, K. (2017). *This artificial sixth sense helps humans orient themselves in the world.* Smithsonian Magazine. https://www.smithsonianmag.com/innovation/artificial-sixth-sense-helps-humans-orient-themselves-world-180961822/. Last accessed 31 January 2020.

Engel, A. K., Maye, A., Kurthen, M., & König, P. (2013). Where's the action? The pragmatic turn in cognitive science. *Trends in Cognitive Sciences, 17*(5), 202–209.

European Commission. (2016). Regulation (EU) 2016/679 of the European Parliament and of the Council of 27 April 2016 on the protection of natural persons with regard to the processing of personal data and on the free movement of such data, and repealing the Directive 95/46/EC (General Data Protection Regulation). *Official Journal of the European Union, L119*, 1–88.

FDA. (2019). *How to determine if your product is a medical device.* https://www.fda.gov/medical-devices/classify-your-medical-device/how-determine-if-your-product-medical-device. Last accessed 31 January 2020.

Fish, W. (2010). *Philosophy of perception: A contemporary introduction.* Routledge.

Fitz, N. S., & Reiner, P. B. (2015). The challenge of crafting policy for do-it-yourself brain stimulation. *Journal of Medical Ethics, 41*(5), 410–412.

Flaxman, S., Goel, S., & Rao, J. M. (2016). Filter bubbles, echo chambers, and online news consumption. *Public Opinion Quarterly*, *80*(S1), 298–320.
Fuller, S., & Lipinska, V. (2014). *The proactionary imperative: A foundation for transhumanism.* Springer.
Furness, D. (2017). *A look inside the budding battle between cyborg supply startups.* DigitalTrends. https://www.digitaltrends.com/cool-tech/inside-budding-battle-between-cyborg-supply-startups/. Last accessed 31 January 2020.
Gavish, N., Gutiérrez, T., Webel, S., Rodríguez, J., Peveri, M., Bockholt, U., et al. (2015). Evaluating virtual reality and augmented reality training for industrial maintenance and assembly tasks. *Interactive Learning Environments*, *23*(6), 778–798.
Goldstein, E. B. (Ed.), (2008). *Blackwell handbook of sensation and perception.* John Wiley & Sons.
Gorton, W. A. (2016). Manipulating citizens: How political campaigns' use of behavioral social science harms democracy. *New Political Science*, *38*(1), 61–80.
Graham, J. (2020). *Once again, iPhones top tech seller of the year, but AirPods the hit.* USA Today. https://eu.usatoday.com/story/tech/2019/12/31/iphone-sales-dip-but-still-no-1-2019-airpods-and-watch-rise/2775193001/. Last accessed 31 January 2020.
Guarniero, G. (1974). Experience of tactile vision. *Perception*, *3*(1), 101–104.
Hameed, J., Harrison, I., Gasson, M. N., & Warwick, K. (2010, September). A novel human-machine interface using subdermal magnetic implants. In *2010 IEEE 9th international conference on cybernetic intelligent systems* (pp. 1–5). IEEE.
Hamilton-Fletcher, G., Mengucci, M., & Medeiros, F. (2016). Synaestheatre: Sonification of coloured objects in space. In *Proceedings of the international conference on live interfaces 2016* (pp. 252–256).
Hansson, S. O. (2015). Ethical implications of sensory prostheses. In J. Clausen & N. Levy (Eds.), *Handbook of neuroethics* (pp. 785–797). Netherlands: Springer.
Harbisson, N. (2018). Hearing colors: My life experience as a cyborg. In X. Pavie (Ed.), *Creativity, imagination and innovation.* World Scientific Publishing.
Harrison, I. (2015). *Sensory enhancement, a pilot perceptual study of subdermal magnetic implants.* (Doctoral dissertation). University of Reading.
Heater, B. (2020). *Mojo Vision's AR contact lenses are very cool, but many questions remain.* TechChrunch. https://techcrunch.com/2020/01/16/mojo-visions-ar-contact-lenses-are-very-cool-but-many-questions-remain/. Last accessed 31 January 2020.
Hillenbrand, S., Raveh, D., & Amedi, A. (2018). What can sensory substitution tell us about the organization of the brain? In F. Macpherson (Ed.), *Sensory substitution and augmentation* (pp. 94–121). Oxford: Oxford University Press.
Hofmann, B., Haustein, D., & Landeweerd, L. (2017). Smart-glasses: Exposing and elucidating the ethical issues. *Science and Engineering Ethics*, *23*(3), 701–721.
Holmes, K. (2016). *An artificial cyborg implant has been fitted on humans for the first time.* Vice. https://www.vice.com/en_us/article/wnpvb5/cyborg-implant-north-sense-fitted-on-humans. Last accessed 31 January 2020.
Jebari, K. (2015). Sensory enhancement. In J. Clausen & N. Levy (Eds.), *Handbook of neuroethics* (pp. 827–838). Netherlands: Springer.
Jewitt, C., Price, S., Leder Mackley, K., Yiannoutsou, N., & Atkinson, D. (2020). *Interdisciplinary insights for digital touch communication.* SpringerOpen.
Johnson, M. (1987). *The body in the mind: The bodily basis of meaning, imagination, and reason.* University of Chicago Press.
Judge, E. F., & Brown, T. E. (2018). A right not to be mapped? Augmented reality, real property, and zoning. *Laws*, 7(2), 23.
Jwa, A. (2015). Early adopters of the magical thinking cap: A study on do-it-yourself (DIY) transcranial direct current stimulation (tDCS) user community. *Journal of Law and the Biosciences*, *2*(2), 292–335.
Koestler, F. A. (2004). *The unseen minority: A social history of blindness in the United States.* American Foundation for the Blind.

Koetsier, J. (2019). *Augmented reality is the operating system of the future. AR cloud is how we get there.* Forbes, Online. https://www.forbes.com/sites/johnkoetsier/2019/02/21/augmented-reality-is-the-operating-system-of-the-future-ar-cloud-is-how-we-get-there/. Last accessed 31 January 2020.

Komesaroff, L., Komesaroff, P. A., & Hyde, M. (2015). Ethical issues in cochlear implantation. In J. Clausen & N. Levy (Eds.), *Handbook of neuroethics* (pp. 815–826). Netherlands: Springer.

Kreitmair, K. V. (2019). Dimensions of ethical direct-to-consumer neurotechnologies. *AJOB Neuroscience*, *10*(4), 152–166.

Kuiken, T. (2016). Governance: Learn from DIY biologists. *Nature*, *531*(7593), 167–168.

Lakoff, G., & Núñez, R. (2000). *Where mathematics comes from (Vol. 6).* New York: Basic Books.

Larratt, S. (2004). *The gift of magnetic vision.* Modblog. https://news.bme.com/2004/02/06/the-gift-of-magnetic-vision-the-publishers-ring/. Accessed 31 January 2020.

Lebeck, K., Ruth, K., Kohno, T., & Roesner, F. (2017, May). Securing augmented reality output. In *2017 IEEE symposium on security and privacy (SP)* (pp. 320–337). IEEE.

Legerton, J., Segre, L., & Marsh, J. (2018). Reality check: Protecting ocular health from headset hazards. *Review of Optometry*, *155*(9), 52–59.

Lehtiniemi, T., & Kortesniemi, Y. (2017). Can the obstacles to privacy self-management be overcome? Exploring the consent intermediary approach. *Big Data & Society*, *4*(2), 2053951717721935.

Levy, S. (1984). Hackers: Heroes of the computer revolution. In *Vol. 14.* Garden City, NY: Anchor Press/Doubleday.

Macpherson, F. (2007). Synaesthesia, functionalism and phenomenology. In *Cartographies of the mind* (pp. 65–80). Dordrecht: Springer.

Macpherson, F. (Ed.), (2018). *Sensory substitution and augmentation.* Oxford: Oxford University Press.

Mann, S. (2013). *My "augmediated" life.* IEEE Spectrum. https://spectrum.ieee.org/geek-life/profiles/steve-mann-my-augmediated-life. Last accessed 31 January 2020.

Mann, S., Lo, R. C., Ovtcharov, K., Gu, S., Dai, D., Ngan, C., et al. (2012). Realtime HDR (high dynamic range) video for eyetap wearable computers, FPGA-based seeing aids, and glasseyes (EyeTaps). In *2012 25th IEEE Canadian conference on electrical and computer engineering (CCECE)* (pp. 1–6).

Mehta, D., Siddiqui, M. F. H., & Javaid, A. Y. (2018). Facial emotion recognition: A survey and real-world user experiences in mixed reality. *Sensors*, *18*(2), 416.

Meijer, P. B. (1992). An experimental system for auditory image representations. *IEEE Transactions on Biomedical Engineering*, *39*(2), 112–121.

Milgram, P., & Kishino, F. (1994). A taxonomy of mixed reality visual displays. *IEICE Transactions on Information and Systems*, 77(12), 1321–1329.

Mitchell, R. (2016). Pokémon Go-es directly to court: How Pokémon Go illustrates the issue of virtual trespass and the need for evolved tort Laws. *Texas Tech Law Review*, *49*, 959.

Mittelstadt, B. D., Allo, P., Taddeo, M., Wachter, S., & Floridi, L. (2016). The ethics of algorithms: Mapping the debate. *Big Data & Society*, *3*(2), 2053951716679679.

Mordor Intelligence. (2019). *Smart wearable market—Growth, trends and forecasts (2020–2025).* https://www.mordorintelligence.com/industry-reports/smart-wearables-market. Accessed 31 January 2020.

Morvan, L., Ovanesoff, A., Carrell-Billiard, M., & Hintermann, F. (2019). Waking up to a new reality: Building a responsible future for immersive technologies. *Accenture.* https://www.accenture.com/_acnmedia/accenture/redesign-assets/dotcom/documents/global/1/accenture-g20-yea-report.pdf. Accessed 24 March 2020.

Müller, O., & Rotter, S. (2017). Neurotechnology: Current developments and ethical issues. *Frontiers in Systems Neuroscience*, *11*, 93.

Nagel, S. K., Carl, C., Kringe, T., Märtin, R., & König, P. (2005). Beyond sensory substitution—Learning the sixth sense. *Journal of Neural Engineering*, *2*(4), R13.

Naughton, J. (2017). *The rebirth of Google Glass shows the merit of failure. The Guardian.* https://www.theguardian.com/commentisfree/2017/jul/23/the-return-of-google-glass-surprising-merit-in-failure-enterprise-edition. Accessed 31 January 2020.

Nechushtai, E., & Lewis, S. C. (2019). What kind of news gatekeepers do we want machines to be? Filter bubbles, fragmentation, and the normative dimensions of algorithmic recommendations. *Computers in Human Behavior*, *90*, 298–307.

Noë, A., & Noë, A. (2004). *Action in perception.* MIT Press.

Norton, Q. (2006). *A sixth sense for a wired world.* Wired, Online. https://www.wired.com/2006/06/a-sixth-sense-for-a-wired-world/. Accessed 31 January 2020.

Nouwens, M., Liccardi, I., Veale, M., Karger, D., & Kagal, L. (2020). Dark patterns after the GDPR: Scraping consent pop-ups and demonstrating their influence. *arXiv* preprint arXiv:2001.02479. https://arxiv.org/abs/2001.02479.

Novich, S. D. (2015). *Sound-to-touch sensory substitution and beyond.* (Doctoral dissertation).

Nye, C. (2018). *Biohacker: Meet the people 'hacking' their bodies.* BBC. https://www.bbc.com/news/technology-46442519. Last accessed 31 January 2020.

O'Connell, M. (2018). *To be a machine: Adventures among cyborgs, utopians, hackers, and the futurists solving the modest problem of death.* Anchor.

O'Regan, J. K., & Noë, A. (2001). A sensorimotor account of vision and visual consciousness. *Behavioral and Brain Sciences*, *24*(5), 939–973.

Pariser, E. (2011). *The filter bubble: What the internet is hiding from you.* UK: Penguin.

Pegg, S. (2017). *Is extreme body modification even legal? The conversation.* https://theconversation.com/is-extreme-body-modification-even-legal-73242. Last accessed 31 January 2020.

Pegg, S. (2019). *Body modification—When consent is not a defence. The Conversation.* https://theconversation.com/body-modification-when-consent-is-not-a-defence-87816. Last accessed 31 January 2020.

Pfeifer, R., & Bongard, J. (2006). *How the body shapes the way we think: A new view of intelligence.* MIT Press.

Platoni, K. (2015). *We have the technology: How biohackers, foodies, physicians, and scientists are transforming human perception, one sense at a time.* Hachette UK.

Popper, B. (2012). *Cyborg America: Inside the strange new world of basement body hackers.* The Verge. https://www.theverge.com/2012/8/8/3177438/cyborg-america-biohackers-grinders-body-hackers. Last accessed 31 January 2020.

Reichert, C. (2020). *Apple shipped 10M more watches than entire Swiss watch industry in 2019, analyst says.* C|Net. https://www.cnet.com/news/apple-shipped-10m-more-watches-than-the-entire-swiss-watch-industry-in-2019-analyst-says/. Last accessed 5 February 2020.

Renier, L., & De Volder, A. G. (2013). Sensory substitution devices: Creating "artificial synesthesias" In J. Simner & E. Hubbard (Eds.), *Oxford handbook of synesthesia* (pp. 853–868). Oxford University Press.

Robertson, A. (2017). *I hacked my body for a future that never came.* The Verge. https://www.theverge.com/2017/7/21/15999544/biohacking-finger-magnet-human-augmentation-loss. Accessed 31 January 2020.

Robertson, A. (2019). *Facebook says it will build AR glasses and map the world.* The Verge. https://www.theverge.com/2019/9/25/20883706/facebook-ar-glasses-prototypes-live-maps-announce-oc6. Last accessed 31 January 2020.

Roesner, F., Kohno, T., & Molnar, D. (2014). Security and privacy for augmented reality systems. *Communications of the ACM*, *57*(4), 88–96.

Romanus, T., Frish, S., Maksymenko, M., Frier, W., Corenthy, L., & Georgiou, O. (2020). Mid-air haptic bio-holograms in mixed reality. *arXiv* preprint arXiv:2001.01441. https://arxiv.org/abs/2001.01441.

Royal Society. (2019). *iHuman: Blurring lines between mind and machine.* Royal Society.

Sabelman, E., & Lam, R. (2015). The real-life dangers of augmented reality. *IEEE Spectrum, 52*, 48–53. https://doi.org/10.1109/MSPEC.2015.7131695.

Sampaio, E., Maris, S., & Bach-y-Rita, P. (2001). Brain plasticity: Visual acuity of blind persons via the tongue. *Brain Research, 908*, 204–207.

Sarpu, B. A. (2014). Google: The endemic threat to privacy. *Journal of High Technology Law, 15*(1), 97–134.

Schwartz, P. H. (2005). Defending the distinction between treatment and enhancement. *The American Journal of Bioethics, 5*(3), 17–19.

Schwartzman, M. (2011). *See yourself sensing. Redefining human perception.* London: Black Dog Publishing.

Sedenberg, E., & Chuang, J. (2017). Smile for the camera: Privacy and policy implications of emotion AI. *arXiv* preprint arXiv:1709.00396. https://arxiv.org/abs/1709.00396.

Sedenberg, E., Wong, R., & Chuang, J. (2018). A window into the soul: Biosensing in public. In *Surveillance, privacy and public space* (pp. 87–110). Routledge.

Skeide, M. A., Kumar, U., Mishra, R. K., Tripathi, V. N., Guleria, A., Singh, J. P., et al. (2017). Learning to read alters cortico-subcortical cross-talk in the visual system of illiterates. *Science Advances, 3*(5), e1602612.

Sparrow, R. (2005). Defending deaf culture: The case of cochlear implants. *Journal of Political Philosophy, 13*(2), 135–152.

Spence, C. (2014). The skin as a medium for sensory substitution. *Multisensory Research, 27*(5–6), 293–312.

Spence, C. (2015). Multisensory flavor perception. *Cell, 161*(1), 24–35.

Spence, C. (2018). Sensory substitution: Unfulfilled promises and fundamental limitations. In F. Macpherson (Ed.), *Sensory substitution and augmentation.* Oxford: Oxford University Press.

Statt, N. (2020). *Pokémon Go never went away—2019 was its most lucrative year ever.* The Verge. https://www.theverge.com/2020/1/10/21060877/pokemon-go-record-revenue-2019-niantic-labs-ar-growth. Accessed 31 January 2020.

Stone, S. A., & Tata, M. S. (2017). Rendering visual events as sounds: Spatial attention capture by auditory augmented reality. *PLos One, 12*(8), e0182635. https://doi.org/10.1371/journal.pone.0182635.

Thaddeus-Johns, J. (2017). *Meet the first humans to sense where north is.* The Guardian. https://www.theguardian.com/technology/2017/jan/06/first-humans-sense-where-north-is-cyborg-gadget. Last accessed 31 January 2020.

ThinkMobiles (n.d.) What is augmented reality (AR) and how does it work? https://thinkmobiles.com/blog/what-is-augmented-reality/. Last accessed 31 January 2020.

Tucker, B. P. (1998). Deaf culture, cochlear implants, and elective disability. *Hastings Center Report, 28*(4), 6–14.

UK Government (n.d.) Tattoo, piercing and electrolysis licence (England and Wales). https://www.gov.uk/skin-piercing-and-tattooing. Last accessed 31 January 2020

Van Boven, R. W., & Johnson, K. O. (1994). The limit of tactile spatial resolution in humans: Grating orientation discrimination at the lip, tongue, and finger. *Neurology, 44*(12), 2361.

Varela, F. J., Thompson, E., & Rosch, E. (1991). *The embodied mind: Cognitive science and human experience.* MIT Press.

Veale, M. (2020). A critical take on the policy recommendations of the EU high-level expert group on artificial intelligence. European Journal of Risk Regulation, 1–10. Cambridge University Press, https://doi.org/10.1017/err.2019.65.

Vincent, J. (2019). *Apple reportedly plans 2022 release for first AR headset, followed by AR glasses in 2023.* The Verge. https://www.theverge.com/2019/11/11/20959066/apple-augmented-reality-ar-headset-glasses-rumors-reported-release-date. Accessed 31 January 2020.

Wachter, S. (2019). Data protection in the age of big data. *Nature Electronics, 2*(1), 6–7.

Wachter, S., & Mittelstadt, B. (2019). A right to reasonable inferences: Re-thinking data protection law in the age of big data and AI. *Columbia Business Law Review, 2*, 494–620.

Wade, G. (2005). *Seeing things in a different light*. BBC. http://www.bbc.co.uk/devon/news_features/2005/eyeborg.shtml. Last accessed 31 January 2020.

Ward, A. F., Duke, K., Gneezy, A., & Bos, M. W. (2017). Brain drain: The mere presence of one's own smartphone reduces available cognitive capacity. *Journal of the Association for Consumer Research*, *2*(2), 140–154.

Ward, J., & Meijer, P. (2010). Visual experiences in the blind induced by an auditory sensory substitution device. *Consciousness and Cognition*, *19*(1), 492–500. https://doi.org/10.1016/j.concog.2009.10.006.

Warwick, K. (2016). Transhumanism: Some practical possibilities. *FifF Kommunikation*, *2016*(2), 24–27.

Welch, R. B. (1978). *Perceptual modification: Adapting to altered sensory environments*. Elsevier.

Wexler, A. (2016). The practices of do-it-yourself brain stimulation: Implications for ethical considerations and regulatory proposals. *Journal of Medical Ethics*, *42*(4), 211–215.

Wexler, A. (2017). The social context of "do-it-yourself" brain stimulation: Neurohackers, biohackers, and lifehackers. *Frontiers in Human Neuroscience*, *11*, 224.

Wexler, A., & Reiner, P. B. (2019). Oversight of direct-to-consumer neurotechnologies. *Science*, *363*(6424), 234–235.

Wilmer, H. H., Sherman, L. E., & Chein, J. M. (2017). Smartphones and cognition: A review of research exploring the links between mobile technology habits and cognitive functioning. *Frontiers in Psychology*, *8*, 605.

Wilson, D. A., & Stevenson, R. J. (2006). *Learning to smell: Olfactory perception from neurobiology to behavior*. Johns Hopkins University Press.

Wolbring, G. (2013). Hearing beyond the normal enabled by therapeutic devices: The role of the recipient and the hearing profession. *Neuroethics*, *6*(3), 607–616.

Wolinsky, H. (2009). Kitchen biology. *EMBO Reports*, *10*(7), 683–685.

Wong, W. L., Su, X., Li, X., Cheung, C. M. G., Klein, R., Cheng, C. Y., et al. (2014). Global prevalence of age-related macular degeneration and disease burden projection for 2020 and 2040: A systematic review and meta-analysis. *The Lancet Global Health*, *2*(2), e106–e116.

World Health Organization. (2019). *Deafness and hearing loss*. https://www.who.int/news-room/fact-sheets/detail/deafness-and-hearing-loss. Accessed 31 January 2020.

Wright, T. D., & Ward, J. (2018). Sensory substitution devices as advanced sensory tools. In F. Macpherson (Ed.), *Sensory substitution and augmentation*. Oxford: Oxford University Press.

Wurzman, R., Hamilton, R. H., Pascual-Leone, A., & Fox, M. D. (2016). An open letter concerning do-it-yourself users of transcranial direct current stimulation. *Annals of Neurology*, *80*(1), 1–4.

Yetisen, A. K. (2018). Biohacking. *Trends in Biotechnology*, *36*(8), 744–747.

Yeung, K. (2017). 'Hypernudge': Big data as a mode of regulation by design. *Information, Communication & Society*, *20*(1), 118–136. https://doi.org/10.1080/1369118X.2016.1186713.

Zabels, R., Osmanis, K., Narels, M., Gertners, U., Ozols, A., Rūtenbergs, K., et al. (2019). AR displays: Next-generation technologies to solve the vergence—Accommodation conflict. *Applied Sciences*, *9*(15), 3147.

Zuboff, S. (2019). *The age of surveillance capitalism: The fight for a human future at the new frontier of power*. Profile Books.

Zuiderveen Borgesius, F. J., Möller, J., Kruikemeier, S., Fathaigh, Ó. R., Irion, K., Dobber, T., et al. (2018). Online political microtargeting: Promises and threats for democracy. *Utrecht Law Review*, *14*(1), 82–96.

CHAPTER FIVE

Do-it-yourself and direct-to-consumer neurostimulation

Anna Wexler*
Department of Medical Ethics & Health Policy, Perelman School of Medicine, University of Pennsylvania, Philadelphia, PA, United States
*Corresponding author: e-mail address: awex@pennmedicine.upenn.edu

Contents

1. Introduction

Researchers, policymakers and clinicians have become concerned with how unintended audiences are co-opting neuroscience knowledge for use outside conventional medical and academic settings. One prominent example surrounds the do-it-yourself (DIY) and direct-to-consumer (DTC)

Developments in Neuroethics and Bioethics, Volume 3
ISSN 2589-2959
https://doi.org/10.1016/bs.dnb.2020.03.005

use of transcranial direct current stimulation (tDCS), a technique that has shown promise in clinical populations, particularly for the treatment of depression and chronic pain (Lefaucheur et al., 2017), and in healthy populations, where studies have suggested that the technique may have cognitive enhancement effects (Buch et al., 2017; Coffman, Clark, & Parasuraman, 2014; Santarnecchi et al., 2015). Due in part to the ease of construction of a tDCS device, in late 2011 individuals began building devices at home and applying stimulation for the purposes of enhancement and/or treatment (Jwa, 2015; Wexler, 2015, 2017). Today, however, it is more common for individuals to purchase direct-to-consumer tDCS devices (Wexler, 2018).

The home use of tDCS[1] has not been well received by scientists and ethicists who have warned against the practice in both media outlets and academic journals. Researchers have voiced concerns about the potential harms of the home use of tDCS, worrying that it may potentially have deleterious effects on cognition (see, e.g., Brem, Fried, Horvath, Robertson, & Pascual-Leone, 2014; Iuculano & Cohen Kadosh, 2013). In 2015, the International Federation of Clinical Neurophysiology became the first professional organization to issue a formal position paper opposing the home use of tDCS, warning "against the use of DIY devices and methods unless they have shown both efficacy and safety" (International Federation of Clinical Neurophysiology, 2015). Regulatory bodies and professional organizations, such as the Food and Drug Administration (FDA), the National Academy of Medicine (formerly the Institute of Medicine), and the American Academy of Arts and Sciences have all held meetings in recent years that have, in part, discussed the ethical issues surrounding the home use of brain stimulation and the regulation of commercially available neurostimulation devices (American Academy of Arts and Sciences, 2016; Institute of Medicine, 2015).

It has now been nearly a decade since the rise of DIY brain stimulation, and the practice has yet to gain a foothold among the general public. Rather, it has remained a subculture fuelled, at least in part, by media coverage in major outlets like *Radiolab*, *Wired* and *The New Yorker* (Batuman, 2015; Miller, 2014; Radiolab, 2014). Still, due to the lack of action from government regulatory bodies, new companies continue to bring devices to the market, and advertise them with claims of treating medical conditions

[1] I use the term "home use" (and "home users") of tDCS to refer to the self-administration of electrical stimulation outside of academic or medical settings.

and enhancing brain function (Coates McCall, Lau, Minielly, & Illes, 2019). From the academic perspective, the ethical and legal concerns expressed by scholars have largely shifted from those regarding the implications of widespread uptake of cognitive enhancers to those regarding misleading advertising and the commercialization of neurostimulation devices for non-medical uses (Coates McCall et al., 2019; Ienca, Haselager, & Emanuel, 2018; Kreitmair, 2019).

In this chapter, I synthesize the latest empirical research on DIY and DTC neurostimulation, and provide an updated review of relevant ethical and legal issues. In addition, I argue that the focus on DIY tDCS in the academic literature has obscured attention to another potentially problematic use of the technique: the provision of tDCS to clients in private clinics (i.e., those not operating under the auspices of a hospital or university), despite the fact that in the United States (USA), there is no tDCS device approved by the FDA for any indication. In Europe and Canada, where there are fewer barriers to bringing a new medical device to market, several companies have recently received approval to market tDCS devices for major depressive disorder, fibromyalgia, and migraine. The ethical issues regarding the private provision of tDCS, particularly in the USA, may be more fraught than those regarding DIY tDCS, as rather than individuals choosing to administer stimulation to themselves, practitioners are administering an experimental brain stimulation technique to others for an out-of-pocket fee.

This chapter begins with an overview of tDCS technology and a historical characterization of the rise of DIY and DTC neurostimulation. Next, I review empirical research on the home use of brain stimulation, focusing on demographics, usage practices, and safety and efficacy. The ensuing section presents previously unpublished pilot data on the provision of tDCS by practitioners in private clinics in the USA. I then outline the regulatory issues surrounding both the sale of DTC neurostimulation devices as well as the provision of neurostimulation services. I conclude with a brief discussion of the ethical issues surrounding both the home use of tDCS and the provision of tDCS in private practice.

2. Transcranial direct current stimulation (tDCS)

Twenty years ago, two neurophysiologists showed that passing a weak electrical current through the brain could modify cortical excitability (Nitsche & Paulus, 2000). Since the publication of their research, thousands of papers that utilize the technique that has come to be known as transcranial direct current stimulation (tDCS) have appeared in science journals. There are two main lines of tDCS research. The first comprises studies in clinical populations that have claimed to show the beneficial effects of tDCS for treating a variety of diseases and disorders, such as depression, chronic pain, schizophrenia, stroke, Parkinson's disease, and eating disorders (Brunoni et al., 2012; Ferrucci, Mameli, Ruggiero, & Priori, 2016; Knotkova, Greenberg, Leuschner, Soto, & Cruciani, 2013; Shiozawa et al., 2014). The second line of tDCS research is comprised of studies claiming to show that tDCS can enhance learning and cognition in healthy individuals, on tasks requiring working memory, motor skills, mathematical ability, motor dexterity, perception, and creative problem solving (for reviews, see Coffman et al., 2014; Hill, Fitzgerald, & Hoy, 2016; Mancuso, Ilieva, Hamilton, & Farah, 2016). Even though researchers have raised questions about tDCS studies—criticizing the proliferation of small sample-size studies, the lack of larger, longitudinal work (see, e.g., Horvath, Forte, & Carter, 2015; Mancuso et al., 2016; Price & Hamilton, 2015; Waltz, 2019), and whether a significant amount of current is even entering the skull (Jwa, Goodman, Law, & Glover, 2019; Vöröslakos et al., 2018)—it has remained a popular technique that continues to be used in both neuroscience investigations and clinical studies.

At its core, a tDCS device consists of a current-providing component such as a 9-V battery, wires that carry the current, and electrodes that interface between the wire and the head. When both electrodes are connected to the scalp, the electrical circuit closes, and the device is thought to deliver a current to the brain. The placement of the wires is considered to be of importance: when the anodal (positive) electrode is placed over a specific brain area, current is thought to flow into that region and stimulate neuronal activity. The cathodal (negative) electrode is the exit point for current, and is thought to inhibit neuronal activity. A common electrode configuration (known as a montage) for depression, for example, involves stimulating the left dorsolateral prefrontal cortex and having the cathode rest a few inches above the right eyebrow.

In comparison to other methods of electric or magnetic brain stimulation, tDCS is unique in that it is both non-invasive (i.e., no surgical implantation is required) and relatively inexpensive to acquire or create. By contrast, transcranial magnetic stimulation (TMS), which is used to treat certain kinds of depression, is non-invasive, but the stimulation device is not simple to replicate. In addition, compared to other non-invasive techniques that stimulate the brain with electricity, such as electroconvulsive therapy (ECT), the level of current used in tDCS is relatively low: most tDCS studies use 0.5–2 mA, whereas ECT utilizes 500–900 mA (American Psychiatric Association, 2001).

3. The rise of DIY and direct-to-consumer tDCS

The DIY brain stimulation movement began in late 2011, when a Yahoo group and a Reddit forum dedicated to DIY tDCS were formed (Wexler, 2016a). By early 2012, there were a number of blogs and websites dedicated exclusively to the topic, indicating that individuals were utilizing tDCS for both self-treatment and cognitive enhancement purposes. While some individuals constructed their own devices, others purchased direct-to-consumer products, such as the $379 device from Hong Kong-based Trans Cranial Technologies. Other tDCS devices offered at the time were more affordable tDCS kits that varied in both price and level of sophistication, but usually consisted of a battery enclosure, wires, electrodes, and a headband to facilitate electrode placement (Wexler, 2015). In addition, some individuals began purchasing and repurposing iontophoresis devices, which are current-providing machines used for transdermal drug delivery (Food and Drug Administration, 2017a) that provide similar levels of current to that used in tDCS.

The consumer tDCS device market underwent a shift in the summer of 2013, when the first wearable direct-to-consumer device, known as the Foc.us version 1, came to market. With its sleek, ready-to-wear headset, advertising campaign featuring photos of an attractive woman, and promised smartphone integration, the product was a clear step apart from the kits sold by small-scale vendors. The release of the Foc.us version 1 heralded the advent of a new chapter in the home use of tDCS, one in which companies with significant financial resources—not just individuals selling devices out of their basement garages—entered the consumer neurostimulation market. Though the Foc.us version 1 device was initially marketed to gamers, there is no empirical evidence that it was ever meaningfully used by this population.

In 2014, two companies, Thync and Halo Neuroscience, announced the receipt of millions of dollars in venture capital funding for consumer brain stimulation products (Biggs, 2014; Stone, 2014). In 2015, Thync released a wearable neurostimulation device ($299), the Thync Edition One, that was controlled via smartphone and provided a form of non-invasive brain stimulation for mood alteration purposes (either a calm vibe or an energy vibe). The company, however, soon encountered financial difficulties, cutting much of its staff and auctioning off its assets (Huet, 2016). It recovered, and in 2017 released another consumer neurostimulation device, known as the Thync Relax Pro (Hartmans, 2017). At present (mid-2019), neither device is available, and the company's website states that it is pursuing clinical trials involving the use of bioelectronics to treat psoriasis (Thync, n.d.).

Halo Neuroscience proved to have greater longevity. The company, whose board includes former chairman of the Federal Communications Commission Reed Hundt, markets a brain stimulator called Halo Sport for athletic enhancement, and has advertised that the product "accelerates gains in strength, explosiveness, and dexterity." Because Halo Sport only claims to stimulate the motor cortex—which lies beneath the area of the head where a pair of headphones might sit—the product does not utilize stray wires or a futuristic headset, but instead takes the recognizable shape of headphones (Wexler, 2016b). The product has reportedly been used by National Basketball Association (NBA) teams such as the Golden State Warriors (Hutchinson, 2015) and a number of Olympic athletes (Attah, 2016), and in 2016 the company was awarded a contract with the experimental arm of the United States Department of Defense (Mehta, 2016).

At present, there are over a dozen companies selling neurostimulation products directly to consumers. Some companies, such as PlatoWork, have turned to crowdfunding to finance the production of their devices (PlatoWork, 2019). Others sell products via their own websites, Amazon, eBay, or Caputron.com, an online retailer specializing in neurostimulation devices and accessories. As I will discuss in detail in the regulatory section, while some companies, such as Flow (Flow Neuroscience, n.d.) make claims about the ability of their product to treat medical indications, others make more ambiguous claims regarding wellness, enhancement, and health (Coates McCall et al., 2019). It also appears that the market is slowly broadening out to encompass other forms of transcranial electrical stimulation: Humm, for example, sells a product that utilizes alternating instead of direct current (Humm, n.d.), and the Xen stimulator from Neuvana claims to target the vagus nerve directly (Neuvana, n.d.).

4. Home users of brain stimulation devices: Demographics and prevalence

Although there have been significant concerns raised in the neuroscience and neuroethics literature regarding the DIY use of brain stimulation, to date only three empirical studies have examined the practices of home users. Two were conducted in late 2013 and early 2014, soon after the first direct-to-consumer wearable tDCS devices were marketed to the general public: Jwa (2015) conducted a content analysis of postings to the online tDCS subreddit forum (www.reddit.com/r/tdcs), a survey ($n = 121$), and in-depth interviews; and Wexler (2016a) presented a preliminary sketch of the practices of home users, based on interviews and reviews of tDCS-related forums, websites, and blogs. Both of these studies relied on the Internet for recruitment. More recently, Wexler (2018) partnered with seven companies selling direct-to-consumer tDCS devices to survey their user base. The resulting sample ($n = 327$) was comprised of individuals who had purchased consumer neurostimulation devices, not just those who had visited online forums.

Both Jwa (2015) and Wexler (2018) found that approximately three-quarters of respondents were based in North America, mostly in the USA. The consistency of this result across studies could reflect a lower prevalence of home use of tDCS outside North America. All three studies found that the home use of tDCS was almost exclusively by men, though Wexler (2018) found a higher proportion of use by women (15.3%) than the 4% previously reported by Jwa (2015). With regard to age, although both Wexler (2016a) and Jwa (2015) had found that the typical home user was in their 20s and 30s, Wexler (2018) more recently reported an older mean age (45.3). As discussed in Wexler (2018), it is unlikely that the variation in mean age is due to differences in sampling methods. Rather, the extensive media coverage of tDCS in 2014 and 2015 may have attracted a group of slightly older individuals, especially those interested in the restoration of cognitive abilities.

When comparing demographic data to population-level data, Wexler (2018) found that those who own consumer tDCS devices tend to be highly educated, politically liberal, and early adopters of technology who frequently read scientific articles. Most respondents in Wexler (2018) reported not attending religious services, which is a particularly salient finding given that a recent Pew report found that Americans who were more religious were less comfortable with the use of enhancement technologies (Pew Research Center, 2016). Indeed, other evidence suggests that faith may play a role in individual willingness to adopt enhancement technologies: a study comparing Swiss student users of cognitive enhancement drugs to non-users found that a higher percentage of users were not affiliated with any religion, and that more users than non-users rated religion as not important (Ott & Biller-Andorno, 2014).

One of the most important unanswered questions regarding the home use of non-invasive brain stimulation is the prevalence and popularity of the phenomenon. Previously, the closest measure of the popularity of tDCS was the number of subscribers to the tDCS subreddit forum, which increased from roughly 1000 subscribers in October 2012 to approximately 12,000 as of August 2019. However, subscribing to the tDCS subreddit is relatively meaningless, as one can visit and post to the tDCS subreddit without subscribing. Thus, the growth in number of subscribers to the tDCS subreddit should be taken as a proxy for increasing interest in the home use of tDCS, rather than an increasing use of tDCS.

A more accurate—though still highly imperfect—measure of the extent of the home use of tDCS can be gleaned from the total number of consumer tDCS devices sold. Indeed, purchasing a device is a financial commitment: it requires not just clicking subscribe on the tDCS subreddit, but actually spending money on a stimulation device: on average, participants in Wexler (2018) paid $177 for their device. The sample in Wexler (2018) included seven companies' customer lists, which totaled approximately 10,000 individuals. Since then, the online retailer Caputron has reported selling tens of thousands of devices a year (Waltz, 2019), and several new companies have brought consumer neurostimulation products to market. It should be emphasized, however, that these sales figures do not reflect the total number of active users of tDCS, as even among the sample in Wexler (2018), which itself may not be representative of all users of consumer tDCS devices, approximately 40% of respondents had quit using tDCS or had never used it.

Despite the lack of data on the prevalence of the home use of tDCS, media reports have portrayed that the home use of brain stimulation is increasing (Denejkina, 2016). This characterization is not unique to tDCS: one study found that 94% of media articles related to the use of cognitive enhancement drugs characterized the phenomenon as either common or growing, despite there being little evidence to justify such claims (Partridge, Bell, Lucke, Yeates, & Hall, 2011). Similar to Jwa (2015) who speculated that there "may be some hype" around the estimates of the size of DIY tDCS, Wexler (2018) argued that there is little evidence to support the notion that the home use of tDCS is increasing. Indeed, while there was certainly more attention paid to the movement in 2014 and 2015 by the media, academia, regulatory authorities, and others, this should not be taken to reflect an increase in the use of consumer tDCS devices, but rather an increase in attention toward those who use them.

5. Home users of brain stimulation devices: Usage practices

Wexler (2018) found that nearly three-quarters of respondents reported using tDCS for cognitive enhancement (i.e., to improve cognitive abilities), and approximately one-quarter had used tDCS for restoration (i.e., to restore diminished cognitive abilities). Among both enhancers and restorers, the two most common indications were to improve focus/concentration and memory. However, the order was flipped for each group: memory was the most common target of restorers, whereas focus/concentration was the most common target of enhancers. Jwa (2015) found a similarly high percentage of users utilizing tDCS for enhancement, with the same top indications as found in Wexler (2018): attention, learning, and working memory.

In Wexler (2018), 43% of respondents reported using tDCS for treatment of a medical/psychological disease or condition; Jwa (2015) found a similar percentage (35%). Across both studies, depression was, by far, the most common treatment indication. Put in perspective, of all individuals who purchased and used a tDCS device on themselves at least once in Wexler (2018), approximately one-third (32%; $n = 97$) had used it specifically to treat depression—despite the fact that most consumer tDCS devices are marketed for enhancement, not treatment. Wexler (2018) also found that treaters were more likely than non-treaters (i.e., those who use tDCS for only enhancement and/or restoration) to report tDCS as being successful. Similarly, Jwa (2015) reported that anecdotally, users reported a high effectiveness of tDCS for depression.

One explanation for this finding is that tDCS actually is more effective for treatment than for enhancement. This possibility is in line with current reports in the tDCS literature, which suggest that the effects of tDCS may be stronger for depression than for other disease or cognitive enhancement indications (Horvath et al., 2015; Lefaucheur et al., 2017). Another possibility is that the effects of stimulation may be more salient for treaters than non-treaters (i.e., a small percentage increase in working memory capabilities may not be noticeable to the average user). A third possibility is that tDCS is not effective for either purpose, but that there is a greater placebo effect for those who use it for treatment than for those who use it for only enhancement and/or restoration.

Wexler (2018) and Jwa (2015) both found overlap in terms of purpose of use: in Jwa (2015), 24% of users reported using tDCS for both enhancement and treatment; and in Wexler (2018), approximately 43% reported using tDCS for multiple indications (i.e., some combination of treatment, enhancement, and/or restoration). Furthermore, on an open-ended initial question in Wexler (2018), approximately one-quarter of participants reported purchasing tDCS out of curiosity and interest in experimentation. Thus, some individuals appear not to be driven by a specific purpose such as treatment or enhancement, but rather by curiosity about the latest technology.

All three previous studies on home users (Jwa, 2015; Wexler, 2016a, 2018) found that they generally adhere to the current levels (1–2 mA) and typical length of stimulation session (20 min) used in scientific studies. Home users did, however, depart from established protocols with regard to frequency of stimulation: Wexler (2018) found that 40% of respondents had self-administered over 21 sessions of tDCS, and among them, 8% were super-users who reported over 100 sessions of tDCS. This chronic use of tDCS is particularly notable, as the effects of frequent stimulation have largely gone unstudied in the scientific literature.

Finally, the methods employed in Wexler (2018) allowed for an examination of the population of users who stopped using tDCS. Results showed that approximately a third of users (34.9%) had stopped using tDCS primarily due to lack of efficacy. Thus, it is not side effects or concern about long-term effects that are driving individuals away from tDCS, rather it is the fact that they are not experiencing expected benefits.

6. Efficacy and safety of consumer brain stimulation devices

Both Wexler (2018) and Jwa (2015) found that approximately 40% of users rated tDCS as effective for its intended use (i.e., either medical treatment or cognitive enhancement). However, the studies of both authors also found a significant percentage of users who were not sure of the efficacy or who found tDCS ineffective.

Only a few studies have attempted to independently examine the efficacy of consumer neurostimulation devices. In a single-blind, sham-controlled, within-subject design of the Foc.us version 1, researchers found that in both online and off-line stimulation protocols, active stimulation with the device, as compared to sham, decreased performance in a subset of working memory tasks (Steenbergen et al., 2016). Furthermore, a small study ($n = 9$) conducted by researchers on the Halo Sport found that stimulation, as opposed to sham, enhanced aspects of sprint cycling ability and cognitive performance (Huang, Deng, Zheng, & Liu, 2019). In addition to these two independent studies, a handful of companies have conducted research using their own devices. For example, employees of Thync published a peer-reviewed study with a finding of significantly decreased tension and anxiety (Tyler et al., 2015). In addition, Halo Neuroscience released several non-peer-reviewed white papers on its website, detailing in-house studies of gains in athletic function following use of the Halo Sport (Halo Neuroscience, n.d.).

All three empirical studies on home users found that users experience side effects that are consistent with those typically found in the tDCS literature (Antal et al., 2017). For example, Jwa (2015) found that 46% of respondents reported side effects such as headache, pain, tingling, itching or burning under the electrodes, fatigue, nervousness, and visual perceptual changes, among others. However, participants consistently rated these effects as very mild (Jwa, 2015). In Wexler (2018), approximately one-third (35%) of all participants reported experiencing skin irritation (i.e., redness, tingling, itching). Less common side effects were headaches (10%), flashes of light in the visual field (8%) and dizziness (2%). In Wexler (2018), 17% of respondents used the word "burn" to describe an unwanted side effect, though most mentions were in reference to heat sensations or varying degrees of skin redness. A small number of participants ($n = 10$), however, reported more serious burns, as determined by mentions of scarring or other language that indicated severity. Reports of burns are also in line with several reports of skin lesions in the tDCS literature that have occurred while tDCS is being administered by trained professionals in laboratory settings (Matsumoto & Ugawa, 2017; Rodríguez, Opisso, Pascual-Leone, & Soler, 2014).

7. The provision of tDCS services to clients

The focus on DIY tDCS in the academic literature has largely obscured attention to another use of the technique: practitioners who provide tDCS to clients in private clinics and who are not operating under the auspices of a hospital or university. To date there has been no published empirical data on those who administer tDCS to others, their practices, and which devices they utilize for which indications. Although the use of tDCS in this manner is not quite DIY or direct-to-consumer—as individuals must visit a provider or practitioner to receive services—it is also distinct from academic and mainstream medical uses. Indeed, as I will show in this section, tDCS has been adopted by complementary and alternative medicine (CAM) practitioners, many of whom lack medical training and provide tDCS for indications for which there is little evidence.

To identify clinics that provide tDCS, I conducted online keyword searches (using the terms "tDCS" or "transcranial direct current stimulation" and "clinic" or "practice" or "brain" or "brain stimulation") in March 2016. Clinics and/or individual practitioners were included if they (a) had a webpage offering the provision of tDCS for the purposes of treatment and/or enhancement (i.e., for non-research purposes) and (b) were not operating under the auspices of a hospital or university, as such institutions generally implement a greater degree of oversight over experimental procedures. Over 40 clinics providing tDCS were found globally, including 19 in the USA.

All 19 US-based practitioners providing tDCS were contacted to set up a telephone interview on the condition of anonymity. The interviews were limited to practitioners based in the USA, where tDCS is still experimental. In total, 12 of the 19 practitioners agreed to an interview, which was conducted by telephone between March and June 2016. The open-ended, semi-structured interviews probed practitioners' backgrounds, treatment protocols, usage indications, pricing structures, and views on regulation. Data were extracted from the 12 interviews along a number of themes.

Based on self-reported estimates from practitioners, approximately 1000 clients received tDCS across all 12 practitioners interviewed. The length of time providing tDCS ranged from 3 months to 10 years, with an average of approximately 4 years. All clients reportedly paid for the stimulation out-of-pocket, with pricing structures ranging from a set hourly rate to a $3000 tDCS package that included a take-home modified iontophoresis device, combined with unlimited Skype consultations. The most common indications for the administration of tDCS were depression, anxiety, and pain; however, practitioners reported utilizing tDCS for a wide array of diseases (e.g., Parkinson's, autism, epilepsy, migraine, schizophrenia, traumatic brain injury, and post-traumatic stress disorder). Three practitioners reported using tDCS for enhancement purposes such as "brain brightening" or improving focus. Three practitioners reported having used tDCS in children: two for autism; one for epilepsy.

Stimulation protocols varied widely, with practitioners reporting everything from daily to weekly sessions; some reported using tDCS twice daily for particularly acute cases. Stimulation duration ranged from 15 to 30 min with 20 min being the most common duration, and reported intensities were generally 1.5–2 mA. Some practitioners had clients perform a cognitive task during stimulation, such as a brain-training game, whereas others administered stimulation without a task.

Practitioners had diverse backgrounds, with a mix of MDs ($n=3$), PhDs ($n=6$), and those without doctoral-level in training in medicine or psychology ($n=3$). Training for tDCS also varied considerably, with practitioners reporting either attending tDCS courses in the USA and abroad, learning from mentors/colleagues, training under direct supervision, taking online webinars from device manufacturers, or various combinations thereof. Most practitioners felt that tDCS was relatively safe when used in the supervised setting of the clinic. Although most practitioners reported not recommending that clients utilize tDCS at home, four had lent or rented tDCS devices to patients, and three actively recommended the home use of tDCS, advising patients on where to buy tDCS devices.

Four practitioners reported using iontophoresis devices. The remainder utilize direct-to-consumer tDCS devices sold by either neurofeedback or cranial electrotherapy stimulation (CES) device manufacturers, which lack FDA approval or clearance despite being sold explicitly for clinical use. Of the four practitioners using iontophoresis devices, three reported deliberately using the device in an off-label manner. In recognition of the unusual nature of off-label use in the tDCS context, one of these four practitioners remarked that it was a "curious loophole" and another called it "worse than off-label" because iontophoresis are contraindicated for use on the head.

Half of all practitioners interviewed were uninformed or misinformed about the regulatory status of tDCS: four thought it was FDA-approved and two were unsure if it had been approved. In some sense, this is not surprising, as prior work has found physicians to be relatively uninformed about the regulatory status of common drugs (Chen, Wynia, Moloney, & Alexander, 2009). Practitioners were mostly concerned about liability: that is, whether they could be held responsible for medical malpractice in the case that a client experienced an adverse event from stimulation.

Other concerns included their comfort level in using the device, and whether they felt tDCS was effective compared to the other therapeutic methods they utilize.

Eighteen of the 19 US clinics identified offered a therapy that is considered alternative or of insufficiently established efficacy by mainstream medicine, most commonly neurofeedback ($n=15$), cranial electrotherapy stimulation (CES; $n=10$), or audio-visual entrainment (AVE; $n=5$) therapy. The use of tDCS in the CAM context appears to have arisen by 2011: the International Society for Neurofeedback and Research (ISNR) conference has featured a talk or workshop about tDCS every year since at least 2011, and in July of that year, a neurofeedback practitioner began a Yahoo group dedicated to tDCS (Yahoo, n.d.). Postings on the Yahoo group seemed to have occurred within an established community, with members—using their real identities—often referring to each other by first name. Furthermore, most practitioners learned about tDCS from the neurofeedback community, including colleagues, conferences, and device manufacturers. This finding is in stark contrast to the home use reports about tDCS, for which most participants learn about tDCS from the Internet, media, or both (Wexler, 2018).

In the CAM context, tDCS is conceptualized as a technique that can be integrated into neurofeedback practice—not a tool to be used on its own. MindAlive and Neurofield, the two main manufacturers of tDCS devices for the neurofeedback community, market the devices as add-ons that can be integrated into the companies' existing CES and neurofeedback systems, respectively. Many practitioners spoke of using neurofeedback as a first-line treatment, but when participants became "stuck," or their brain patterns became "phase-locked," tDCS could be useful to help "move" them. Some practitioners referred to tDCS as being useful for "underactive" or "overactive" brains: one practitioner related how he utilized anodal stimulation on a "real underactive spot on the right hemisphere," and an article by the owner of MindAlive recommended using anodal stimulation for "excessive" delta, theta and alpha activity, and cathodal stimulation for "excessive" beta activity (Siever, 2013).

Overall, the use of tDCS in private clinics in this manner is distinct both from its use among lay individuals at home (i.e., DIY tDCS) and neuromodulation researchers. Thus, tDCS is currently utilized in different ways by at least three separate groups, as shown in Table 1, largely operating in independent domains and with minimal interaction with one other.

Table 1 tDCS in the domain of research, clinic and the home.

	Research use of tDCS	**Private clinical use of tDCS**	**Home or "DIY" use of tDCS**
Population of users	Neuromodulation researchers; clinicians	Practitioners with varying backgrounds	Lay individuals
Location of use	Laboratory	Clinic	Home
Subject of stimulation	Research subjects	Clients	Self
Primary purpose of stimulation	Research (i.e., creation of generalizable knowledge); experiments in both clinical and healthy populations	Treatment of individual clients (mostly for mood disorders); occasional uses for neuroenhancement	Self-improvement (treatment and/or cognitive enhancement)
Degree of regulatory oversight in the USA	Stringent; all usage requires prior approval by Institutional Review Board (IRB)	Moderate; practitioners subject to state laws (e.g., scope of practice and medical malpractice)	None; though consumer tDCS companies marketing unapproved devices for medical indications may be in violation of federal law

8. Regulation of consumer brain stimulation devices

In the USA, where the vast majority of consumer neurostimulation companies are based, products intended for use in the treatment or diagnosis of disease or other medical conditions are considered medical devices (Maslen, Douglas, Kadosh, Levy, & Savulescu, 2015; Wexler, 2015) and regulated by the FDA. Historically, regulators have relied upon the content of the product's advertising and labeling to determine the intended use of a product. As detailed in Wexler (2015) and Coates McCall et al. (2019), some consumer neurostimulation devices are advertised for medical indications, such as treating depression. In the USA, although these products would clearly be considered illegally marketed medical devices—because they do not have prior approval as a medical device—the FDA has not taken any enforcement action against their manufacturers. To date, the only regulatory action against a tDCS device company has come at a state level, from the California Department of Public Health (CDPH), which shut down a California based-firm in 2013 for unlawfully selling a device that made implied medical claims (Wexler, 2015).

It is less clear, however, whether consumer neurostimulation devices that are marketed for non-medical indications—such as improving memory, mood, or cognitive function—would fall under FDA jurisdiction (Coates McCall et al., 2019; Wexler, 2015). Although the FDA stated in a recent guidance that it does not intend to regulate low risk products that are marketed for general wellness, it noted that neurostimulation products, such as those marketed for memory enhancement, would not be considered low risk "due to the risks to a user's safety from electrical stimulation" (Food and Drug Administration, 2019). However, while the FDA has seemingly left open the possibility of regulating neurostimulation products marketed for cognitive enhancement, it has not yet taken regulatory enforcement action in this domain.

Because of the patchwork of marketing strategies and the dearth of enforcement action, at least one group of scholars have suggested that there is a regulatory gap for consumer cognitive enhancement devices (Dubljević, Saigle, & Racine, 2014). Others have considered the appropriate level of regulation that might be applied to such devices (Dubljević, 2015). However, the lack of enforcement action should not be conflated with the lack of regulation (Wexler, 2015) as there exists a comprehensive regulatory framework for both medical devices and consumer products. Indeed, the Federal Trade Commission (FTC) has recently taken enforcement action against a number of companies making unsubstantiated cognitive enhancement claims (Federal Trade Commission, 2015a, 2015b, 2016a, 2016b).

When neurostimulation devices marketed for non-medical indications first appeared on the market in the early 2010s, several European scholars, as well as the Nuffield Council for Bioethics, advocated extending medical device legislation in Europe to cover these products (Maslen, Douglas, Cohen Kadosh, Levy, & Savulescu, 2014; Nuffield Council on Bioethics, 2013). In 2017, new legislation was passed that would encompass these devices under the European legal framework for medical devices (European Union, 2017). The new Medical Device Regulation (MDR), which will go into effect on May 26, 2020, specifically states under Annex XVI that "equipment intended for brain stimulation that apply electrical currents or magnetic or electromagnetic fields that penetrate the cranium to modify neuronal activity in the brain" and marketed for a non-medical purpose will need to comply with certain MDR requirements, such as a risk management plan and periodic safety update reports (European Union, 2017; Medicines & Healthcare Products Regulatory Agency, 2017). While the comprehensive set of requirements for manufacturers of neurostimulation devices marketed for non-medical purposes has not yet been published, the new EU regulations—which encompass these products within the medical device framework—represent a clear departure from the regulatory schema in the USA.

Some scholars have pointed out that focusing on regulatory enforcement might be impractical or infeasible (Jwa, 2015; Wexler, 2015) and instead have supported the open engagement approach to the DIY community first proposed by Fitz and Reiner (2015). In 2016, researchers wrote an open letter to the DIY community, outlining the potential risks of the home use of tDCS (Wurzman, Hamilton, Pascual-Leone, & Fox, 2016). Along these same lines, Wexler and Reiner (2019) recently argued that oversight from FTC alone is insufficient for DTC neurotechnologies given the sheer number of products, the dynamic nature of software applications which can change with each update, the flexibility required to oversee them, and the potential ethical issues involved. They proposed the formation of an independent working group that would survey the main domains of DTC neurotechnology and provide succinct appraisals of potential harms and probable efficacy. Rather than evaluating each and every product or providing overarching framing questions, the appraisals of the group would outline the evidence base, potential risks, and identify future research directions.

9. Regulation of neurostimulation services

While the regulatory issues surrounding the sale of direct-to-consumer neurostimulation devices in the USA have been the subject of much discussion, as reviewed above, the use of such devices in clinical settings presents a distinct set of legal considerations. As the FDA does not regulate the practice of medicine (Food and Drug Administration, 2017), the regulation of tDCS services for the purposes of medical treatment largely falls to individual states, which define the practice of medicine and limit who (i.e., individuals with which types of licensing) can offer medical services for which indications (Cohen, 1998). A similar regulatory framework exists in both Europe and Canada, where individual member nations and provinces, respectively, regulate the practice of medicine. At least some practitioners in the pilot study described above, who lack advanced degrees or licensing in relevant fields, appear to be operating outside their legal scope of practice.

In the USA, the provision of tDCS for clinical (i.e., non-research) purposes has sometimes been referred to as off-label tDCS (Bikson, Paneri, & Giordano, 2016; Fregni et al., 2015). However, the term "off-label tDCS" is problematic for a number of reasons. First, there is no on-label, approved, use of tDCS; thus there cannot be true off-label tDCS. Second, only one-third of practitioners interviewed in the pilot study reported using an iontophoresis device to provide tDCS; the others reported using unapproved devices, which would not be considered off-label use. Third, using an iontophoresis device to provide tDCS requires that significant modifications be made, in terms of the stimulation apparatus where different electrodes need to be used, site of stimulation with transcranial instead of transdermal electrodes, and indication from hyperhidrosis/drug delivery to mood disorders. One major iontophoresis device manufacturer even lists transcranial use as a contraindication (Chattanooga, 2012). Given these extensive modifications, the utilization of iontophoresis devices to provide tDCS arguably stretches the boundaries of the term off-label. Overall, the term off-label tDCS appears to be a misnomer both in terminology and in its description of the phenomenon itself.

While state laws in the US govern who can prescribe certain drugs—for example, medical doctors have broad prescribing privileges, but naturopathic practitioners generally do not (American Medical Association, 2017)—there are no analogous restrictions governing the use of medical devices. In other words, no special licensing or certification is necessary to utilize medical devices on others in a clinical setting. Thus, practitioners are free to utilize any device: whether FDA-approved and being used for the approved indication, whether FDA-approved and being used in an off-label manner, or entirely unapproved. Practitioners' main legal considerations center not on the regulatory status of a given device, but around issues of liability.

A number of legal cases in the USA—mostly related to orthopedic surgery—have addressed how the off-label nature of a medical device may play a role in medical malpractice cases. In *Femrite v. Abbott Northwestern Hospital* (1997), the court found that off-label use of a device in itself does not constitute medical malpractice, and in *Alvarez v. Smith* (1998) the court found that physicians are not legally required to inform the patient of the regulatory status of a device, provided that the patient is clearly informed of relevant procedural risks (Brown, 2012; Sheha, Hammouri, Snyder, Campbell

Jr., & Vitale, 2014; Tomaszewski, 2006). Thus, in the case of tDCS, a theoretical medical malpractice case would not rest solely on the regulatory status of the device, but rather would consider a number of contextual factors, such as negligence, informed consent, scope of practice, standard of care.

10. Ethical considerations with regard to the home use of tDCS

Contemporary discussions of the ethics of the home use of tDCS have largely paralleled earlier literature on pharmacological enhancement, with debates arising about whether there are meaningful differences between various methods of enhancement (Farah et al., 2004; Fitz, Nadler, Manogaran, Chong, & Reiner, 2014; Greely et al., 2008). The main ethical concerns, elaborated below, have surrounded safety, coercion and/or peer pressure, and distributive justice; more recent concerns have been raised about misleading advertising.

10.1 Safety

In medical and research contexts, safety is generally defined in relation to the probability of an adverse event—a short-term, quantifiable, health-related effect. One review paper (Antal et al., 2017) found that across 18,000 subjects who had received various forms of transcranial electrical brain stimulation, no serious adverse events were reported, although mild adverse events (e.g., headache, tingling, burning sensations) were relatively common. Skin burns, considered a moderate adverse event, were rare. As described above, this characterization of the safety profile of tDCS is largely in line with those reported by home users of neurostimulation.

In addition to adverse events, safety has also been discussed in the context of the unknown cognitive effects of neurostimulation. Along these lines, some researchers have speculated that enhancement of one cognitive ability via tDCS may come at the cost of others, and at least one study has found initial evidence to support this claim (Iuculano & Cohen Kadosh, 2013; Sarkar, Dowker, & Cohen Kadosh, 2014). However, it may also be argued that deeming tDCS risky due to unknown long-term cognitive effects is holding the technique to an unreasonably high standard: as a matter of practice, regulatory bodies do not test pharmaceuticals—or even consumer products—for potential downstream effects on cognition. Still, it is true that the effects of long-term use of tDCS are largely unknown, which may be particularly concerning given that a subset of recreational users of neurostimulation devices utilize the device much more frequently than in scientific protocols (Wexler, 2018).

10.2 Coercion

Some scholars have raised concerns about the potential for coerced use of neurotechnology, fearing that if a neurotechnology becomes widespread in educational, military or occupational contexts, individuals may feel compelled to adopt such technologies, either explicitly, via regulations and policies, or implicitly, via social pressure. However, as pointed out by Voarino, Dubljević, and Racine (2017) and Wexler (2018), coercion may only be a practical concern if a given neurotechnology is both effective and has demonstrated a propensity for widespread social uptake. The lack of overwhelming efficacy reported in Jwa (2015) and Wexler (2018), particularly for cognitive enhancement, combined with the fact that to date no contemporary pharmacological or neurotechnological enhancement intervention has achieved widespread social uptake, suggest that peer pressure or coercion may not be a particularly pressing concern in the near future.

10.3 Distributive justice

If cognitive enhancement neurotechnologies become disproportionately available to those in higher socioeconomic classes, they might work to exacerbate current gaps in inequality. Indeed, there is empirical evidence that those who purchase direct-to-consumer neurostimulation devices have higher incomes relative to the general US population (Wexler, 2018). It is important to note, however, that concerns would only be relevant if a technology were to be both effective for cognitive enhancement and achieve widespread uptake.

10.4 Misleading advertising

It is unclear whether DTC neurostimulation technologies work as advertised (i.e., to ameliorate symptoms or enhance cognition). Many neurotechnology companies have conducted little to no research on the effectiveness of their products (Coates McCall et al., 2019; Wexler & Thibault, 2018), and some rely solely on studies conducted in laboratory settings on research-grade devices (Wexler & Reiner, 2019). Furthermore, as described earlier, there is a lack of consensus in the scientific literature itself regarding the effects of tDCS. Given that it is difficult for consumers to assess the validity of claims made by DTC neurotechnology companies, more care should be taken by companies to ensure that their claims are truthful and consistent with the scientific literature (Coates McCall et al., 2019).

11. Ethical considerations with regard to the provision of tDCS to clients

Is it ethical for practitioners to provide tDCS to patients in private clinics? On the one hand, it can be argued that tDCS has been shown to be relatively low risk (Antal et al., 2017), and a medical practitioner, who is bound by a duty to relieve patient suffering, is well within professional boundaries to utilize emerging therapies that may be beneficial (and not overtly harmful). Along these same lines, it can be argued that an individual should have the freedom to try a novel therapy, regardless of scientific consensus. On the other hand, it may be argued that because there is little evidence for tDCS efficacy, the technology should not be employed in clinical settings. From this standpoint, the relative strength of scientific evidence plays a major role in the determining the ethical permissibility of tDCS use, regardless of the legal status of the tDCS device itself.

American Medical Association guidelines, which state that off-label use is appropriate when based on "sound scientific evidence and sound medical opinion" suggest scientific evidence is necessary to ethically justify the use of tDCS (Dresser & Frader, 2009). Although there is no widespread agreement regarding the efficacy of tDCS, one consensus paper found probable efficacy for the use of tDCS in the treatment of fibromyalgia, addiction/craving, and a major depressive episode without drug resistance (Lefaucheur et al., 2017). Practitioners in the pilot study reported above, however, mentioned using tDCS for a wide variety of indications for which there has been a lack of properly powered studies and/or mixed evidence (Szymkowicz, McLaren, Suryadevara, & Woods, 2016).

In principle, the ethical issues raised by providing tDCS in clinical settings parallel those raised by other emerging therapies, CAM techniques, and off-label uses of pharmaceuticals (Dresser & Frader, 2009; Largent, Miller, & Pearson, 2009; Smith, Ernst, Colquhoun, & Sampson, 2016), where scientific consensus is lacking. While the ethical issues introduced by the use of tDCS in clinical settings are not necessarily novel—studies have found that scientific support is lacking for many of the off-label indications for which physicians prescribe drugs (Eguale et al., 2016; Radley, Finkelstein, & Stafford, 2006; Walton et al., 2008; Wong et al., 2017)—the ways that tDCS is currently being used are particularly concerning. Below are three main ethical recommendations with regard to the clinical provision of tDCS; recently published guidelines for tDCS personnel highlight similar ethical issues (Antal, Woods, & Knotkova, 2019).

11.1 Practitioner competency both with tDCS and the treatment indication

At minimum, practitioners should have an in-depth understanding of—and ideally, considerable experience with—both the tool they are using (i.e., tDCS) and the indication(s) they are treating; this consideration overlaps with legal requirements regarding scope of practice. Additionally, given that tDCS is a novel technique, practitioners have a heightened responsibility to be broadly informed both about the device's regulatory status and the evidence base supporting its usage.

11.2 Awareness and minimization of conflicts of interest

Because the provision of tDCS requires a practitioner to invest in a stimulation device, as well as administer multiple in-office procedures, there is a likelihood of conflicts of interest (COIs) between the financial interests of the practitioner and the best interests of the patient. Although these potential COIs are not unique to tDCS and are frequently encountered in other areas of medicine, it is nonetheless important for practitioners to both maintain an awareness of them and work to minimize them. For example, practitioners could specify intervention protocols a priori to prevent seeking a particular treatment response. In addition, guidelines from professional associations can help minimize the risks of COI and should therefore be encouraged.

11.3 Adequate disclosures to patients

Patients should always be informed of the experimental nature of tDCS—that tDCS is still in the research phase—the evidence base for its use, and its concomitant risks. It is not sufficient for practitioners to use the term "experimental" when describing clinical uses of tDCS to patients, as researchers and patients may understand the term differently (see, e.g., Sankar, 2004). Furthermore, while referring to tDCS as a "novel but promising" technique may be appropriate for an indication such as depression, for which there exists a body of scientific evidence, it may be deceptive when describing neurofeedback-related uses, as there is little evidence to support this application.

12. Conclusion

tDCS represents the case of an emerging therapy being adopted at home and in CAM settings before the scientific consensus and regulatory apparatus have caught up. The emergence of tDCS in this manner is likely due to its affordability, ease of construction and administration, and reputation for being low risk. From a regulatory perspective, the ambiguity surrounding the legal status of non-invasive neurostimulation devices marketed for enhancement, combined with the lack of extant regulatory enforcement action, have signaled to manufacturers that bringing a new product to market is a low-risk endeavor. In the CAM setting, the relatively lax restrictions regarding the use of medical devices in clinical practice as compared to pharmaceuticals, combined with its adoption among neurofeedback practitioners, have likely contributed to its flourishing. Although the regulatory considerations differ in each of those contexts, the ethical considerations remain relevant and, in principle, are broadly applicable to the clinical use of any emerging therapeutic technology.

Acknowledgement

I thank Dr. Alvaro Pascual-Leone for valuable inputs on the sections relating to the provision of tDCS in private clinics.

References

American Academy of Arts and Sciences. (2016). *The regulatory and ethical dimensions of human performance enhancement.* Bulletin of the American Academy of Arts and Sciences. Retrieved from https://www.amacad.org/multimedia/pdfs/HPE-workshop-summary-Bulletin.pdf.

American Medical Association. (2017). *State law chart: Naturopath licensure and scope of practice.* American Medical Association. Retrieved from https://www.ama-assn.org/sites/default/files/media-browser/specialty%20group/arc/ama-chart-naturopath-scope-practice-2017.pdf.

American Psychiatric Association, (Ed.), (2001). *The practice of electroconvulsive therapy: Recommendations for treatment, training, and privileging (a task force report of the American Psychiatric Association).* (2nd ed.). American Psychiatric Association.

Antal, A., Alekseichuk, I., Bikson, M., Brockmöller, J., Brunoni, A. R., Chen, R., et al. (2017). Low intensity transcranial electric stimulation: Safety, ethical, legal regulatory and application guidelines. *Clinical Neurophysiology, 128*(9), 1774–1809. https://doi.org/10.1016/j.clinph.2017.06.001.

Antal, A., Woods, A. J., & Knotkova, H. (2019). *Transcranial direct current stimulation ethics and professional conduct.* In H. Knotkova, M. A. Nitsche, M. Bikson, & A. J. Woods (Eds.), *Practical guide to transcranial direct current stimulation* (pp. 407–427): *Vol. 16.* Cham: Springer International Publishing. https://doi.org/10.1007/978-3-319-95948-1_14.

Attah, T. (2016, August 1). *Rio 2016 Olympic athletes using halo sport headset*. Sports Illustrated. Retrieved from https://www.si.com/tech-media/2016/08/01/halo-sport-halo-neuroscience-rio-2016-olympics.

Batuman, E. (2015, April 6). Adventures in transcranial direct current stimulation. *The New Yorker*. Retrieved from http://www.newyorker.com/magazine/2015/04/06/electrified.

Biggs, J. (2014, May 1). *Halo, the brain-improving wearable, raises $1.5 million*. TechCrunch. Retrieved from https://techcrunch.com/2014/05/01/halo-the-brain-improving-wearable-raises-1-5-million/.

Bikson, M., Paneri, B., & Giordano, J. (2016). The off-label use, utility and potential value of tDCS in the clinical care of particular neuropsychiatric conditions. *Journal of Law and the Biosciences*, *3*(3), 642–646. https://doi.org/10.1093/jlb/lsw044.

Brem, A. K., Fried, P. J., Horvath, J. C., Robertson, E. M., & Pascual-Leone, A. (2014). Is neuroenhancement by noninvasive brain stimulation a net zero-sum proposition? *NeuroImage*, *85*(3), 1058–1068. https://doi.org/10.1016/j.neuroimage.2013.07.038.

Brown, O. W. (2012). Legal implications of pushing the endovascular envelope. *Journal of Vascular Surgery*, *56*(1), 273–274. https://doi.org/10.1016/j.jvs.2012.03.264.

Brunoni, A., Nitsche, M., Bolognini, N., Bikson, M., Wagner, T., Merabet, L., et al. (2012). Clinical research with transcranial direct current stimulation (tDCS): Challenges and future directions. *Brain Stimulation*, *5*(3), 175–195. https://doi.org/10.1016/j.brs.2011.03.002.

Buch, E. R., Santarnecchi, E., Antal, A., Born, J., Celnik, P. A., Classen, J., et al. (2017). Effects of tDCS on motor learning and memory formation: A consensus and critical position paper. *Clinical Neurophysiology*, *128*(4), 589–603. https://doi.org/10.1016/j.clinph.2017.01.004.

Chattanooga. (2012). *Iontophoresis 2013*. 1–8. Retrieved from https://international.chattgroup.com/sites/default/files/ionto_catalogue_2013_en.pdf.

Chen, D. T., Wynia, M. K., Moloney, R. M., & Alexander, G. C. (2009). U.S. physician knowledge of the FDA-approved indications and evidence base for commonly prescribed drugs: Results of a national survey. *Pharmacoepidemiology and Drug Safety*, *18*(11), 1094–1100. https://doi.org/10.1002/pds.1825.

Coates McCall, I., Lau, C., Minielly, N., & Illes, J. (2019). Owning ethical innovation: Claims about commercial wearable brain technologies. *Neuron*, *102*(4), 728–731. https://doi.org/10.1016/j.neuron.2019.03.026.

Coffman, B. A., Clark, V. P., & Parasuraman, R. (2014). Battery powered thought: Enhancement of attention, learning, and memory in healthy adults using transcranial direct current stimulation. *NeuroImage*, *85*(3), 895–908. https://doi.org/10.1016/j.neuroimage.2013.07.083.

Cohen, M. H. (1998). *Complementary & alternative medicine: Legal boundaries and regulatory perspectives*. Baltimore, MD: John Hopkins University Press.

Denejkina, A. (2016, April 10). *The promise and peril of DIY electrical brain stimulation*. The Daily Dot. Retrieved from http://kernelmag.dailydot.com/issue-sections/headline-story/16371/diy-electrical-brain-stimulation-tdcs-promise-and-peril/.

Dresser, R., & Frader, J. (2009). Off-label prescribing: A call for heightened professional and government oversight. *The Journal of Law, Medicine & Ethics*, *37*(3), 476–486. https://doi.org/10.1111/j.1748-720X.2009.00408.x.

Dubljević, V. (2015). Neurostimulation devices for cognitive enhancement: Toward a comprehensive regulatory framework. *Neuroethics*, *8*(2), 115–126. https://doi.org/10.1007/s12152-014-9225-0.

Dubljević, V., Saigle, V., & Racine, E. (2014). The rising tide of tDCS in the media and academic literature. *Neuron*, *82*(4), 731–736. https://doi.org/10.1016/j.neuron.2014.05.003.

Eguale, T., Buckeridge, D. L., Verma, A., Winslade, N. E., Benedetti, A., Hanley, J. A., et al. (2016). Association of off-label drug use and adverse drug events in an adult population. *JAMA Internal Medicine, 176*(1), 55–63. https://doi.org/10.1001/jamainternmed.2015.6058.

European Union. (2017). Council Regulation (EU) 2017/745 of 5 April 2017 on medical devices, amending Directive 2001/83/EC, Regulation (EC) No 178/2002 and Regulation (EC) No 1223/2009 and repealing Council Directives 90/385/EEC and 93/42/EEC. *Official Journal of the European Union, L117*, 1–93. Retrieved from https://eur-lex.europa.eu/legal-content/EN/TXT/PDF/?uri=CELEX:32017R0745.

Farah, M., Illes, J., Cook-Deegan, R., Gardner, H., Kandel, E., King, P., et al. (2004). Neurocognitive enhancement: What can we do and what should we do? *Nature Reviews Neuroscience, 5*(5), 421–425. https://doi.org/10.1038/nrn1390.

Federal Trade Commission. (2015, Jnury 20a). *Makers of jungle rangers computer game for kids settle FTC charges that they deceived consumers with baseless "brain training" claims.* Federal Trade Commission. Retrieved from https://www.ftc.gov/news-events/press-releases/2015/01/makers-jungle-rangers-computer-game-kids-settle-ftc-charges-they.

Federal Trade Commission. (2015, Septemer 17b). *FTC charges marketers of "vision improvement" app with deceptive claims.* Federal Trade Commission. Retrieved from https://www.ftc.gov/news-events/press-releases/2015/09/ftc-charges-marketers-vision-improvement-app-deceptive-claims.

Federal Trade Commission. (2016, Jnury 5a). *Lumosity to pay $2 million to settle FTC deceptive advertising charges for its "brain training" program.* Federal Trade Commission. Retrieved from https://www.ftc.gov/news-events/press-releases/2016/01/lumosity-pay-2-million-settle-ftc-deceptive-advertising-charges.

Federal Trade Commission. (2016, May 18b). *Marketers of one-on-one "brain training" programs settle FTC charges that claims about ability to treat severe cognitive impairments are unsupported.* Federal Trade Commission. Retrieved from https://www.ftc.gov/news-events/press-releases/2016/05/marketers-one-one-brain-training-programs-settle-ftc-charges.

Ferrucci, R., Mameli, F., Ruggiero, F., & Priori, A. (2016). Transcranial direct current stimulation as treatment for Parkinson's disease and other movement disorders. *Biochemical Pharmacology, 6*(1), 53–61. https://doi.org/10.1016/j.baga.2015.12.002.

Fitz, N., Nadler, R., Manogaran, P., Chong, E., & Reiner, P. (2014). Public attitudes toward cognitive enhancement. *Neuroethics*, 7(2), 173–188. https://doi.org/10.1007/s12152-013-9190-z.

Fitz, N. S., & Reiner, P. B. (2015). The challenge of crafting policy for do-it-yourself brain stimulation. *Journal of Medical Ethics, 41*(5), 410–412. https://doi.org/10.1136/medethics-2013-101458.

Flow Neuroscience (n.d.). Flow: Medication-free depression treatment. Flow Neuroscience. Retrieved from https://flowneuroscience.com/

Food and Drug Administration. (2017). *FDA's role in regulating medical devices.* Food and Drug Administration. Retrieved from https://www.fda.gov/MedicalDevices/ProductsandMedicalProcedures/HomeHealthandConsumer/HomeUseDevices/ucm204884.htm.

Food and Drug Administration. (2017a). *CFR Title 21, part 890: Physical medicine devices, subpart F—Physical medicine therapeutic devices, sec. 890.5525 iontophoresis devices.* Food and Drug Administration. Retrieved from https://www.accessdata.fda.gov/scripts/cdrh/cfdocs/cfcfr/CFRSearch.cfm?fr=890.5525.

Food and Drug Administration. (2019). *General wellness: Policy for low risk devices—Guidance for industry and Food and Drug Administration staff.* Food and Drug Administration, pp. 1–13. Retrieved from https://www.fda.gov/media/90652/download.

Fregni, F., Nitsche, M. A., Loo, C. K., Brunoni, A. R., Marangolo, P., Leite, J., et al. (2015). Regulatory considerations for the clinical and research use of transcranial direct current stimulation (tDCS): Review and recommendations from an expert panel. *Clinical Research and Regulatory Affairs, 32*(1), 22–35. https://doi.org/10.3109/10601333.2015.980944.

Greely, H., Sahakian, B., Harris, J., Kessler, R., Gazzaniga, M., Campbell, P., et al. (2008). Towards responsible use of cognitive-enhancing drugs by the healthy. *Nature, 456*(7223), 702–705. https://doi.org/10.1038/456702a.

Halo Neuroscience (n.d.). tDCS research. Halo Neuroscience. Retrieved from https://www.haloneuro.com/pages/researchpapers

Hartmans, A. (2017, March 23). *This strange-looking device sticks to your neck to help you de-stress and sleep better.* Business Insider. Retrieved from https://www.businessinsider.com/thync-relax-pro-device-stress-sleep-photos-2017-3.

Hill, A., Fitzgerald, P., & Hoy, K. (2016). Effects of anodal transcranial direct current stimulation on working memory: A systematic review and meta-analysis of findings from healthy and neuropsychiatric populations. *Brain Stimulation: Basic, Translational, and Clinical Research in Neuromodulation, 9*(2), 197–208. https://doi.org/10.1016/j.brs.2015.10.006.

Horvath, J. C., Forte, J. D., & Carter, O. (2015). Quantitative review finds no evidence of cognitive effects in healthy populations from single-session transcranial direct current stimulation (tDCS). *Brain Stimulation: Basic, Translational, and Clinical Research in Neuromodulation, 8*(3), 535–550. https://doi.org/10.1016/j.brs.2015.01.400.

Huang, L., Deng, Y., Zheng, X., & Liu, Y. (2019). Transcranial direct current stimulation with Halo Sport enhances repeated sprint cycling and cognitive performance. *Frontiers in Physiology, 10*, 118. https://doi.org/10.3389/fphys.2019.00118.

Huet, E. (2016, May 29). *How Thync, startup behind brain-zapping gadget, almost died.* Bloomberg News. Retrieved from https://www.bloomberg.com/news/articles/2016-05-23/how-thync-startup-behind-brain-zapping-gadget-almost-died.

Humm. (n.d.). How Humm helps you think better. Humm. Retrieved from https://thinkhumm.com/science

Hutchinson, A. (2015, June 15). For the golden state warriors, brain zapping could provide an edge. *The New Yorker.* Retrieved from https://www.newyorker.com/tech/annals-of-technology/for-the-golden-state-warriors-brain-zapping-could-provide-an-edge.

Ienca, M., Haselager, P., & Emanuel, E. (2018). Brain leaks and consumer neurotechnology. *Nature Biotechnology, 36*(9), 805–810. https://doi.org/10.1038/nbt.4240.

Institute of Medicine. (2015). *Non-invasive neuromodulation of the central nervous system: opportunities and challenges: workshop summary.* Retrieved from https://www.ncbi.nlm.nih.gov/books/NBK305684/.

International Federation of Clinical Neurophysiology. (2015). *Position statement: Transcranial electric stimulation in do-it-yourself applications.* International Federation of Clinical Neurophysiology. Retrieved from http://www.ifcn.info/uploadfiles/documents/2015/Using_tES_devices_as_DIY_FINAL_13Dec15.pdf.

Iuculano, T., & Cohen Kadosh, R. (2013). The mental cost of cognitive enhancement. *The Journal of Neuroscience, 33*(10), 4482–4486. https://doi.org/10.1523/JNEUROSCI.4927-12.2013.

Jwa, A. (2015). Early adopters of the magical thinking cap: A study on do-it-yourself (DIY) transcranial direct current stimulation (tDCS) user community. *Journal of Law and the Biosciences, 2*(2), 292–335. https://doi.org/10.1093/jlb/lsv017.

Jwa, A., Goodman, J. S., Law, C. S. W., & Glover, G. H. (2019, October). In vivo current mapping in transcranial direct current stimulation (tDCS): A cautionary tale of premature application of tDCS in and out of clinical context. In *Poster session presented at the international neuroethics society annual meeting, Chicago, IL.*

Knotkova, H., Greenberg, A., Leuschner, Z., Soto, E., & Cruciani, R. (2013). P 127. Evaluation of outcomes from transcranial direct current stimulation (tDCS) for the treatment of chronic pain. *Clinical Neurophysiology, 124*(10), e125–e126. https://doi.org/10.1016/j.clinph.2013.04.205.

Kreitmair, K. (2019). Dimensions of ethical direct-to-consumer neurotechnologies. *AJOB Neuroscience*, *10*(4), 152–166. https://doi.org/10.1080/21507740.2019.1665120.
Largent, E. A., Miller, F. G., & Pearson, S. D. (2009). Going off-label without venturing off-course: Evidence and ethical off-label prescribing. *Archives of Internal Medicine*, *169*(19), 1745–1747. https://doi.org/10.1001/archinternmed.2009.314.
Lefaucheur, J.-P., Antal, A., Ayache, S. S., Benninger, D. H., Brunelin, J., Cogiamanian, F., et al. (2017). Evidence-based guidelines on the therapeutic use of transcranial direct current stimulation (tDCS). *Clinical Neurophysiology*, *128*(1), 56–92. https://doi.org/10.1016/j.clinph.2016.10.087.
Mancuso, L. E., Ilieva, I. P., Hamilton, R. H., & Farah, M. J. (2016). Does transcranial direct current stimulation improve healthy working memory?: A meta-analytic review. *Journal of Cognitive Neuroscience*, *28*(8), 1063–1089. https://doi.org/10.1162/jocn_a_00956.
Maslen, H., Douglas, T., Cohen Kadosh, R., Levy, N., & Savulescu, J. (2014). The regulation of cognitive enhancement devices: Extending the medical model. *Journal of Law and the Biosciences*, *1*(1), 68–93. https://doi.org/10.1093/jlb/lst003.
Maslen, H., Douglas, T., Kadosh, R. C., Levy, N., & Savulescu, J. (2015). Do-it-yourself brain stimulation: A regulatory model. *Journal of Medical Ethics*, *41*(5), 413–414. https://doi.org/10.1136/medethics-2013-101692.
Matsumoto, H., & Ugawa, Y. (2017). Adverse events of tDCS and tACS: A review. *Clinical Neurophysiology Practice*, *2*, 19–25. https://doi.org/10.1016/j.cnp.2016.12.003.
Medicines & Healthcare Products Regulatory Agency. (2017). *Guidance for products without an intended medical purpose (annex XVI) under the new medical device regulation (EU 2017/745)*. Medicines & Healthcare Products Regulatory Agency. Retrieved from https://assets.publishing.service.gov.uk/government/uploads/system/uploads/attachment_data/file/748131/Guidance_leaflet_on_Annex_XVI_products_.pdf.
Mehta, A. (2016, October 14). *DIUx offers $36 million in FY16 contracts*. Defense News. Retrieved from https://www.defensenews.com/pentagon/2016/10/14/diux-offers-36-million-in-fy16-contracts/.
Miller, G. (2014, May 5). *Inside the strange new world of DIY brain stimulation*. Wired. Retrieved from http://www.wired.com/2014/05/diy-brain-stimulation/.
Neuvana. (n.d.). The science behind xen: Meet the vagus nerve. Neuvana. Retrieved from https://neuvanalife.com/science/
Nitsche, M. A., & Paulus, W. (2000). Excitability changes induced in the human motor cortex by weak transcranial direct current stimulation. *The Journal of Physiology*, *527*(Pt. 3), 633–639. https://doi.org/10.1111/j.1469-7793.2000.t01-1-00633.x.
Nuffield Council on Bioethics. (2013). *Novel neurotechnologies: Intervening in the brain* (1st ed.). Nuffield Council on Bioethics. Retrieved from http://nuffieldbioethics.org/wp-content/uploads/2013/06/Novel_neurotechnologies_report_PDF_web_0.pdf.
Ott, R., & Biller-Andorno, N. (2014). Neuroenhancement among Swiss students: A comparison of users and non-users. *Pharmacopsychiatry*, *47*(01), 22–28. https://doi.org/10.1055/s-0033-1358682.
Partridge, B. J., Bell, S. K., Lucke, J. C., Yeates, S., & Hall, W. D. (2011). Smart drugs "as common as coffee": Media hype about neuroenhancement. *PLoS One*, *6*(11), e28416. https://doi.org/10.1371/journal.pone.0028416.
Pew Research Center. (2016). *U.S. public wary of biomedical technologies to "enhance" human abilities*. Pew Research Center. Retrieved from http://www.pewinternet.org/2016/07/26/u-s-public-wary-of-biomedical-technologies-to-enhance-human-abilities/.
PlatoWork. (2019). *PlatoWork brain stimulator*. PlatoWork. Retrieved from https://www.indiegogo.com/projects/platowork-brain-stimulator.
Price, A. R., & Hamilton, R. H. (2015). A re-evaluation of the cognitive effects from single-session transcranial direct current stimulation. *Brain Stimulation*, *8*(3), 663–665. https://doi.org/10.1016/j.brs.2015.03.007.

Radiolab. (2014, June 26). *9-volt nirvana*. Radiolab. Retrieved from http://www.radiolab.org/story/9-volt-nirvana.

Radley, D. C., Finkelstein, S. N., & Stafford, R. S. (2006). Off-label prescribing among office-based physicians. *Archives of Internal Medicine*, *166*(9), 1021–1026. https://doi.org/10.1001/archinte.166.9.1021.

Rodríguez, N., Opisso, E., Pascual-Leone, A., & Soler, M. D. (2014). Skin lesions induced by transcranial direct current stimulation (tDCS). *Brain Stimulation: Basic, Translational, and Clinical Research in Neuromodulation*, 7(5), 765–767. https://doi.org/10.1016/j.brs.2014.06.005.

Sankar, P. (2004). Communication and miscommunication in informed consent to research. *Medical Anthropology Quarterly*, *18*(4), 429–446. https://doi.org/10.1525/maq.2004.18.4.429.

Santarnecchi, E., Brem, A.-K., Levenbaum, E., Thompson, T., Kadosh, R. C., & Pascual-Leone, A. (2015). Enhancing cognition using transcranial electrical stimulation. *Current Opinion in Behavioral Sciences*, *4*, 171–178. https://doi.org/10.1016/j.cobeha.2015.06.003.

Sarkar, A., Dowker, A., & Cohen Kadosh, R. (2014). Cognitive enhancement or cognitive cost: Trait-specific outcomes of brain stimulation in the case of mathematics anxiety. *The Journal of Neuroscience*, *34*(50), 16605–16610. https://doi.org/10.1523/JNEUROSCI.3129-14.2014.

Sheha, E. D., Hammouri, Q., Snyder, B. D., Campbell, R. M., Jr., & Vitale, M. G. (2014). Off-label use of pediatric orthopaedic devices. *The Journal of Bone & Joint Surgery*, *96*(3), e21–e27. https://doi.org/10.2106/JBJS.M.00288.

Shiozawa, P., Fregni, F., Benseñor, I., Lotufo, P., Berlim, M., Daskalakis, J., et al. (2014). Transcranial direct current stimulation for major depression: An updated systematic review and meta-analysis. *International Journal of Neuropsychopharmacology*, *17*(9), 1443–1452. https://doi.org/10.1017/s1461145714000418.

Siever, D. (2013). *Transcranial DC stimulation*. NeuroConnections., pp. 33–40. Spring 2013. Retrieved from https://www.isnr.org/neuroconnections.

Smith, K., Ernst, E., Colquhoun, D., & Sampson, W. (2016). "Complementary & alternative medicine" (CAM): Ethical and policy issues. *Bioethics*, *30*(2), 60–62. https://doi.org/10.1111/bioe.12243.

Steenbergen, L., Sellaro, R., Hommel, B., Lindenberger, U., Kühn, S., & Colzato, L. S. (2016). "Unfocus" on foc.us: Commercial tDCS headset impairs working memory. *Experimental Brain Research*, *234*(3), 637–643. https://doi.org/10.1007/s00221-015-4391-9.

Stone, B. (2014, October 8). *Thync lets you give your mind a jolt*. Bloomberg News. Retrieved from https://www.bloomberg.com/news/articles/2014-10-08/thync-raises-13-million-for-its-brain-stimulating-electrodes.

Szymkowicz, S. M., McLaren, M. E., Suryadevara, U., & Woods, A. J. (2016). Transcranial direct current stimulation use in the treatment of neuropsychiatric disorders: A brief review. *Psychiatric Annals*, *46*(11), 642–646. https://doi.org/10.3928/00485713-20161006-01.

Thync. (n.d.). Using the nervous system to change medicine. Thync. Retrieved from https://www.thync.com/

Tomaszewski, C. (2006). Off-label: Just what the doctor ordered. *Journal of Medical Toxicology*, *2*(3), 87–88. https://doi.org/10.1007/BF03161014.

Tyler, W. J., Boasso, A. M., Mortimore, H. M., Silva, R. S., Charlesworth, J. D., Marlin, M. A., et al. (2015). Transdermal neuromodulation of noradrenergic activity suppresses psychophysiological and biochemical stress responses in humans. *Scientific Reports*, *5*, 13865. https://doi.org/10.1038/srep13865.

Voarino, N., Dubljević, V., & Racine, E. (2017). tDCS for memory enhancement: Analysis of the speculative aspects of ethical issues. *Frontiers in Human Neuroscience, 10*(678), 1–13. https://doi.org/10.3389/fnhum.2016.00678.

Vöröslakos, M., Takeuchi, Y., Brinyiczki, K., Zombori, T., Oliva, A., Fernández-Ruiz, A., et al. (2018). Direct effects of transcranial electric stimulation on brain circuits in rats and humans. *Nature Communications, 9*(1), 483. https://doi.org/10.1038/s41467-018-02928-3.

Walton, S. M., Schumock, G. T., Lee, K.-V., Alexander, G. C., Meltzer, D., & Stafford, R. S. (2008). Prioritizing future research on off-label prescribing: Results of a quantitative evaluation. *Pharmacotherapy, 28*(12), 1443–1452. https://doi.org/10.1592/phco.28.12.1443.

Waltz, E. (2019). The brain hackers. *Nature Biotechnology, 37*, 983–987. https://doi.org/10.1038/s41587-019-0238-4.

Wexler, A. (2015). A pragmatic analysis of the regulation of consumer transcranial direct current stimulation (tDCS) devices in the United States. *Journal of Law and the Biosciences, 2*(3), 669–696. https://doi.org/10.1093/jlb/lsv039.

Wexler, A. (2016a). The practices of do-it-yourself brain stimulation: Implications for ethical considerations and regulatory proposals. *Journal of Medical Ethics, 42*(4), 211–215. https://doi.org/10.1136/medethics-2015-102704.

Wexler, A. (2016, June 14b). *Consumer neurotechnology: New products, more regulatory complexity*. The Neuroethics Blog. Retrieved from http://www.theneuroethicsblog.com/2016/06/consumer-neurotechnology-new-products.html.

Wexler, A. (2017). The social context of "do-it-yourself" brain stimulation: Neurohackers, biohackers, and lifehackers. *Frontiers in Human Neuroscience, 11*(224), 1–6. https://doi.org/10.3389/fnhum.2017.00224.

Wexler, A. (2018). Who uses direct-to-consumer brain stimulation products, and why? A study of home users of tDCS devices. *Journal of Cognitive Enhancement, 2*(1), 114–134. https://doi.org/10.1007/s41465-017-0062-z.

Wexler, A., & Reiner, P. B. (2019). Oversight of direct-to-consumer neurotechnologies. *Science, 363*(6424), 234–235. https://doi.org/10.1126/science.aav0223.

Wexler, A., & Thibault, R. (2018). Mind-reading or misleading? Assessing direct-to-consumer electroencephalography (EEG) devices marketed for wellness and their ethical and regulatory implications. *Journal of Cognitive Enhancement, 3*(1), 131–137. https://doi.org/10.1007/s41465-018-0091-2.

Wong, J., Motulsky, A., Abrahamowicz, M., Eguale, T., Buckeridge, D. L., & Tamblyn, R. (2017). Off-label indications for antidepressants in primary care: Descriptive study of prescriptions from an indication based electronic prescribing system. *BMJ, 356*, j603–j608. https://doi.org/10.1136/bmj.j603.

Wurzman, R., Hamilton, R. H., Pascual-Leone, A., & Fox, M. D. (2016). An open letter concerning do-it-yourself users of transcranial direct current stimulation. *Annals of Neurology, 80*(1), 1–4. https://doi.org/10.1002/ana.24689.

Yahoo! (n.d.). Transcranial direct current stimulation. Yahoo! Groups Transcranial Direct Current Stimulation. Retrieved from https://groups.yahoo.com/neo/groups/tDCStim/info

CHAPTER SIX

Neuroenhancement using transcranial electrical brain stimulation in adolescence: Ethical and social concerns

Maya Willms, Naznin Virji-Babul*
University of British Columbia, Vancouver, BC, Canada
*Corresponding author: e-mail address: naznin.virji-babul@ubc.ca

Contents

Developments in Neuroethics and Bioethics, Volume 3
ISSN 2589-2959
https://doi.org/10.1016/bs.dnb.2020.03.006

1. Introduction

Neuroenhancement (NE) refers to the use of neurotechnologies to improve basic brain functions such as memory, learning and mood in healthy individuals (Maier, Liechti, Herzig, & Schaub, 2013). While the use of pharmaceuticals for neuroenhancement in healthy adults has dominated the academic literature, the rise of do-it-yourself (DIY) of non-invasive brain stimulation raises new ethical and societal issues, particularly with respect to adolescents.

Current neurobiological models highlight that the adolescent brain shows asynchronous and distinct developmental trajectories. Across cultures, adolescents show an increase in sensation seeking and risk-taking behavior, the desire to experience new and rewarding stimuli and to take risks. These behaviors peak during the late teen years and decrease in the 20s. In contrast, the ability to plan, regulate decision making and modulate thoughts and actions, gradually increases over adolescence and plateaus in the mid-20s (Steinberg et al., 2017). This leads to an imbalance between systems, as the reward system is easily aroused creating a tendency toward sensation seeking, while the self-regulation system, which regulates these tendencies, matures later (Casey, Jones, & Hare, 2008; Steinberg, 2007). This developmental asynchrony, manifested as an increase in risk-taking and exploration, is now interpreted as critical for successful maturation and a drive toward independence (Crone & Dahl, 2012).

The unique characteristics of this developmental period prime adolescents to be highly receptive to global trends and changes in new technologies (Dahl, Allen, Wilbrecht, & Suleiman, 2018). In fact, adolescents are often the early adopters of new products and technologies, and in the Western world, reportedly have the highest prevalence of using pharmaceutical enhancers to manage the pressures of academia, sports and school life (Atkinson, 2017; Canadian Centre on Substance Use and Addiction, 2019). According to the 2019 Prescription Drug Summary, 5.4% of Canadian youth aged 15–19 use pharmaceutical stimulants. Adolescents are also more easily influenced than adults to utilize these products as a result of social peer pressure (Juvonen, Nishina, & Graham, 2000; Steinberg & Monahan, 2007).

The entrance of easily accessible and available tools for neurostimulation in the market is situated within a complex web of interconnected developmental issues related to the effects on the developing brain. This calls for clear guidelines based on a constant updating of knowledge. Dynamic changes in adolescent brain-behavior relationships must be considered within the context of the changing needs, roles and priorities of the adolescent that pertain to their autonomy and self-development, as well as the influences of parents and peers. The aims of this chapter therefore, are to: (i) briefly review the basics of transcranial electrical stimulation (TES) and its known effects on the developing brain, highlighting the gaps in knowledge in this area, (ii) present a neurodevelopmental framework to situate neuroenhancement during the adolescent period of development, (iii) discuss the ethical, and societal issues of using TES for enhancement in typically developing youth that emerge from a neurodevelopmental framework, and finally (iv) propose some guiding principles to consider in using TES-based enhancement in adolescence. In this chapter, we adopt the definition provided by Bard et al. (2018) and consider only healthy adolescents who do not have any clinical diagnoses and who voluntarily choose to use neurotechnology to enhance their cognitive, affective or behavioral functions.

2. Transcranial electrical stimulation: An overview

Non-invasive TES involves the application of weak/low electric currents to the scalp via electrodes. Although the underlying mechanisms of TES are not fully understood, researchers propose that the electric current either depolarizes or hyperpolarizes the neuronal resting membrane potential and changes cortical excitability. TES does not directly result in activation or inhibition of an action potential but rather is associated with shifts in the resting membrane potential that either increase or decrease the likelihood of an action potential (Stagg et al., 2009; Stagg & Nitsche, 2011). The effects of TES begin a few seconds after the start of the stimulation and continue throughout the application and can last for several hours or longer depending on a number of factors including the size and arrangement of the electrodes, the intensity of the stimulation, as well as the arousal and attentional state of the individual (see Polanía, Nitsche, & Ruff, 2018 for review). The arrangement of the electrodes, the electrode size, current intensity and location of the applied stimulus varies depending on the area of study or treatment (Heinrichs, 2012). Most commonly, the current is typically 1–2 mA and the electrode size 25–35 cm^2 (Kadosh, 2013). The pattern in which the stimulus is applied also varies, and includes direct current (tDCS), random noise (tRNS) and alternating current stimulation (tACS).

Transcranial direct current stimulation (tDCS) involves the application of a weak direct electrical current (e.g., 1–2 mA) through two or more electrodes placed on the scalp. The two electrodes are referred to as the anode and the cathode due to the current being positive (anode) or negative (cathode). In tACS, instead of a direct current, a sinusoidal current at a specific frequency is applied to the scalp with the aim of influencing brain oscillations. In tRNS an alternating current is applied with a stimulation frequency that changes between 0.1 and 640 Hz (see Woods et al., 2016).

TES is non-invasive, highly portable, and can be used with an active/sham stimulation control condition, in which no stimulation is applied, but reportedly feels similar to actual stimulation to inexperienced subjects (Gandiga, Hummel, & Cohen, 2006). Based on human trial testing and reporting, animal models and computational modeling, reports indicate that brain injury occurs at a significantly higher magnitude of current stimulation than that used in traditional TES. Nitsche and Bikson (2017) have suggested that up to 4 mA is safe and tolerable in humans, expanding the conventional limits of 1–2 mA used in TES. There have been no reports of brain injury associated with the application of TES (Bikson et al., 2016; Jackson et al., 2017).

2.1 Applications of TES

Due to the simplicity and portability of TES, the effects of this tool have been investigated in both healthy adults and in a range of neurocognitive disorders, such as major depressive disorder, bipolar disorder and Alzheimer's disease (Bersani et al., 2017; Bikson et al., 2008; Liu et al., 2016) with varying degrees of success. In healthy adults, TES has shown improvements in memory (Fregni et al., 2005; Marshall, Mölle, Hallschmid, & Born, 2004), motor learning (Dockery, Liebetanz, Birbaumer, Malinowska, & Wesierska, 2011; Reis & Fritsch, 2011), and planning ability (Dockery, Hueckel-Weng, Birbaumer, & Plewnia, 2009). Thus far, the majority of studies have reported significant but short-term changes in cognition, memory and mood only when combined with behavioral or cognitive training in both animal and human studies (e.g., Dockery et al., 2011; Elmasry, Loo, & Martin, 2015). Horvath, Forte, and Carter (2015) showed that a single session of tDCS did not show any significant changes in cognitive function. Elmasry et al. (2015) conducted a systematic review involving 465 healthy participants and determined that studies that paired tDCS with cognitive training, showed enhanced performance on a variety of cognitive functions such as working memory and cognitive control.

Within the pediatric population, the effect of TES has generally been studied in children with specific learning, neurological or behavioral disorders such as learning disabilities, attention-deficit/hyperactivity disorder

(ADHD), depression and eating disorders (Bandeira et al., 2016; Costanzo et al., 2017; Kirton et al., 2015; Looi et al., 2017; Salehinejad, Wischnewski, Nejati, Vicario, & Nitsche, 2019). These studies primarily focused on minors between 3 and 18 years of age. Looi et al. (2017) recently showed that cognitive training combined with tRNS in children aged 8–11 with mathematical learning disabilities, showed improvements in accuracy, response time and a steeper learning slope in comparison to matched children that received only the sham throughout the cognitive training. Other studies using TES as treatment in adolescents have also demonstrated positive results such as decreases in depression and anxiety scores in young patients with major depressive disorder (Kirton et al., 2015).

These and other studies show that researchers and clinicians have begun to trial the effects of TES—either independently or in combination with various interventions in a wide range of neurological conditions; however, there is a significant gap in research studies that have studied TES for neuroenhancement in the adolescent population and none that we are aware of that report on the use of TES for self-treatment in adolescence.

2.2 Anatomical and physiological factors that may modulate the effects of neurostimulation

There is ongoing debate on the effectiveness of TES and its applications, and much of this is due to the lack of empirical evidence, as well as the subjective and state-dependent nature of neurostimulation (Horvath et al., 2015). There are several anatomical and physiological factors which influence how stimulation will affect the brain and how it will be tolerated. These factors vary not only on an individual basis, but also depend on the level of growth, maturation and age (Beauchamp et al., 2011; Lasiene, Matsui, Sawa, Wong, & Horner, 2009). Variability in individual response makes it challenging to predict the effects of neurostimulation on the brain and behavior (Tremblay et al., 2014).

With age, both the thickness of the scalp and calvarium changes, as does the scalp to brain distance and the total volume of cerebrospinal fluid (CSF) (Beauchamp et al., 2011; Fillmore, Phillips-Meek, & Richards, 2015; Herting, Gautam, Spielberg, Dahl, & Sowell, 2015; Sierawska et al., 2019). These factors all modulate the conduction of electrical current in the brain. In a recent model of electrical fields in the human brain during TES (Huang et al., 2017) it was predicted that the highly conductive CSF pulls significant current through the inter-hemispheric fissure and may be producing stronger electrical fields in deeper brain structures. Findings from this study demonstrated that electrical field intensity was greatest underneath stimulation electrodes and in brain tissue neighboring CSF containing ventricles (Huang et al., 2017). Similarly, the thickness of the scalp and skull affects how deeply the electrical stimulus is able to penetrate, and this factor changes throughout life and also varies between sexes (Herting et al., 2015).

Structural neuroimaging studies show decreases in gray matter volume in specific regions, from adolescence into adulthood, accompanied by increases in white matter myelination (Mills, Goddings, Herting, et al., 2016; Uhlhaas & Singer, 2011). Myelination plays a critical role in the conduction of action potentials (Hodler, Stämpfli, & Tasaki, 1952) and increases functional connectivity between distant brain regions, however, the effect of the increase in myelin in mediating electrical current in youth is still unknown (Davis, 2014).

Due to the many structural and functional differences between adolescent and adult brains, the magnitude and spatial distribution of electrical fields are essentially unknown (Davis, 2014; Huang et al., 2017). Yet, in spite of these known physiological differences, the research evidence to date indicates that TES is safe and tolerated quite well by adolescents in short-term applications within the controlled environment of a laboratory (Poreisz, Boros, Antal, & Paulus, 2007).

2.3 TES: From the laboratory to at-home use

Non-invasive neurostimulation devices are moving from the realm of the laboratory to the home. TES devices can now easily be built with instructions available on do-it-yourself (DIY) websites or purchased online, and once purchased, can be applied anywhere on the scalp (Kadosh, Levy, O'Shea, Shea, & Savulescu, 2012). Yet, little is known about the demographics of adolescent consumers and the purposes for which TES is used in the adolescent population.

Jwa (2015) conducted a survey among Reddit users who subscribed to the DIY tDCS subreddit forum. The 119 respondents were individuals who had either built or purchased tDCS devices and had used them for cognitive enhancement. They reported that in their sample of 119 survey respondents, 6% were between 16 and 20 years of age and 16% were within the 21–25 age range. Youth under the age of 18 years also reported using commercially available TES devices. A recent study by Wexler (2017) used the emails of individuals who had bought TES devices from some of the companies selling directly to the public. The study found a higher age demographic than the previous report by Jwa (2015), though Wexler (2017) also reported that some users (n = 6) bought devices for their children. These findings are limited by low response rates (Wexler, 2017) and non-random sampling (Jwa, 2015) that weaken the representativeness of the samples. Given the uncertainties of TES related to efficacy, individual variability and the limited information about the short- and long-term effects on the developing brain, how can parents and adolescents navigate this terrain? What framework can be used to disentangle the issues related to the use of NE in adults vs. use in adolescents? How could such a framework help to then develop guidelines for the use of NE in adolescents? We discuss these issues further in the section on guidelines for practice.

3. Neurodevelopmental framework of neuroenhancement in adolescents

In this section, we propose and describe the components of a neurodevelopmental framework that can guide our approach to NE in adolescents (Fig. 1). We begin by describing three concepts that are inherent to this framework: multiple maturities (in relation to brain-behavior interactions), sensitive periods and brain connectivity.

Over the past decade, increasing evidence has shown that the brains of children, adolescents and adults are different in very complex ways. Across development, the pattern of changes in both the brain and in behavior may be linear or nonlinear, as a result of the structural, functional and network level changes that occur across the lifespan (Casey, 2014). These changes are a consequence of interactions with the environment, hormones and genetic influences, to name a few. A neurodevelopmental informed policy for the use of NE should, at the foundational level, recognize that the brains of youth are still developing and that they differ in critical ways from a mature brain. However, the complexities and the multi-layered process of neurodevelopment poses challenges for establishing a baseline reference that indicates when the brain is mature (Somerville, 2016).

A prime example is risk-taking behavior in adolescents. Adolescent behavior is highly contextual and influenced by emotion, social evaluation, rewards and threats. Steinberg et al. (2017) have shown that adolescents are more likely to engage in risky behaviors when they are with their peers than when they are alone. Somerville (2016) has elegantly referred to this contextual dependency of adolescent behavior as "waves of maturity" that shape and influence behavior. A neurodevelopmental-informed policy therefore,

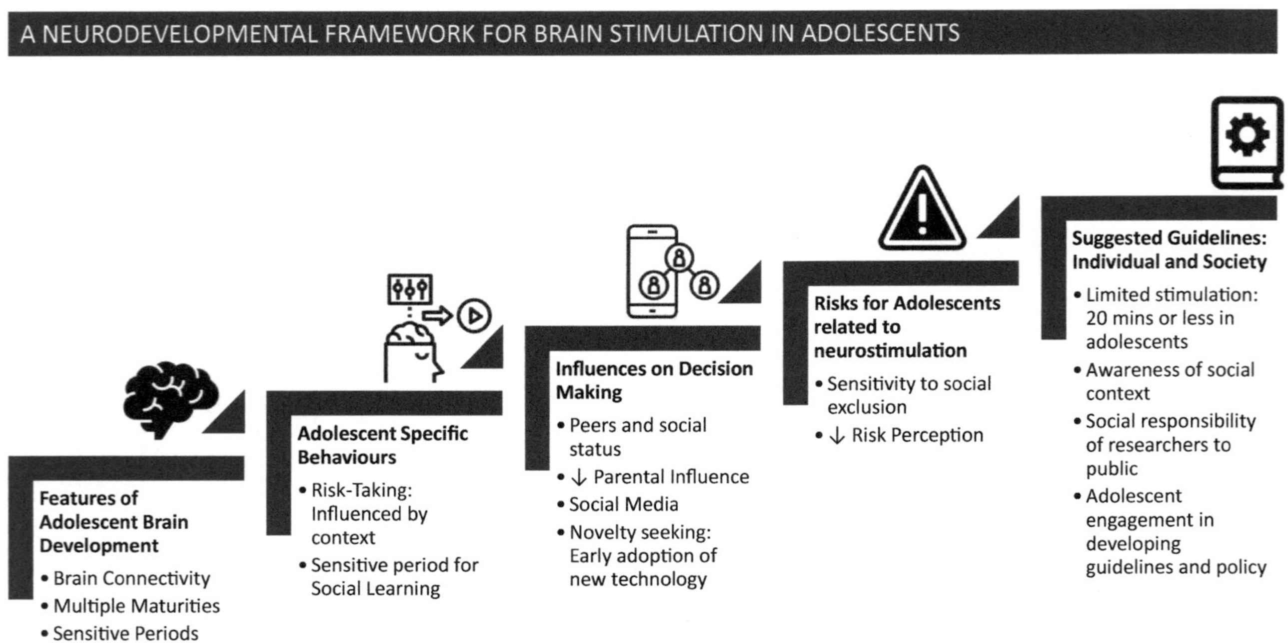

Fig. 1 A neurodevelopmental framework with issues and suggested guidelines that emerge from this framework that are specific to the adolescent period of development.

cannot simply be based on a single age criterion of the age of 18 years as a cut off for determining adult maturity. Understanding the nuances in the progression of brain and behavior within the context of environmental cues will require more careful and thoughtful discussion of the conditions and contexts under which NE is situated. As neuroscientists are still grappling with the issue of how to define and quantify brain maturation, a framework policy that recognizes the importance of emotional and social contexts on adolescent goal-directed behavior is critical.

A related concept to multiple maturities is sensitive periods. Knudsen (2004) defines sensitive periods as a window of development when the effects of particular experiences have a very strong influence on brain and behavior. During this time, experiences interact with neural circuitry to process or represent information that is adaptive for the individual. Identifying sensitive periods is critical for understanding when particular behaviors are most flexible (Casey, 2014). It is well established that infancy is a sensitive period for sensory-motor, visual and language learning. Emerging evidence highlights that adolescence is a sensitive period for social-emotional learning and is associated with shifts in processing affective and social information that enhance the importance of social status and social relationships (see Dahl et al., 2018 for review). This sensitive period during adolescence must inform NE policy related to adolescents and importantly, be leveraged for designing interventions that will capture the attention of adolescents to be engaged in the process of developing policy.

The third concept relates to functional brain connectivity. The brain is a complex entity comprising widely distributed but highly interconnected regions. Probing the brain during the resting state has revealed that even when the brain is not involved in active cognitive tasks, it exhibits a highly structured organization of activity characterized by synchronous neural activity over spatially distributed networks (Raichle et al., 2001). These networks play a critical role in mediating complex functions such as memory, language, and emotional states (Raichle et al., 2001; Raichle & Mintun, 2006). Several key features of functional connectivity are particularly relevant to a neurodevelopmental framework for NE. First, resting-state functional networks are not independent; there is a high degree of interaction between brain-state networks (Hutchison, Womelsdorf, Gati, Everling, & Menon, 2013). Second, there are very strong links between resting-state networks and task-based networks and the brain appears organized to achieve a balance between these networks and accordingly shifts cognitive resources between networks depending on the context (Brem, Fried, Horvath, Robertson, & Pascual-Leone, 2014). Third, functional networks are not stationary but rather switch dynamically between different brain states, even in the resting state (e.g., see Muller & Virji-Babul, 2018). The implication of these complex, interacting features is that the effects of TES will not be constrained simply to specific brain regions underneath the electrodes, but rather that there will be widespread effects as a result of the inherent interactions between the networks. There is an emerging literature on the effects of TES on brain connectivity in adults (see Brem et al., 2014 for review), but to date, we know very little about how network connectivity will be impacted by applying TES in a developing brain. Nevertheless, we feel this is a critical concept that should be included in the framework to help guide future research and policy. But what are the ethical implications relevant to adolescent development that should be taken into account in light of introducing this framework?

Below we describe the value of using a neurodevelopmental lens to understand the specific ethical issues that emerge during adolescence.

4. Ethical issues relevant to adolescents

If the goal of NE in typically developing adolescents is to boost cognitive/physical performance above an average level, there are a number of inter-related issues that arise as a result of using a neurodevelopmental framework. Some overlap with issues reported in the adult literature such as safety and efficacy (Chatterjee, 2013). Others such as motivation/social learning, risk-taking behavior and authenticity are specifically discussed in relation to adolescent development. Below we consider these issues.

4.1 Safety and efficacy

Within the general medical context, it is largely accepted that children and youth are not simply small adults. Despite this, when medical interventions are translated to pediatrics, both children and youth are often treated as though there are no differences in translating the effects from adults to children (Davis, 2014). In fact, most medical devices are rarely tested in a pediatric population before approval for use in this population (Hwang, Kesselheim, & Bourgeois, 2014). While adult brain stimulation is recognized as a safe practice when application guidelines are observed (Bikson et al., 2016; Davis, 2014), there remain several gaps in our knowledge of neurostimulation, which complicate its translation to adolescents.

Similarly, in a research context, there are no policies in place controlling the use of brain stimulation and no standard dosing guidelines in relation to adolescents. This is in part due to the lack of empirical evidence supporting the efficacy of brain stimulation, as the underlying mechanisms are still unknown (Sierawska et al., 2019). This creates a cyclical effect in which research on brain stimulation in adolescence may not be justified as there is no real benefit to the participant and the long-term risks are not known. But "absence of evidence is not evidence of absence."

The lack of clear dosing guidelines is likely an issue of greater concern to adolescents with psychological/mental health issues. TES is thought to be a generally safe form of neurostimulation when conservative guidelines are followed (Davis, 2014); however, there is a risk of misleading public perceptions that could "lead to negative implications for ethical and regulatory oversight" (Dubljević, Saigle, & Racine, 2014). When adult TES home users were assessed on their attitudes toward DIY tDCS in the media, 42.7% of responders agreed that media coverage "blows the issue out of proportion" (Wexler, 2017). Currently, there are no data indicating what percentage of the adolescent population has used TES devices nor that there are any trends in the use of the devices. Without adequate communication of standard safety and dosing guidelines, adolescents may be unaware of how to manage risks associated with equipment failure, inappropriate dosage or application of electrodes (i.e., burns on the scalp). Guidelines and principles for use in adults are emerging (cf. Wexler, 2016; Wurzman, Hamilton, Pascual-Leone, & Fox, 2016), however, no specific guidelines for use in adolescents have yet been developed.

4.2 Motivation and social learning

Recent evidence is highlighting that there is a significant shift in how the brain attends to processes and integrates information during adolescence. Adolescents have a heightened sensitivity to socially relevant cues that bias attention and actions toward social stimuli. This increased sensitivity to social cues is perfectly matched for navigating the social challenges of this period of development. However, the increased priority placed on social status means that learning from, and having the respect of their peers, can exert a powerful influence on motivational learning. In particular, decision making processes during this developmental stage are highly influenced by a desire for enhanced social status. Within this context, adolescents may be motivated to try novel technologies to enhance their cognitive or physical performance, thereby raising their social status within their peer groups, without a complete understanding of the risks involved.

The increased emphasis on social and motivational processes also means that many adolescents become particularly sensitive to social exclusion and social rejection. There is a risk that this hypersensitivity may trigger mental health issues, particularly in youth who are subjected to chronic social rejection or other significant negative social experiences, or those who are most vulnerable and are unable to manage such negative social experiences (Silk et al., 2014). Cohen et al. (2016) showed that older adolescents between the ages of 18 and 21 exhibit diminished cognitive performance during negative emotional arousal in comparison to adults over the age of 21 years. Adolescents also show increased reactivity to negative social situations, which is especially heightened in teens with depression (Silk et al., 2014). This combination of increased reactivity and decreased cognitive performance in negative social situations has implications for adolescents making decisions and experimenting with novel neurotechnologies in highly emotionally charged contexts such as in sport or when under pressure to perform in academic contexts.

4.3 Risk-taking

While sensation seeking behavior peaks in adolescents, as highlighted by the higher incidence of motor vehicle accidents, drug and alcohol abuse, unprotected sex and criminal behavior (Steinberg, 2007), emerging evidence suggests that adolescents are more averse to clearly stated risks than older adults. This nuanced approach appears to be driven by their acceptance of uncertainty and ambiguity (Tymula et al., 2012; Van den Bos & Hertwig, 2017). Adolescents are far more willing to accept situations where the outcomes of their decisions are unclear or unknown, and they have a greater ability to consider more than one interpretation of a situation. Importantly, however, adolescents are more likely to engage in risky behaviors when they are with their peers than when they are alone (Steinberg et al., 2017).

In terms of decision making, the two most important sources of social influence for adolescents are parents and peers. The relative importance of these two sources shift over the period of adolescence (Blakemore & Mills, 2014). Knoll, Magis-Weinberg, Speekenbrink, and Blakemore (2015) showed that young adolescents (12–14 years of age) are more strongly influenced by the opinions of other teenagers in comparison with the opinions of adults during a risk perception task. They also observed that the influence of peers decreases with age. More recent work suggests that the influence of these two sources may be more nuanced, with parents having more influence in decision making related to values or morals and peers having more influence in relation to social relationships (Welborn et al., 2016) and risk taking (Steinberg et al., 2017). Overall, the increased influence of peers may impact both the decision to try novel neurotechnologies and to perhaps downplay the mitigating factors related to safety. These factors should be considered and further work needs to be done to investigate the relative influences of peers and parents on the decision to use NE.

In addition, Crone and Konijn (2018) have shown that the interaction between the increased saliency of social-emotional information during early adolescence combined with increased social media use may make adolescents more susceptible to media related influences of risk perception and increase sensitivity to media that is saturated with emotionally laden messaging. Web based posts that advertise statements such as "hack your brain for learning and focus" may be particularly attractive to adolescents and may peak their interest in trying novel untested tools and technologies.

4.4 Authenticity

There is much discussion in the literature about the concepts of authenticity in relation to neuroenhancement, particularly in relation to whether neuroenhancement alters the true self (see Schuijer, Van Atteveldt, & Kupper, 2017). We argue that the notion of self is not a static concept but should rather be viewed as an emerging and dynamic process, particularly in relation to the developing brain. The interaction between the myriad life experiences, the influence of the physical, social and emotional environment, as well as genetics, continually mature and shape the self. Within this context, there is essentially no end point of when the authentic self is realized and achieved. In addition, to date, TES has only targeted the enhancement of isolated and highly specialized skills or behaviors such as motor learning, working memory and mood. While it can be argued that these isolated skills and behaviors are essential components of the self, enhancing one or more of these isolated components cannot impact the developing identity of the adolescent. The brain is a highly inter-connected, dynamic organ with tremendous potential for plasticity (Fuhrmann, Knoll, & Blakemore, 2015), particularly during the adolescent phase. All experiences—whether internally driven and motivated or externally derived—will create change. TES should be considered as one aspect of an externally derived mechanism of change.

5. Guidelines for oversight and practice

In order to contextualize the issues of governance, oversight and guidelines for practice, Wexler (2017) highlights two important points related to the home-use of brain stimulation (in adults). First, in contrast to predictions, the home-use of brain stimulation has not become widespread and has remained a subculture. Second, there appears to be a high attrition rate (based on informal observations) suggesting that many users do not use brain stimulation for long periods. Whether this attrition is due to lack of observed changes in brain function, difficulties with the physical set up, lack of knowledge about use or other factors remains unknown and further research is needed to probe the perception and experiences of users.

The safety and efficacy of TES is clearly not been well established. In addition, there are no industry standards or regulations and companies sell directly to consumers with unsubstantiated claims. Given this current state, how can the public make decisions about the efficacy of TES? Wexler (2017) proposes that an independent working group be developed that draws on the expertise of scientists, health professionals, consumer groups, ethicists, industry representatives, regulators and funders to evaluate the evidence as well as outline the potential risks. It is still early to determine whether this suggestion has been accepted or acted on by the larger community.

To our knowledge, only one group composed of scientists and clinicians has taken up the challenge of providing general guidelines for use of tDCS for professionals as well as to DIY users. Wurzman et al. (2016) succinctly summarize the most relevant issues so that the public may assess the level of risk of using tDCS. While they do not apply this information to children or adolescents, they acknowledge that children require special consideration due to neurodevelopmental issues.

Given the gaps in the field, and the current lack of regulation of TES devices, we propose several guiding principles for the oversight and use of TES in adolescents, based on our review of the current literature on governance. Below we discuss the issues related to the influence of brain state and the need to engage adolescents in developing guidelines and policy.

5.1 tDCS dose

In their review, Bikson, Datta, and Elwassif (2009) highlight that safety limits for tDCS based on standard montages show that using relatively large wet sponges (25–35 cm^2) with currents of 1–2 mA for a duration of 20 min are considered safe. However, there is extremely limited research on how increasing the duration, current or electrode size will impact brain function—particularly in adolescence. For this reason, we urge companies to adhere to these established protocols when designing systems for home-use and limit stimulation duration to no more than 20 min per day, until further research has established the optimal stimulation parameters for adolescents. We also urge consumers to be aware of this gap in the literature and to place limits on dose.

5.2 Influence of brain state

The main foundations of the neurodevelopmental framework that we presented are that (1) adolescent behavior is highly contextual and influenced by emotion, social evaluation and rewards/threats and (2) the adolescent period is a sensitive period for social-emotional learning and experiences. This implies that the effects of NE in adolescents will likely be much more strongly influenced by the brain state in comparison with children and adults. Wurzman et al. (2016) highlight that simultaneous cognitive or behavioral activity combined with applying tDCS could result in quite different changes in the brain in comparison with the resting state. While some of these effects are known in adults (i.e., Gill, Shah-Basak, & Hamilton, 2015), to our knowledge, the effects of emotional and motivational contexts during the application of tDCS on the adolescent brain has not been studied. Thus, highly charged environmental contexts may modulate the effects of TES on behavior in unknown ways. This consideration is likely further complicated by the type of device used, dosage, current propagation and other issues that are specific to adolescent brain development. Thus, we caution the use and application of TES in highly charged environmental contexts, such as in preparation for sports competition, a critical test/examination, or under emotional contexts, until further research can provide explanations of how the effects of TES are modulated under such conditions.

5.3 Social responsibility of researchers

Scientists and researchers have a responsibility to ensure that the long-terms effects of TES on the developing brain are investigated and communicated to the public in the timeliest fashion. Grassroots initiatives to establish research clusters that have a focus on the developmental effects of TES can be launched to provide a clear focus on developmental questions. These initiatives can be used to leverage funding for targeted research studies and for communication of results. Incorporating citizen science initiatives with specialized subgroups of researchers is a valuable addition and eventually may be supported by policy makers to develop clear, evidence-based guidelines.

5.4 Adolescent engagement and empowerment

We acknowledge that currently, most adolescents globally are most likely still unaware of TES; however, rather than waiting to develop guidelines when the demand for this novel technology increases, we encourage an anticipatory strategy to prepare for how best to manage knowledge and information about the use and risks of TES. What are the most effective methods for a public health-based intervention for delivering information to adolescents? Yeager, Dahl, and Dweck (2018) conducted a review showing that many traditional behavioral interventions directed at adolescents, fail when information is delivered top-down from adults to adolescents. Adolescents often interpret this type of instruction as their inability to make choices on their own. Given that there is a biological drive for autonomy and independence, albeit with parental support and guidance, adolescents should be involved and engaged in the process of considering the risks and benefits of neuroenhancement. In fact, the recent Lancet Commission on Adolescent Health and Wellbeing strongly advocates that adolescents be encouraged to contribute to the discussion of programs that affect them and their peers, and that this contribution should be structured in a way that helps to unravel key questions (Patton et al., 2016). The rise of citizen science initiatives, such as the web-based project to investigate the relationship between chronic pain and weather (https://www.cloudywithachanceofpain.com) is a prime example of how the interested public can contribute to data collection to answer key questions. Such approaches could be adapted to build frameworks for the use of TES in adolescents.

As early adopters of new technologies and tools, adolescents should play a key role in developing flexible, ethical guidelines for the use of this emerging neurotechnology. We endorse the views put forward by Wexler (2017) of forming an independent working group and encourage the inclusion of adolescent representatives who can actively be involved in evaluating the effects of emerging neurotechnologies, monitoring advertising claims and more importantly, can lead efforts to communicate and disseminate up-to-date information to their peers.

6. Conclusion

Many critical questions remain largely unanswered about how this emerging neurotechnology will impact both brain and behavior during adolescence. There is a need for well-designed, longitudinal studies to evaluate the long-term effects of TES on the developing brain. There is a need to understand the influences, motivation and conditions under which typical adolescents would be willing to try tools for neuroenhancement. Scientists, educators, industry representatives and a diverse group of children and adolescents should be engaged in creating awareness and policy that reflects the values of discovery, innovation and autonomy, yet minimizes risks and ensures that informed decisions are made with respect to the use of TES. Until these steps have been taken, we urge users to reflect on and be aware of the social context when making the decision to use TES and to limit the stimulation dose to 20 min or less per day.

Acknowledgment

We would like to thank Rebecca Kenny and Ella Weik for their feedback and to the editors for their thoughtful review of this chapter.

References

Atkinson, L. D. (2017). Prescription stimulant abuse on college campuses. *Journal of the American Academy of Child & Adolescent Psychiatry*, *56*(10), S79.

Bandeira, I. D., Guimarães, R. S. Q., Jagersbacher, J. G., Barretto, T. L., de Jesus-Silva, J. R., Santos, S. N., et al. (2016). Transcranial direct current stimulation in children and adolescents with attention-deficit/hyperactivity disorder (ADHD): A pilot study. *Journal of Child Neurology*, *31*(7), 918–924.

Bard, I., Gaskell, G., Allansdottir, A., da Cunha, R. V., Eduard, P., Hampel, J., et al. (2018). Bottom up ethics—Neuroenhancement in education and employment. *Neuroethics*, *11*(3), 309–322. https://doi.org/10.1007/s12152-018-9366-7.

Beauchamp, M. S., Beurlot, M. R., Fava, E., Nath, A. R., Parikh, N. A., Saad, Z. S., et al. (2011). The developmental trajectory of brain-scalp distance from birth through childhood: Implications for functional neuroimaging. *PLoS One*, *6*(9), e24981.

Bersani, F. S., Minichino, A., Bernabei, L., Spagnoli, F., Corrado, A., Vergnani, L., et al. (2017). Prefronto-cerebellar tDCS enhances neurocognition in euthymic bipolar patients. Findings from a placebo-controlled neuropsychological and psychophysiological investigation. *Journal of Affective Disorders*, *209*, 262–269.

Bikson, M., Bulow, P., Stiller, J. W., Datta, A., Battaglia, F., Karnup, S. V., et al. (2008). Transcranial direct current stimulation for major depression: A general system for quantifying transcranial electrotherapy dosage. *Current Treatment Options in Neurology*, *10*(5), 377–385.

Bikson, M., Datta, A., & Elwassif, M. (2009). Establishing safety limits for transcranial direct current stimulation. *Clinical Neurophysiology*, *120*(6), 1033–1034.

Bikson, M., Grossman, P., Thomas, C., Zannou, A., Jiang, J., Adnan, T., et al. (2016). Safety of transcranial direct current stimulation: Evidence based update 2016. *Brain Stimulation, 9*(5), 641–661.
Blakemore, S., & Mills, K. L. (2014). Is adolescence a sensitive period for sociocultural processing? *Annual Review of Psychology, 65*(1), 187–207.
Brem, A., Fried, P. J., Horvath, J. C., Robertson, E. M., & Pascual-Leone, A. (2014). Is neuroenhancement by noninvasive brain stimulation a net zero-sum proposition? *Neuroimage, 85*(3), 1058–1068.
Canadian Centre on Substance Use and Addiction. (2019). *Canadian prescription stimulants (Canadian drug summary).* https://www.ccsa.ca/prescription-stimulants-canadian-drug-summary.
Casey, B. J. (2014). A neurodevelopmental perspective on the research domain criteria (RDoC) framework. *Biological Psychiatry, 76*(5), 350–353. https://doi.org/10.1016/j.biopsych.2014.01.006.
Casey, B. J., Jones, R. M., & Hare, T. A. (2008). The adolescent brain. *Annals of the New York Academy of Sciences, 1124*(1), 111–126.
Chatterjee, A. (2013). The ethics of neuroenhancement. *Handbook of Clinical Neurology, 118*, 323–334. https://doi.org/10.1016/B978-0-444-53501-6.00027-5.
Cohen, A. O., Breiner, K., Steinberg, L., Bonnie, R. J., Scott, E. S., Taylor-Thompson, K., et al. (2016). When is an adolescent an adult? Assessing cognitive control in emotional and nonemotional contexts. *Psychological Science, 27*(4), 549–562.
Costanzo, F., Maritato, A., Menghini, D., Castiglioni, M. C., Zanna, V., & Vicari, S. (2017). Non-invasive brain stimulation treatment in a group of adolescents with anorexia. *European Psychiatry, 41*, S548.
Crone, E. A., & Konijn, E. A. (2018). Media use and brain development during adolescence. *Nature Communications, 9*(1), 588–510.
Crone, E. A., & Dahl, R. E. (2012). Understanding adolescence as a period of social–affective engagement and goal flexibility. *Nature Reviews Neuroscience, 13*, 636–650.
Dahl, R. E., Allen, N. B., Wilbrecht, L., & Suleiman, A. B. (2018). Importance of investing in adolescence from a developmental science perspective. *Nature, 554*(7693), 441–450.
Davis, N. J. (2014). Transcranial stimulation of the developing brain: A plea for extreme caution. *Frontiers in Human Neuroscience, 8*, 600.
Dockery, C. A., Hueckel-Weng, R., Birbaumer, N., & Plewnia, C. (2009). Enhancement of planning ability by transcranial direct current stimulation. *Journal of Neuroscience, 29*, 7271–7277.
Dockery, C. A., Liebetanz, D., Birbaumer, N., Malinowska, M., & Wesierska, M. J. (2011). Cumulative benefits of frontal transcranial direct current stimulation on visuospatial working memory training and skill learning in rats. *Neurobiology of Learning and Memory, 96*(3), 452–460.
Dubljević, V., Saigle, V., & Racine, E. v (2014). The rising tide of tDCS in the media and academic literature. *Neuron, 82*(4), 731–736.
Elmasry, J., Loo, C., & Martin, D. (2015). A systematic review of transcranial electrical stimulation combined with cognitive training. *Restorative Neurology and Neuroscience, 33*(3), 263–278.
Fillmore, P. T., Phillips-Meek, M. C., & Richards, J. E. (2015). Age-specific MRI brain and head templates for healthy adults from 20 through 89 years of age. *Frontiers in Aging Neuroscience, 7*, 44.
Fregni, F., Boggio, P. S., Nitsche, M., Bermpohl, F., Antal, A., Feredoes, E., et al. (2005). Anodal transcranial direct current stimulation of prefrontal cortex enhances working memory. *Experimental Brain Research, 166*, 23–30.
Fuhrmann, D., Knoll, L., & Blakemore, S. J. (2015). Adolescence as a sensitive period of brain development. *Trends in Cognitive Sciences, 19*(10), 558–566.

Gandiga, P. C., Hummel, F. C., & Cohen, L. G. (2006). Transcranial DC stimulation (tDCS): A tool for double-blind sham-controlled clinical studies in brain stimulation. *Clinical Neurophysiology*, *117*(4), 845–850.

Gill, J., Shah-Basak, P. P., & Hamilton, R. (2015). It's the thought that counts: Examining the task-dependent effects of transcranial direct current stimulation on executive function. *Brain Stimulation*, *8*(2), 253–259.

Heinrichs, J. (2012). The promises and perils of non-invasive brain stimulation. *International Journal of Law and Psychiatry*, *35*(2), 121–129.

Herting, M. M., Gautam, P., Spielberg, J. M., Dahl, R. E., & Sowell, E. R. (2015). A longitudinal study: Changes in cortical thickness and surface area during pubertal maturation. *PLoS One*, *10*(3), e0119774.

Hodler, J., Stämpfli, R., & Tasaki, I. (1952). Role of potential wave spreading along myelinated nerve fiber in excitation and conduction. *American Journal of Physiology-Legacy Content*, *170*(2), 375–389.

Horvath, J. C., Forte, J. D., & Carter, O. (2015). Quantitative review finds no evidence of cognitive effects in healthy populations from single-session transcranial direct current stimulation (tDCS). *Brain Stimulation*, *8*(3), 535–550.

Huang, Y., Liu, A., Lafon, B., Friedman, D., Dayan, M., Wang, X., et al. (2017). Measurements and models of electric fields in the in vivo human brain during transcranial electric stimulation. *Brain Stimulation*, *10*(4), e25–e26.

Hutchison, R. M., Womelsdorf, T., Gati, J. S., Everling, S., & Menon, R. S. (2013). Resting-state networks show dynamic functional connectivity in awake humans and anesthetized macaques. *Human Brain Mapping*, *34*(9), 2154–2177.

Hwang, T. J., Kesselheim, A. S., & Bourgeois, F. T. (2014). Postmarketing trials and pediatric device approvals. *Pediatrics*, *133*(5), e1197–e1202.

Jackson, M. P., Truong, D., Brownlow, M. L., Wagner, J. A., McKinley, R. A., Bikson, M., et al. (2017). Safety parameter considerations of anodal transcranial direct current stimulation in rats. *Brain, Behavior, and Immunity*, *64*, 152–161.

Juvonen, J., Nishina, A., & Graham, S. (2000). Peer harassment, psychological adjustment, and school functioning in early adolescence. *Journal of Educational Psychology*, *92*(2), 349–359.

Jwa, A. (2015). Early adopters of the magical thinking cap: A study on do-it-yourself (DIY) transcranial direct current stimulation (tDCS) user community. *Journal of Law and the Biosciences*, *2*(2), 292–335.

Kadosh, C. (2013). Using transcranial electrical stimulation to enhance cognitive functions in the typical and atypical brain. *Translational Neuroscience*, *4*(1), 20–33.

Kadosh, R. C., Levy, N., O'Shea, J., Shea, N., & Savulescu, J. (2012). The neuroethics of non-invasive brain stimulation. *Current Biology*, *22*(4), R108–R111.

Kirton, A., Langevin, L. M., Wilkes, T. C., Jasaui, Y., Sembo, M., & MacMaster, F. (2015). Brain stimulation for adolescent depression. *European Psychiatry*, *30*, 689.

Knoll, L. J., Magis-Weinberg, L., Speekenbrink, M., & Blakemore, S. (2015). Social influence on risk perception during adolescence. *Psychological Science*, *26*(5), 583–592.

Knudsen, E. I. (2004). Sensitive periods in the development of the brain and behavior. *Journal of Cognitive Neuroscience*, *16*(8), 1412–1425.

Lasiene, J., Matsui, A., Sawa, Y., Wong, F., & Horner, P. J. (2009). Age-related myelin dynamics revealed by increased oligodendrogenesis and short internodes. *Aging Cell*, *8*(2), 201–213.

Liu, A., Bryant, A., Jefferson, A., Friedman, D., Minhas, P., Barnard, S., et al. (2016). Exploring the efficacy of a 5-day course of transcranial direct current stimulation (TDCS) on depression and memory function in patients with well-controlled temporal lobe epilepsy. *Epilepsy and Behavior*, *55*, 11–20.

Looi, C. Y., Lim, J., Sella, F., Lolliot, S., Duta, M., Avramenko, A. A., et al. (2017). Transcranial random noise stimulation and cognitive training to improve learning and cognition of the atypically developing brain: A pilot study. *Scientific Reports*, 7(1), 4633.
Maier, L. J., Liechti, M. E., Herzig, F., & Schaub, M. P. (2013). To dope or not to dope: Neuroenhancement with prescription drugs and drugs of abuse among swiss university students. *PLoS One*, *8*(11), e77967.
Marshall, L., Mölle, M., Hallschmid, M., & Born, J. (2004). Transcranial direct current stimulation during sleep improves declarative memory. *Journal of Neuroscience*, *24*, 9985–9992.
Mills, K. L., Goddings, A. L., Herting, M. M., et al. (2016). Structural brain development between childhood and adulthood: Convergence across four longitudinal samples. *NeuroImage*, *141*, 273–281.
Muller, A. M., & Virji-Babul, N. (2018). Stuck in a state of inattention? Functional hyperconnectivity as an indicator of disturbed intrinsic brain dynamics in adolescents with concussion: A pilot study. *ASN Neuro*, *10*, 1759091417753802.
Nitsche, M. A., & Bikson, M. (2017). Extending the parameter range for tDCS: Safety and tolerability of 4 mA stimulation. *Brain Stimulation*, *10*(3), 541–542.
Patton, G. C., Sawyer, S. M., Santelli, J. S., Ross, D. A., Afifi, R., Allen, N. B., et al. (2016). Our future: A Lancet commission on adolescent health and wellbeing. *Lancet (London, England)*, *387*(10036), 2423–2478. https://doi.org/10.1016/S0140-6736(16)00579-1.
Polanía, R., Nitsche, M. A., & Ruff, C. C. (2018). Studying and modifying brain function with non-invasive brain stimulation. *Nature Neuroscience*, *21*(2), 174–187.
Poreisz, C., Boros, K., Antal, A., & Paulus, W. (2007). Safety aspects of transcranial direct current stimulation concerning healthy subjects and patients. *Brain Research Bulletin*, *72*(4), 208–214.
Raichle, M. E., & Mintun, M. A. (2006). Brain work and brain imaging. *Annual Review of Neuroscience*, *29*(1), 449–476.
Raichle, M. E., MacLeod, A. M., Snyder, A. Z., Powers, W. J., Gusnard, D. A., & Shulman, G. L. (2001). A default mode of brain function. *Proceedings of the National Academy of Sciences of the United States of America*, *98*(2), 676–682.
Reis, J., & Fritsch, B. (2011). Modulation of motor performance and motor learning by transcranial direct current stimulation. *Current Opinion in Neurology*, *24*, 590–596.
Salehinejad, M. A., Wischnewski, M., Nejati, V., Vicario, C. M., & Nitsche, M. A. (2019). Transcranial direct current stimulation in attention-deficit hyperactivity disorder: A meta-analysis of neuropsychological deficits. *PLoS One*, *14*(4), e0215095.
Schuijer, J. W., Van Atteveldt, N. M., & Kupper, J. F. H. (2017). Transcranial electrical stimulation to enhance cognitive performance of healthy minors: A complex governance challenge. *Frontiers in Human Neuroscience*, *11*, 1–15.
Sierawska, A., Prehn-Kristensen, A., Moliadze, V., Krauel, K., Nowak, R., Freitag, C. M., et al. (2019). Unmet needs in children with attention deficit hyperactivity disorder-can transcranial direct current stimulation fill the gap? Promises and ethical challenges. *Frontiers in Psychiatry*, *10*, 334.
Silk, J. S., Siegle, G. J., Lee, K. H., Nelson, E. E., Stroud, L. R., & Dahl, R. E. (2014). Increased neural response to peer rejection associated with adolescent depression and pubertal development. *Social Cognitive and Affective Neuroscience*, *9*(11), 1798–1807.
Somerville, L. H. (2016). Searching for signatures of brain maturity: What are we searching for? *Neuron*, *92*(6), 1164–1167.
Stagg, C. J., Best, J. G., Stephenson, M. C., O'Shea, J., Wylezinska, M., Kincses, Z. T., et al. (2009). Polarity-sensitive modulation of cortical neurotransmitters by transcranial stimulation. *Journal of Neuroscience*, *29*(16), 5202–5206.
Stagg, C. J., & Nitsche, M. A. (2011). *Physiological basis of transcranial direct current stimulation*. Los Angeles, CA: SAGE Publications.

Steinberg, L. (2007). Risk taking in adolescence: New perspectives from brain and behavioral science. *Current Directions in Psychological Science*, *16*(2), 55–59.

Steinberg, L., Icenogle, G., Shulman, P., Breiner, K., Chein, J., Bacchini, D., et al. (2017). Around the world, adolescence is a time of heightened sensation seeking and immature self-regulation. *Developmental Science*, *21*(2), e12532.

Steinberg, L., & Monahan, K. C. (2007). Age differences in resistance to peer influence. *Developmental Psychology*, *43*(6), 1531–1543.

Tremblay, S., Lepage, J., Latulipe-Loiselle, A., Fregni, F., Pascual-Leone, A., & Théoret, H. (2014). The uncertain outcome of prefrontal tDCS. *Brain Stimulation*, 7(6), 773–783.

Tymula, A., Lior, A., Belmaker, R., Roy, A. K., Ruderman, L., Manson, K., et al. (2012). Adolescents' risk-taking behavior is driven by tolerance to ambiguity. *Proceedings of the National Academy of Sciences of the United States of America*, *109*(42), 17135–17140.

Uhlhaas, P. J., & Singer, W. (2011). The development of neural synchrony and large-scale cortical networks during adolescence: Relevance for the pathophysiology of schizophrenia and neurodevelopmental hypothesis. *Schizophrenia Bulletin*, *37*, 514–523.

Van den Bos, W., & Hertwig, R. (2017). Adolescents display distinctive tolerance to ambiguity and to uncertainty during risky decision making. *Scientific Reports*, 7(1), 40962.

Welborn, B. L., Lieberman, M. D., Goldenberg, D., Fuligni, A. J., Galván, A., & Telzer, E. H. (2016). Neural mechanisms of social influence in adolescence. *Social Cognitive and Affective Neuroscience*, *11*(1), 100–109.

Wexler, A. (2016). The practices of do-it-yourself brain stimulation: Implications for ethical considerations and regulatory proposals. *Journal of Medical Ethics*, *42*, 211–215.

Wexler, A. (2017). The social context of "do-it-yourself" brain stimulation: Neurohackers, biohackers, and lifehackers. *Frontiers in Human Neuroscience*, *11*, 224. https://doi.org/10.3389/fnhum.2017.00224.

Woods, A. J., Antal, A., Bikson, M., Boggio, P. S., Brunoni, A. R., Celnik, P., et al. (2016). A technical guide to tDCS, and related non-invasive brain stimulation tools. *Clinical Neurophysiology*, *127*(2), 1031–1048, 2015/2016.

Wurzman, R., Hamilton, R. H., Pascual-Leone, A., & Fox, M. D. (2016). An open letter concerning do-it-yourself users of transcranial direct current stimulation. *Annals of Neurology*, *80*, 1–4. https://doi.org/10.1002/ana.24689.

Yeager, D. S., Dahl, R. E., & Dweck, C. S. (2018). Why interventions to influence adolescent behavior often fail but could succeed. *Perspectives on Psychological Science*, *13*, 101–122.

CHAPTER SEVEN

DIY brain stimulation: On the difficulty of measuring effectiveness and its ethical implications

Ying-Tung Lin*
Institute of Philosophy of Mind and Cognition, National Yang-Ming University, Taipei City, Taiwan
*Corresponding author: e-mail address: linyingtung@gmail.com

Contents

1. Introduction

Humans have long sought to enhance themselves psychologically; for example, there is evidence of ingestion of herbal plants to enhance memory in Ancient Greece (Le Strange, 1977). As mind and brain sciences progress, more and more neurotechnologies for intervening in our brain states,

Developments in Neuroethics and Bioethics, Volume 3
ISSN 2589-2959
https://doi.org/10.1016/bs.dnb.2020.03.007

like pharmaceuticals and brain stimulation, become available. Similar to the use of pharmaceuticals for enhancement (Morein-Zamir & Sahakian, 2011; Smith & Farah, 2011), brain stimulation is not limited to use in experimental or clinical settings; due to its easy accessibility and ease of use, some are increasingly being adopted by individuals in a home setting (Fitz & Reiner, 2014; Wexler, 2016). Among neurotechnologies, transcranial direct current stimulation (tDCS) has drawn the most attention (Fitz & Reiner, 2015; Hildt, 2014; Jwa, 2015; Wexler, 2018). This home-based, self-administered form of brain stimulation is referred to as do-it-yourself (DIY) brain stimulation. In this chapter, I focus on the use of tDCS. Not only is tDCS the most common technique, but also numerous studies have been conducted that provide a characterization of the DIYers' practice of tDCS (Jwa, 2015; Wexler, 2016, 2018; Wurzman, Hamilton, Pascual-Leone, & Fox, 2016). Even though this chapter centers on the use of tDCS, the analysis and discussion can be applied to the contexts of other methods such as the use of pharmaceuticals.

Comparing the use of brain stimulation in a home setting and in a professional setting brings to light a number of substantial differences. First, the former requires users to be equipped with a certain degree of knowledge and information in order to obtain and operate a device on themselves, including finding the tDCS montage and electrode positions, and applying the brain stimulation. One can rely on the professionals in the latter case. Second, home-based brain stimulation involves observation and management of a person's own cognitive activities. During the process, one is required to pay attention to his or her experiences of, and changes in cognitive performance, as well as modify the mode of application based on these observations. Third, the expectations and motives of DIYers are likely to be more diverse than those of users in a professional setting. In addition, the means for measuring the effectiveness of tDCS in improving cognitive abilities in laboratory studies with a larger subject group and a control group cannot be replicated in a home setting with an individual user (Wexler, 2016).

This chapter aims to highlight the problems of measuring the effects of DIY brain stimulation and its ethical implications, given the characteristics of home users described above. The difficulty of assessing these effects risks undermining truthfulness and authenticity. Truthfulness concerns the understanding of the potential effects and uncertainties of using tDCS. It requires subjects being provided with full and honest information about tDCS and to be honest with themselves. Authenticity concerns the users' identification with the actions they chose to perform given their goal, and their understanding of how their goals can be achieved. The problem of uncertain effectiveness will be examined in consideration of these two issues.

The role of metacognition—the ability to monitor and control one's cognitive functioning—will also be examined, as it is crucial for DIY brain stimulation. Whereas the use of brain stimulation to manipulate a person's cognitive functioning can be seen as a kind of control, metacognitive monitoring contributes to users' perceived effects and subjective judgments of the effectiveness of the stimulation. Taking the role of metacognition role into account to consider the ethical implications of the problem of uncertain effectiveness, the analysis highlights the importance of subjects being informed of and being capable to understand the effects and uncertainties of using tDCS in a home setting. I suggest that education and communication are important factors in addition to regulation (Maslen, Douglas, Kadosh, Levy, & Savulescu, 2015). Increasing neuroscience literacy can help to avoid some unwanted normative consequences.

In the following sections, I review the practice of DIY brain stimulation using tDCS, particularly how DIYers consider the effect and effectiveness of brain stimulation. The role of metacognition in DIY tDCS and the problem of evaluating the effects of brain stimulation are analyzed. Finally, I examine the issues of truthfulness and authenticity, considering the difficulty of assessing effectiveness, and discuss practical implications.

It is worth noting that the concept of a DIYer is sometimes used specifically to refer to users who build a brain stimulation device on their own, instead of purchasing from direct-to-customer manufacturers (e.g., Wexler & Reiner, 2017). This distinction is not important for the purposes of this chapter. Therefore, a broader concept is adopted here to include users who use brain stimulation at home rather than in a professional setting. Furthermore, as the normative concerns with regard to the problem of potential ineffectiveness may differ when targeting a medical condition, the discussion in this chapter is limited to DIY brain stimulation used for cognitive enhancement. However, analysis of the difficulty of measuring the effect of an intervention is also applicable to DIY brain stimulation used for therapeutic purposes.

2. Do-it-yourself brain stimulation

This section outlines DIYers' practices, particularly how users consider the effects of brain stimulation using tDCS.

2.1 The practice of DIY tDCS

DIY brain stimulation refers to lay individuals' use of devices to stimulate their brains for treatment or enhancement outside of professional academic or medical settings. One of the most popular brain interventions is tDCS: a neural modulation technique that directs a low amplitude electric current through scalp electrodes to stimulate specific parts of the brain. As it is non-invasive, portable, inexpensive, painless, easy to administer, and relatively safe, tDCS can be used in home and everyday settings and has thus been adopted by DIY communities. There are several ways in which DIYers can obtain a brain stimulation device despite the fact that most DIYers are not qualified to purchase professional devices from medical device companies. One can choose to buy direct-to-consumer products (e.g., foc.us headset), device kits, or an iontophoresis device and modify it. Building a tDCS device from material available in hardware stores is another option. Product reviews and tutorials about how to construct such devices are shared and accessible on forums and websites (Wexler, 2016).

The main reasons for using DIY tDCS are cognitive enhancement and the treatment of medical conditions. Concerning the former, the cognitive functions that are primarily targeted by DIYers are attention, learning (e.g., motor, verbal, and numerical learning), working memory, long-term memory, and perception (e.g., visual, somatosensory, and auditory perception) (Jwa, 2015). Other reasons include curiosity, novelty, and/or self-experimentation (Wexler, 2018). Researchers have noted that there can be more than one reason for a DIYer to use tDCS (Jwa, 2015; Wexler, 2018).

When applying tDCS, DIYers rely on scientific articles and information shared by other DIYers on websites and blogs. Without any standard protocol, DIYers often follow the standardized electrode placement[a] and the maximum of 2 mA of current used conventionally in scientific studies (Wexler, 2016). However, to attempt to increase effectiveness, DIYers experiment on different methods of application (e.g., frequency, montages) beyond those reported in the scientific literature. For instance, while single sessions tend to be used in scientific studies of cognitive enhancement effects on healthy subjects, and 5–15 sessions in studies on clinical effects, Wexler (2018) found that more than 40% of tDCS DIYers had self-administered more than 20 sessions. This raises safety concerns while

[a] The standard 10–20 system is an internationally recognized method used in electroencephalography (EEG) to describe and apply the location of scalp electrodes as well as compare electrode positioning between subjects.

the effects of tDCS are unknown, especially considering the lack of research on its long-term effects.

2.2 Considerations of the effects and effectiveness of DIY tDCS

How does an individual DIYer assess the extent of the achieved effects of tDCS? In scientific studies, researchers have found tDCS to affect working memory (Fregni et al., 2005; Hill, Fitzgerald, & Hoy, 2016; Mancuso, Ilieva, Hamilton, & Farah, 2016; Ohn et al., 2008), episodic memory (Manenti, Brambilla, Petesi, Ferrari, & Cotelli, 2013; Penolazzi et al., 2010), attention (Coffman, Clark, & Parasuraman, 2014), decision-making (Fecteau et al., 2007; Hecht, Walsh, & Lavidor, 2010), and other cognitive abilities. Even though the effectiveness of tDCS has been reported, the same assessment methods cannot be used to measure effectiveness on a single individual, since DIY users at home cannot replicate experimental conditions, which often involve a larger number of subjects, a control group (i.e., sham stimulation in tDCS), and a blind design. Without any scientific method of determining the effectiveness of their tDCS use, some users attempt to quantify the effectiveness by tracking their improvement in cognitive tasks (Wexler, 2016), while other members of the DIY community rely mainly on subjective judgment or feelings of improvement (Jwa, 2015; Wexler, 2016).

The outcome of the use of tDCS has been reported as either positive or neutral by DIYers who seek cognitive enhancement or self-treatment. According to a study by Jwa (2015) that used an online questionnaire survey, when users were asked to rate the success of their use of tDCS on a scale of one to five, from "totally unsuccessful" to "extremely successful," most of the 121 respondents scored four or five (44%) or three (41%). In addition, when asked if they would continue using tDCS, a large proportion of them (92%) replied positively. Nearly half of the respondents (46%) were willing to recommend using tDCS to family or friends with similar conditions or enhancement goals. The results suggest that around half of the respondents found tDCS helpful in improving their conditions or achieving their aims.

Using an online survey, a study by Wexler (2018) targeted users who had purchased devices from manufacturers for treatment, restoration, or enhancement purposes. According to this study, almost half of the participants reported that they did not attempt to measure the effects, while about a fifth (19.5%) reported that they measured the effects based on self-observation and self-reflection. A smaller number of users (12.3%) took measurements using self-developed tests. In the results reported in the study, examples of these personal tests included comparing memory encoding and retrieval tasks with or without brain stimulation, using cognitive tasks to track their performance, and self-evaluation of a skill (e.g., "I pay attention to if my guitar playing sounds better and it's like I've made a breakthrough or leap to another level" (Wexler, 2018:125)). The participants were also asked to rate the extent to which they felt tDCS was successful. The majority of participants (52.4%) reported their impression to be positive (somewhat or totally successful), whereas 29.9% of users were not sure.

In the studies previously reviewed (Jwa, 2015; Wexler, 2018), DIYers' impressions of the success of the use of tDCS was investigated. However, the implications of these results are vague. First, it is unclear, from the DIYer's point of view, what success would consist of; that is, it is not clear what criteria the DIYers who completed the questionnaire used to determine their rating. Many factors—such as users' goals, expectations, criteria, and awareness of improvements—can influence how an individual evaluates the success of the practice. Therefore, it would be interesting to obtain a deeper understanding of what constitutes the success of tDCS administration for DIYers and to understand how the perceived effect is weighed.

Second, Jwa (2015) interpreted DIYers' ratings of success concerning the use of tDCS as their ratings of the "perceived effect" (pp. 310–311). Nevertheless, because of the variety of factors that might be involved in the rating, rating of success cannot be equivalent to the rating of perceived effect. That is, to perceive tDCS as more successful does not necessarily relate to a positive or significant effect. For example, successfully self-administering tDCS may lead to the higher rating of success but not the perceived effect. To date, there has been no study that focuses on the DIYers' perceived effect of the use of tDCS. Empirical studies on the subjectively perceived effects of DIY tDCS will allow researchers to access users' perspectives on the perceived effects as well as learn more about relevant perceived effects. Such studies will help to understand the motivation behind using, or ceasing to use, tDCS, as well as what self-administering brain stimulation means to the users.

3. Metacognition in DIY brain intervention

Having reviewed the use of tDCS and the issue of measuring its effectiveness, this section highlights metacognition and the role it plays in the practice of home-based tDCS.

3.1 Metacognition

Metacognition is defined as thinking about thinking, or the cognition of cognition (Flavell, 1979). Such a conceptualization of metacognition presumes two levels of function (Nelson & Narens, 1990). The first is object- or cognitive-level processing, where cognitive functions—such as memory, attention, and learning—occur. It is the target level of cognitive enhancement. The second is the meta, or metacognitive, level, which governs operations at the object level by monitoring and applying top-down control. For example, if anomalies in cognitive activity occur, the metacognitive monitoring process will detect them, and a control process will be activated to deal with the problem.

Metacognition consists of three components: monitoring, control (or regulation), and metacognitive knowledge and experience (Beran, 2012; Nelson, 1996). First, metacognitive monitoring refers to the evaluation of an individual's cognitive functioning, including the level of confidence and accuracy of judgments, knowledge, and performance (Koriat, 2006; Nelson & Narens, 1990). It comes in various forms. For example, a tip-of-the-tongue phenomenon involves a metacognitive judgment about the possession of a piece of information even though the information is not immediately accessible (e.g., "I know that I know this, but I can't recall it at this moment"). Other forms of monitoring include retrospective monitoring, which evaluates the adequacy of a response, and prospective monitoring, which judges a person's ability to carry out a cognitive task.

The assessment of cognitive performance (including decision-making) is crucial for the guidance of behavior and further decision-making, especially when external feedback is unavailable or only sporadic. However, metacognitive monitoring sometimes fails and is not always reliable (for example, when an individual exhibits overconfidence that he or she will remember more information on a future test than he or she actually does). Furthermore, metacognitive abilities vary across individuals (Fleming, Weil, Nagy, Dolan, & Rees, 2010; Kelemen, Frost, & Weaver, 2000) and can be sensitive to physiological and environmental factors (Nelson et al., 1990; Slife & Weaver, 1992). Neuroimaging studies and evidence from neurological disorders suggest that prospective judgments are supported by medial prefrontal cortex (PFC) function, whereas retrospective judgments depend on lateral PFC (Fleming & Dolan, 2012). Metacognitive monitoring is useful for the subject as it can be used by metacognitive control to facilitate cognitive performance.

Metacognitive control refers to the ability to adjust the cognitive process (Dunlosky & Metcalfe, 2008; Nelson et al., 1990). It appears in several forms, depending on the task and the resources available to the subject. For example, an individual can avoid answering questions or provide more general answers to questions that he or she is likely to get wrong. When engaging in tasks, an individual can strategically manipulate his or her attention to improve overall performance; whereas while learning, cognitive control allows the individual to select more effective learning strategies, such as deciding to continue or stop the current practice.

To achieve a cognitive goal, metacognitive monitoring and control work closely together. Metacognitive monitoring enables an assessment of the effort required for tasks before actually performing them and leads to the strategic manipulation of cognitive resources. That is, the evaluation of current cognitive activity through metacognitive monitoring indicates if the current object-level operation is sufficient for achieving a subject's goals or whether an adjustment is required. For instance, given limited time, an evaluation of the difficulty of tasks can inform cognitive control about time allocation so as to maximize performance. In addition, metacognitive control is guided by the monitoring operation, which informs an individual if the top-down regulation is effective and, if not, how to readjust. For example, monitoring a person's competency to accomplish a task and the improvement of his or her ability in a learning process allows for an assessment of the current learning strategy.

In addition to metacognitive monitoring and control, metacognitive knowledge and experience constitute an individual's descriptive knowledge and experience about the interplay between the characteristics of his or her cognitive ability, the task at hand, and strategies adopted or available (Dunlosky & Metcalfe, 2008; Flavell, 1979). For instance, metacognitive knowledge may be the judgment of the individual's competency at a given task (e.g., memorizing people's names) and available strategy to cope with this task (e.g., adopting mnemonic techniques or practicing repeatedly). In comparison, a feeling of knowing—the experience of being able to retrieve a specific piece of information—is an example of a metacognitive feeling. Other examples of metacognitive knowledge and experience include confidence judgments (judging confidence in a response), ease of learning judgments (perception of the difficulty of items in advance of learning), and knowing judgments (reducing uncertainty about belief accuracy). Metacognitive knowledge and experience allow a subject to reflect and report on the cognitive process.

As discussed earlier, monitoring may fail to correctly assess a person's own cognitive functioning. As such, metacognitive knowledge and experience may be incorrect, since an individual may over- or underestimate his or her ability and the difficulty of the task. Furthermore, recent studies have identified dissociations between cognitive performance and metacognitive judgment (Schwartz & Díaz, 2014; Schwartz & Metcalfe, 1992), and such findings support the view that metacognitive judgment results from heuristic cues instead of direct access to the content of cognitive activities (Koriat, 2000). For example, a tip-of-the-tongue phenomenon is an example of a feeling of knowing without a corresponding retrieval of contents.

How is metacognition relevant in DIYers' practice of stimulating their own brains? The next section discusses the roles that metacognition might play in DIY brain stimulation.

3.2 The role of metacognition in DIY brain stimulation

DIYers' use of tDCS relies on the users' self-regulation of the brain stimulation, and thus metacognitive monitoring and control are heavily involved. The heavy involvement of metacognitive abilities distinguishes the use of brain stimulation in a home setting from that in a professional setting. In the latter, the application of brain stimulation does not necessarily depend on the subject's monitoring, and therefore, compared with DIY brain stimulation, there is less control or less sense of control over the stimulation.

Metacognitive monitoring in DIY tDCS requires users to observe their cognitive activities and performance in cognitive tasks, and such observation provides guidance for the practice. As reviewed in Section 2.2, when stimulating their brains for cognitive enhancement, DIYers rely on their own subjective judgments or their performance on cognitive tasks to determine the effects and effectiveness of the practice (Jwa, 2015; Wexler, 2016, 2018). Some DIYers share information on Reddit about how they observe and record their cognitive activities, performance, strategies, and thoughts. For example, a user wrote a software program to record the application and perceived effect;[b] while another one combined electroencephalography (EEG) to record brain activation.[c] Such observations and judgments demand that an individual monitors his or her cognitive activities, capacities, and

[b] https://www.reddit.com/r/tDCS/comments/bc0jbc/i_made_some_code_to_keep_a_record_of_using_tdcs/.

[c] https://www.reddit.com/r/tDCS/comments/2f6y5k/recorded_the_effects_of_that_tdcs_device_with_a/.

changes, as well as metacognitive knowledge and experience about his or her cognitive condition. Individuals may infer improvement in the cognitive capacities from their metacognitive knowledge and experience concerning the difficulty of tasks, their confidence in their performance outcome, or their ability to perform a cognitive task. For instance, a tDCS DIYer may use a working memory task, an n-back task,[d] to track his or her performance and the effect of using brain stimulation on working memory. In addition to objectively recorded performance scores, judgments about the ease of the task and confidence in their performance may also contribute to individuals' perceived improvement.

With respect to metacognitive control, using brain stimulation provides users with an additional way to manage their cognitive state or ability. Metacognitive control allows individuals to modify their cognitive processing, guided by metacognitive monitoring. For instance, DIYers explore not only how to stimulate their brains to maximize the effect, but also which strategy works better for a particular cognitive task (e.g., maintaining a pattern, or rehearsing or not rehearsing for n-back tasks to track working memory enhancement).[e] They evaluate their strategy and performance, which may lead them to decide to adopt the same strategy, modify it, or switch to a new one.

Furthermore, manipulating a person's cognitive functioning or psychological state through tDCS is comparable to metacognitive control. Similar to the typical metacognitive control, the way in which users adjust their stimulation can be guided by the monitoring process. When an adjustment is made, cognitive functioning and performance are monitored in order to evaluate the adjustment. Using brain stimulation devices to intervene in a person's brain and manipulate cognitive functions can be regarded as an alternative, indirect way to control cognitive processing.

However, given that metacognitive monitoring can be inaccurate, the perceived improvement that results from, or is influenced by, metacognitive knowledge or experience may be misleading. Therefore, the role of metacognition can be an unreliable indicator of a person's cognitive performance. The incorrect judgments of an individual's cognitive functioning and abilities may create illusory perceived effects or prevent the user from being aware of the effects, either of which can undermine the use of brain stimulation.

[d] The n-back task is widely used to assess working memory. Participants are instructed to monitor a series of stimuli and to respond when a stimulus is the same as one previously presented in the trials.

[e] https://www.reddit.com/r/tDCS/comments/3vew3o/touching_5_on_dual_nback/.

In addition to the role of metacognition in managing the cognitive operation in using brain stimulation to enhance cognitive functions, theoretically speaking, metacognition itself can potentially be a target of enhancement. Even though metacognition has not yet been a target of intervention for the tDCS user community, it can potentially be a more effective way to enhance a person's cognitive performance. By enhancing their metacognitive abilities, users could improve their learning abilities, and this could indirectly be considered a kind of cognitive enhancement. To date, only a few studies have investigated the effects on metacognitive abilities caused by using tDCS. For instance, Chua and Ahmed (2016) found that stimulating dorsolateral prefrontal cortex (DLPFC) led to improved memory monitoring accuracy; that is, the subjects' feeling-of-knowing judgments became better predictors of memory performance. However, it is crucial to note that such an effect may vary with the cognitive bases of the judgments (e.g., familiarity with the stimulus) (Gaynor & Chua, 2019). Since there is no sufficient evidence to indicate the effects of tDCS on metacognitive ability—let alone the effects on learning—metacognitive enhancement remains only a distant possibility.

4. The problem of measuring effectiveness in DIY brain stimulation and metacognition

For DIYers who stimulate their brains outside of professional settings with the purpose of enhancing cognitive performance, one of the most important concerns is to know the effect of this practice. However, there are many difficulties in evaluating the effectiveness of this kind of practice. This section lays out these challenges. First, while scientific studies have found tDCS to have positive effects on cognitive functions, its effectiveness has yet to be established. Methodological reasons include the small sample size, short-term trials, and the variety of measures used in the studies (Farah, 2015).

Second, even if a method of stimulating the brain is reported to be effective, it is far from guaranteed to produce the same effect in every individual. Complications increase as the applications are used by individual DIYers at home with the expectation of producing effects similar to those reported in the scientific literature. One of the reasons for this is that no standard protocol is available, and some DIYers who consult the protocols in scientific studies alter them (e.g., by changing current density, total charge, stimulation duration, electrode location) as they experiment on themselves.

Furthermore, individual differences in biological and psychological traits may affect the results, since the specific electrode placement affects brain regions in different ways (Utz, Dimova, Oppenländer, & Kerkhoff, 2010).

The effectiveness of brain stimulation relies on finding the correct stimulation site, and functional magnetic resonance imaging (fMRI) is often employed to identify locations in the most effective studies on the enhancement benefits of tDCS (Dresler et al., 2013). However, fMRI is unlikely to be incorporated into home-based tDCS. As such, it remains a challenge to ensure the correct stimulation location. The difference in handedness is another example. While most of the subjects in previous studies have been right-handed, left-handed DIYers, due to the lateralized brain function, may encounter moderately different effects compared to right-handed individuals (Schade, Moliadze, Paulus, & Antal, 2012). Considering such differences, some left-handed DIYers modified their protocols (e.g., flipping the montages)[f] in the hope of accounting for the lateralization (Fitz & Reiner, 2014). Other factors—such as an individual's familiarity and prior experience with DIY tDCS, expectations of the effect, and subjective assessment—are all likely to influence the effects on an individual.

Third, while there is no guarantee of the reported effects of tDCS on every individual, measuring the effects of the application on an individual subject is challenging. The effects can be subtle and take time to appear. The experimental setting cannot be replicated at home to examine the effects since it requires an appropriate number of subjects. Even if one tries to mimic the setting in a scientific study and compare the effects against the reported performance of the control group, the lack of long-term study will prevent one from being able to examine the effects of the application on oneself over a relatively longer period of time. Some tDCS DIYers pursuing cognitive enhancement track their performance in cognitive tasks in the hope of measuring potential effects. However, the observed improvement in performance does not necessarily result from the brain stimulation: it can also be due to a placebo effect[g] or the repetitive practice of cognitive tasks. Furthermore, it is unclear whether the act of self-administering tDCS facilitates the placebo effect. If it does so, this leads to more variables in comparing an individual home-based tDCS user and controls in scientific studies. It would be difficult to differentiate these on an individual level.

---

[f] https://www.reddit.com/r/tDCS/comments/5fd6w3/left_handed_tdcsers_may_want_to_flip_the_montage/.

[g] What is more, a placebo response may be fortified by tDCS (Schambra, Bikson, Wager, DosSantos, & DaSilva, 2014).

Fourth, as reviewed in Section 2.2, DIYers rely on their subjective judgments of the effects of brain stimulation to determine effectiveness. However, as highlighted earlier, our metacognition may not always be reliable. The self-evaluation of performance or effects may be inaccurate due to individual and environmental factors. Furthermore, even if an individual can accurately judge his or her cognitive progress, determining the causes of such progress is a challenge, as the effects of placebo, cognitive training, and bias still cannot be easily excluded. In conclusion, the adoption of subjective judgments and the lack of effective alternative methods make it impossible to determine whether the application of brain stimulation provides the particular effects that an individual is after.

5. The ethical dimension

The uncertainty of the effectiveness of DIY brain stimulation raises some ethical concerns. In this section, I will consider three issues: truthfulness, authenticity, and the conception of enhancement. This section illustrates the relevance of these issues to the difficulty of measuring the effects and examines what these considerations entail. I suggest that the ethical implications do not simply pertain to the permissibility or regulation of DIY practice, but rather concern something much subtler: DIY users' knowledge of the practice of using neurotechnology, their expectations, their awareness of the potential ineffectiveness, and general scientific literacy.

To illustrate the ethical issues involved in DIY brain stimulation, I introduce four hypothetical tDCS home users: The first is a DIY practitioner (user A), who believes that brain stimulation with tDCS will improve her particular cognitive ability and self-administers brain stimulation at home in pursuit of modifying her cognition through the application. To assess the effects, the user utilizes cognitive tasks to track the effects of the brain stimulation. When an improvement in her performance is observed, it is directly attributed to the use of tDCS without being aware of, or consciously considering, other possibilities. Similarly, the second practitioner (user B), who, with a similar belief and goal, is also using tDCS to improve her cognitive function and is tracking her progress, is, in her case, aware of the complexity of measuring the effects of the brain stimulation. This user, unlike user A, is aware of the possibility that her perceived improvement may not be the result of the application. The third user (user C) is similar to users A and B, except insofar as her goal is to improve her particular cognitive ability through the process of applying the tDCS to herself, which includes the brain stimulation, the act of self-administration, and accompanying activities such as the cognitive tasks—rather than through the brain stimulation alone. This person may or may not understand the problem of measuring the effects of brain stimulation. Finally, the last user (user D) identifies herself as a member of the citizen science movement, or as someone whose intention is to experiment with brain stimulation and acquire the power to change her physiological and psychological states. She uses tDCS at home and also tracks her own performance in cognitive tasks. In the following sections, these cases will be used to examine the issues of truthfulness, authenticity, and the conception of enhancement in the rest of this section. It is noted that the fictitious cases are different from non-fictitious ones (Hildt, 2014), and these cases are simplified for the convenience of the analysis and discussion. A real-life tDCS DIYer is likely to be a combination of these hypothetical users (or of other kinds of user not mentioned here), or they might alternate between these characterizations.

5.1 Truthfulness

It is crucial that DIYers are fully informed before and while using brain stimulation devices on themselves, i.e., they must have a full understanding of how to use the device as well as the potential effects and risks associated with its use. There are two ways in which the issue of truthfulness is of concern in this matter: the first relates to users being informed honestly by the manufacturers of brain stimulation devices for non-medical purposes; the second relates to users being self-aware about and honest with themselves with regard to results.

The concern of users being treated truthfully and honestly is pertinent to users who purchase their devices from manufacturers. In such cases, manufacturers bear the responsibility of providing detailed characteristics of the products and they ought to be cautious of the information provided to market the products since this is one of the main sources of information for users who have purchased their devices (Wexler, 2018). Attributing any effects of brain stimulation to the product without sufficient evidence or failing to warn of side effects will mislead consumers and cultivate false expectations. Evidence-based guidelines will help prevent users from perceiving illusory effects. In addition to manufacturers, the information provided by other parties, such as the media, other users in the DIY brain stimulation community, and professionals in medicine, science, or ethics—even though it can provide information from multiple perspectives and therefore promote truthfulness—can also potentially undermine truthfulness (Dubljević, Saigle, & Racine, 2014). Aside from positive effects and potentially adverse effects, DIYers should also be informed of the potential ineffectiveness of using the device and the difficulties of measuring the effects on themselves to grant users reasonable expectations. However, it is unclear who should bear this responsibility.

The other worry of diminishing truthfulness relates to individuals deceiving themselves or failing to keep themselves well-informed about the effects and effectiveness of brain stimulation. Here, self-deception is regarded as "information-processing biases that give priority to welcome over unwelcome information" (von Hippel & Trivers, 2011:2). What motivates an individual to continue self-administering brain stimulation may be the perceived effects, curiosity, or patience in waiting for the effects to kick in. However, as we have seen in Section 4, the manifold difficulties of measuring the effectiveness of tDCS on an individual DIYer in a home-based setting suggest that an individual rarely has an accurate picture of the effects of brain stimulation. Even when he or she perceives an improvement in cognitive capacity, it may be the result of other factors, rather than the use of brain stimulation. The worry of undermining truthfulness depends on whether the user is able to fully grasp, and make an honest assessment of his or her own condition, including all the possibilities and options available to them. A full awareness of the potential ineffectiveness of brain stimulation entails a user conceiving of all the potential causes of an improvement in cognitive functioning (if there is perceived improvement), and not just attributing improvement to brain stimulation. This can be clearly illustrated by comparing users A and B, introduced earlier. User A is concerning, as she is either ignorant of the complications and simply attributes any perceived effect to brain stimulation, or she comprehends all the possibilities but chooses to ignore them and thus deceives herself into firmly believing in the effectiveness of the practice. In comparison, the effects and the effectiveness of brain stimulation may be equally uncertain for user B. However, she is informed, and open to herself about all the potential underlying causes of improvement, and is truthful as to what can be happening.

Let us now compare users A and B with user C. Their differences lie in the subtle difference in their goals: the former expects to improve her cognitive ability directly by brain stimulation, whereas the latter expects to do so through the whole process of stimulating her own brain, which includes the act of self-administration. For user C, it does not matter whether the perceived improvement results from the brain stimulation, the repetitive practice of cognitive tasks, the placebo effect, or other causes, for it is not relevant to her motive. Therefore, it is not crucial for user C to be aware of the complexity of measuring the effects and effectiveness of brain stimulation.

5.2 Authenticity and autonomy

Another key issue is authenticity, which is one of the central topics of dispute in the cognitive enhancement debate. There are two major conceptions of authenticity: authenticity as self-discovery, and authenticity as self-creation (Bolt, 2007). This section focuses on the latter conception for two reasons. First, understanding authenticity as self-discovery presumes some kind of essential nature or defining characteristics as the self to which an individual ought to be true (Elliott, 1998). However, this static nature or related characteristics are unclear. Second, ethicists endorsing the former tend to argue against individuals using neurotechnologies to modify their cognition. The central concerns for them are the permissibility of using technology to modify a person's cognitive functioning and whether the technology can indeed modify cognitive functioning (e.g., Erler, 2011), instead of the difficulty of measuring the effectiveness of neurotechnologies on an individual DIYer, which is not an issue of direct relevance to their concern of changing the essential nature or defining characteristics of oneself.

How is the concept of authenticity understood in terms of self-creation? According to DeGrazia, "any self-creation project that is autonomous and honest is ipso facto authentic" (DeGrazia, 2005:112); that is, authenticity consists of three key elements: self-creation, autonomy, and honesty. Self-creation refers to "the conscious, deliberate shaping of one's own personality, character, other significant traits [...] or life direction" (DeGrazia, 2005:89–90). Using brain stimulation devices to intervene in the brain in order to manipulate a person's cognitive processing is regarded as a project of self-creation.

This concept of authenticity is intertwined with the concept of autonomy; indeed, they are sometimes defined interchangeably. Take Christman's (2018) generic definition as an example: "to be autonomous is to be one's own person, to be directed by considerations, desires, conditions, and characteristics that are not simply imposed externally upon one, but are part of what can somehow be considered one's authentic self." Autonomy, in addition, requires the subject to prefer the actions he or she performs, the preference to be identified by the subject, and the identification to be carefully reflected on to exclude any influences that might lead to alienation. Therefore, according to DeGrazia (2005), projects of self-creation should involve setting goals and plans about the kind of person an individual identifies as and the way in which one wishes to lead his or her life, as well as taking actions and making decisions to realize plans and achieve goals in a truthful manner. Accordingly, the issues of authenticity and autonomy, beyond the consideration of truthfulness, concern not only whether the subject is in possession of full knowledge about his or her own possible conditions and options (which is addressed in the issue of truthfulness), but also the actions one chooses to perform given the set goal and knowledge of it.

With respect to DIY brain stimulation, the issue of authenticity prompts DIYers to reflect on their self-creation—brain stimulation—plan, including the goals, actions, outcome, and how these are related to one another. It requires the individual to contemplate his or her reasons for using brain stimulation (e.g., to improve cognitive abilities, to experiment with tDCS), as well as to have an understanding of what actions could help to achieve their goals and what the potential outcomes are. This includes understanding the difficulties of measuring the effectiveness of the intervention and considering other possible causes of the effects. Examining the cases of the four users helps to illustrate how the issue is considered. Awareness of the possibility that the brain stimulation is not effective is important for users A and B, as the effectiveness of the intervention is crucial in determining whether the given action (i.e., stimulating the brain in such a way) will help achieve their goals. Suppose user A's self-administration is not effective or less effective than she believes it to be. Before she realizes the issue, she is deceived or self-deceived in the pursuit of her cognitive enhancement goal and fails to reflect on her goals and actions. On the other hand, consider user C; her goal of using tDCS is to improve her cognitive ability in the process of stimulating her own brain—a different kind of goal from that of users A and B. Whether user C is aware of the potential causes of effects does not influence the user with respect to the issue of authenticity as the effectiveness of the brain stimulation alone is not as crucial for her to achieve her goal.

On the other hand, DIYers using brain stimulations to intervene in their brain activities can potentially foster authenticity. Consider user D, if the core value for her to use home-based neurotechnologies is to own and use the tool of manipulating her own cognitive functioning (Wexler, 2017). For this kind of user, the act of using brain stimulation is aligned with the goal which involves whom the users identify themselves to be.

Furthermore, to manage a person's own cognitive functioning using metacognitive monitoring and self-administration of brain stimulation—in addition to metacognitive control—allows users to manifest their mental agency in a new way; that is, the access to brain stimulation devices expands user D's potential to govern her own psychological states and cognitive abilities. However, in user D's case, being aware of the potential ineffectiveness is crucial for her, as failing to do so can result in an illusory sense of agency, of being in control, and therefore risks undermining authenticity.

In summary, it depends not only on how informed the DIY practitioners are—whether they are aware of the potential ineffectiveness—but also on what goals they have set for using their brain stimulation devices. In fact, there is a way of understanding the concept of enhancement which allows the act of using brain stimulation to be considered as enhancement in and of itself, despite its potential ineffectiveness.

5.3 The concept of enhancement

It would be a mistake to conclude from the previous discussion that using tDCS to improve cognitive abilities is not an attainable goal for DIYers using brain stimulation. It merely requires users to be aware of the possibilities concerning their conditions, including potential ineffectiveness and other causes for improvements in cognitive ability. In this sense, tDCS can be used for cognitive enhancement. However, there are other interesting ways of regarding DIY brain stimulation as enhancement. This section briefly reviews the conceptual issue of enhancement and discusses the different perspectives prompted by the conceptual issue.

The concept of cognitive enhancement has received a diverse range of treatments (Parens, 2005; Savulescu, Sandberg, & Kahane, 2011). On the one hand, it can be seen as any intervention that improves a particular cognitive function beyond what is necessary for good health (Parens, 1998). This has been the most common notion used in scientific literature on enhancement studies which focus on a single cognitive function and involve healthy subjects, an intervention, and a particular way of measuring performance and improvement. On the other hand, we can adopt a much broader conception of enhancement, such as *any interventions that manipulate a person's psychological state or capacity based on the self-interests of the individual.* Savulescu et al.' (2011) welfarist definition shares the same spirit: "[a]ny change in the biology or psychology of a person which increases the chances of leading a good life in the relevant set of circumstances" (Savulescu et al., 2011:7).

The first concept is about the condition of a particular cognitive ability, whereas the second concept concerns the overall state of the subject. A major difference between these two conceptions is the way to determine whether the subject is better off after the intervention. For the former, it is determined locally: it is based on the change of a single cognitive ability (e.g., working memory), perhaps through a single type or limited types of assessment method (e.g., n-back task). In contrast, with respect to the second conception, determining whether an intervention is considered as an enhancement requires a more holistic assessment. The interests of the individual are taken into account, and therefore the improvement of a single cognitive ability may not necessarily imply enhancement. Cases in which the improvement of one cognitive faculty may lead to the cost of another (Iuculano & Cohen Kadosh, 2013) are regarded as cognitive enhancement in the first sense, but not necessarily in the second.

The difficulty of measuring the effectiveness of brain stimulation is of concern if cognitive enhancement is understood in the first sense. Whether there exists any effect and whether this effect is a result of the stimulation determines whether this is an instance of cognitive enhancement. However, if endorsing the second conception of cognitive enhancement, this is determined by the subject's self-interests. If a user, like users A and B, demands improvement effects from the application of tDCS, then the difficulty of measuring the effectiveness matters. However, for some users, like user C, it does not matter if the perceived improvement results from the stimulation or placebo,[h] as long as there is an improvement in cognitive performance accompanying the process of DIY brain stimulation.

There can be other senses in which a subject is enhanced by the act of self-administering brain stimulation if the focus is on the subject as a whole rather than a particular cognitive function, and the self-interests of the subject are taken into account. First, the mere act of using a brain stimulation device as a potential tool to manipulate a person's cognition can be regarded as a kind of enhancement if the interest of the subject (like user D) is to gain control over a person's own psychology. Using neurotechnologies to manipulate his or her cognitive processing and abilities provides a sense of control or agency. Such manipulation can be regarded as an unorthodox example of cognitive enhancement if endorsing the second conception, as long as taking control of one's psychological states and cognitive abilities is

[h] https://www.reddit.com/r/tDCS/comments/5x0y6c/experience_with_tdcs_for_enhanced_learning/.

considered a part of leading a good life. Second, as discussed in Section 3.2, manipulating one's cognitive functioning or psychological states through brain stimulation is comparable to metacognitive control. Using brain stimulation devices to intervene in a person's own brain and manipulate cognitive functions can be regarded as an alternative, indirect way to control cognitive processing. A DIYer's gaining access to a DIY brain stimulation device and putting it on themselves enhances their ability to control their own cognitive functioning and accordingly this can be seen as a kind of metacognitive enhancement.

5.4 Practical implications for stakeholders

Examining the problem of measuring the effectiveness of brain stimulation in relation to the normative issues of truthfulness and authenticity and the conceptual issue of cognitive enhancement provides some practical implications for stakeholders. However, such implications are not as straightforward as to simply question the permissibility of DIY brain stimulation or how it should be regulated. Rather, the situation is more nuanced. As I have detailed above, the difficulties of measuring the effects and effectiveness of using brain stimulation are manifold, and deciphering the causes of any perceived improvement is challenging. Whether the potential ineffectiveness compromises the values of truthfulness and authenticity depends on whether the user is aware of the complicated underlying causes, how the user sets his or her goals, and what kind of person the user wants to be. The uncertainty of the effectiveness of brain stimulation on an individual does not amount to supporting or opposing the permissibility of using brain stimulation devices, and considerations about regulation are not sufficient to address these worries. Instead, education is equally—if not more—important (De Ridder, Vanneste, & Focquaert, 2014). The key is to equip the DIYers with relevant background knowledge and neuroscientific literacy to determine their relationship to the neurotechnology, including their goal in using it, how the practice is undertaken, and their expected outcomes. In addition, it is crucial to have responsible and truthful communication about the intervention's effects and the uncertainties between DIYers and other parties—such as the device manufacturers, media, other users in the DIY brain stimulation community, and professionals in medicine, science, or ethics. Thus, a platform is required where these various parties can come together to interact and communicate with one another.

I agree with Wexler (2017) in that the use of brain stimulation in a home setting should be considered within the context of citizen science or the neurohacking movement. Since the main purpose of using tDCS for enhancement may not solely be to improve a person's own cognitive functioning (Wexler, 2018), I suggest that individuals' access to brain intervention can be regarded as an improvement in the control of their own cognitive functioning. This can be regarded as a kind of enhancement, in the senses discussed above. In addition, for DIYers who want to gain control of their psychological states and functioning, brain stimulation technology has the potential to foster authenticity.

6. Conclusion

This chapter focused on the illusory effects that users of brain stimulation may experience and considered the potential concerns arising from these effects. A significant number of DIYers rely on their subjective judgments to determine the effects of the stimulation. However, deciphering the causes of these effects is challenging, since they can result not only from the stimulation but also from placebo effects, the repetitive practice of cognitive tasks, sample bias, and other causes. I highlighted the role of metacognition as it is crucial for brain stimulation DIYers in two ways. First, cognitive monitoring contributes to users' perceived effects and subjective judgment of the effectiveness of the stimulation. Considering the individual differences in metacognitive monitoring and its sensitivity to environmental factors helps us to understand the complexity of evaluating effectiveness. Second, DIYers monitoring their cognitive activities and controlling them through the use of brain stimulation can be seen as gaining mental agency. The access to this new way of managing a person's own cognitive functioning can potentially promote authenticity and be considered an enhancement for individuals who seek to have control over their mental states.

Acknowledgment

I am grateful to Elisabeth Hildt, Imre Bard, and Judy Illes for their helpful comments on earlier versions of this paper.

References

Beran, M. J. (Ed.), (2012). *Foundations of metacognition.* (1st ed.). Oxford University Press.

Bolt, L. L. E. (2007). True to oneself? Broad and narrow ideas on authenticity in the enhancement debate. *Theoretical Medicine and Bioethics, 28*(4), 285–300. https://doi.org/10.1007/s11017-007-9039-8.

Christman, J. (2018). Autonomy in moral and political philosophy. In E. N. Zalta (Ed.), *The Stanford encyclopedia of philosophy*. (Spring 2018 ed.). https://plato.stanford.edu/archives/spr2018/entries/autonomy-moral/.

Chua, E. F., & Ahmed, R. (2016). Electrical stimulation of the dorsolateral prefrontal cortex improves memory monitoring. *Neuropsychologia, 85*, 74–79. https://doi.org/10.1016/j.neuropsychologia.2016.03.008.

Coffman, B. A., Clark, V. P., & Parasuraman, R. (2014). Battery powered thought: Enhancement of attention, learning, and memory in healthy adults using transcranial direct current stimulation. *NeuroImage, 85*, 895–908. https://doi.org/10.1016/j.neuroimage.2013.07.083.

De Ridder, D., Vanneste, S., & Focquaert, F. (2014). Outstanding questions concerning the regulation of cognitive enhancement devices. *Journal of Law and the Biosciences, 1*(3), 316–321. https://doi.org/10.1093/jlb/lsu024.

DeGrazia, D. (2005). *Human identity and bioethics*. Cambridge University Press.

Dresler, M., Sandberg, A., Ohla, K., Bublitz, C., Trenado, C., Mroczko-Wąsowicz, A., et al. (2013). Non-pharmacological cognitive enhancement. *Neuropharmacology, 64*, 529–543. https://doi.org/10.1016/j.neuropharm.2012.07.002.

Dubljević, V., Saigle, V., & Racine, E. (2014). The rising tide of tDCS in the media and academic literature. *Neuron, 82*(4), 731–736. https://doi.org/10.1016/j.neuron.2014.05.003.

Dunlosky, J., & Metcalfe, J. (2008). *Metacognition*. Sage Publications.

Elliott, C. (1998). The tyranny of happiness: Ethics and cosmetic psychopharmacology. In E. Parens (Ed.), *Enhancing human traits: Ethical and social implications* (pp. 177–188). Washington, DC: Georgetown University Press.

Erler, A. (2011). Does memory modification threaten our authenticity? *Neuroethics, 4*, 235–249.

Farah, M. J. (2015). The unknowns of cognitive enhancement. *Science, 350*(6259), 379–380. https://doi.org/10.1126/science.aad5893.

Fecteau, S., Pascual-Leone, A., Zald, D. H., Liguori, P., Theoret, H., Boggio, P. S., et al. (2007). Activation of prefrontal cortex by transcranial direct current stimulation reduces appetite for risk during ambiguous decision making. *Journal of Neuroscience, 27*(23), 6212–6218. https://doi.org/10.1523/JNEUROSCI.0314-07.2007.

Fitz, N. S., & Reiner, P. B. (2014). The perils of using electrical stimulation to change human brains. In R. C. Kadosh (Ed.), *The stimulated brain* (pp. 61–83). Academic Press. https://doi.org/10.1016/B978-0-12-404704-4.00003-X.

Fitz, N. S., & Reiner, P. B. (2015). The challenge of crafting policy for do-it-yourself brain stimulation. *Journal of Medical Ethics, 41*(5), 410–412. https://doi.org/10.1136/medethics-2013-101458.

Flavell, J. H. (1979). Metacognition and cognitive monitoring: A new area of cognitive-developmental inquiry. *American Psychologist, 34*(10), 906–911. https://doi.org/10.1037/0003-066x.34.10.906.

Fleming, S. M., & Dolan, R. J. (2012). The neural basis of metacognitive ability. *Philosophical Transactions of the Royal Society B: Biological Sciences, 367*(1594), 1338–1349. https://doi.org/10.1098/rstb.2011.0417.

Fleming, S. M., Weil, R. S., Nagy, Z., Dolan, R. J., & Rees, G. (2010). Relating introspective accuracy to individual differences in brain structure. *Science, 329*(5998), 1541–1543. https://doi.org/10.1126/science.1191883.

Fregni, F., Boggio, P. S., Nitsche, M., Bermpohl, F., Antal, A., Feredoes, E., et al. (2005). Anodal transcranial direct current stimulation of prefrontal cortex enhances working memory. *Experimental Brain Research, 166*(1), 23–30. https://doi.org/10.1007/s00221-005-2334-6.

Gaynor, A. M., & Chua, E. F. (2019). Transcranial direct current stimulation over the prefrontal cortex alters encoding and judgments of learning based on fluency. *Journal of Cognitive Neuroscience, 31*(11), 1710–1725. https://doi.org/10.1162/jocn_a_01449.

Hecht, D., Walsh, V., & Lavidor, M. (2010). Transcranial direct current stimulation facilitates decision making in a probabilistic guessing task. *Journal of Neuroscience, 30*(12), 4241–4245. https://doi.org/10.1523/JNEUROSCI.2924-09.2010.

Hildt, E. (2014). On the current neuroenhancement use of transcranial direct current stimulation by healthy individuals—A non-fictional snap-shot: Commentary on Lapenta et al. 2014. *Psychology & Neuroscience*, 7(2), 181–182. https://doi.org/10.3922/j.psns.2014.011.

Hill, A. T., Fitzgerald, P. B., & Hoy, K. E. (2016). Effects of anodal transcranial direct current stimulation on working memory: A systematic review and meta-analysis of findings from healthy and neuropsychiatric populations. *Brain Stimulation, 9*(2), 197–208. https://doi.org/10.1016/j.brs.2015.10.006.

Iuculano, T., & Cohen Kadosh, R. (2013). The mental cost of cognitive enhancement. *Journal of Neuroscience, 33*(10), 4482–4486. https://doi.org/10.1523/JNEUROSCI.4927-12.2013.

Jwa, A. (2015). Early adopters of the magical thinking cap: A study on do-it-yourself (DIY) transcranial direct current stimulation (tDCS) user community. *Journal of Law and the Biosciences, 2*(2), 292–335. https://doi.org/10.1093/jlb/lsv017.

Kelemen, W. L., Frost, P. J., & Weaver, C. A. (2000). Individual differences in metacognition: Evidence against a general metacognitive ability. *Memory & Cognition, 28*(1), 92–107. https://doi.org/10.3758/BF03211579.

Koriat, A. (2000). The feeling of knowing: Some metatheoretical implications for consciousness and control. *Consciousness and Cognition, 9*(2), 149–171. https://doi.org/10.1006/ccog.2000.0433.

Koriat, A. (2006). Metacognition and consciousness. In P. Zelazo, M. Moscovitch, & E. Thompson (Eds.), *The Cambridge handbook of consciousness* (pp. 289–325). Cambridge University Press.

Le Strange, R. (1977). *A history of herbal plants*. Angus & Robertson.

Mancuso, L. E., Ilieva, I. P., Hamilton, R. H., & Farah, M. J. (2016). Does transcranial direct current stimulation improve healthy working memory? A meta-analytic review. *Journal of Cognitive Neuroscience, 28*(8), 1063–1089. https://doi.org/10.1162/jocn_a_00956.

Manenti, R., Brambilla, M., Petesi, M., Ferrari, C., & Cotelli, M. (2013). Enhancing verbal episodic memory in older and young subjects after non-invasive brain stimulation. *Frontiers in Aging Neuroscience, 5*. https://doi.org/10.3389/fnagi.2013.00049.

Maslen, H., Douglas, T., Kadosh, R. C., Levy, N., & Savulescu, J. (2015). Do-it-yourself brain stimulation: A regulatory model. *Journal of Medical Ethics, 41*(5), 413–414. https://doi.org/10.1136/medethics-2013-101692.

Morein-Zamir, S., & Sahakian, B. J. (2011). Pharmaceutical cognitive enhancement. In J. Illes & B. J. Sahakian (Eds.), *Oxford handbook of neuroethics* (pp. 229–244). Oxford University Press.

Nelson, T. O. (1996). Consciousness and metacognition. *American Psychologist, 15*, 102–116.

Nelson, T. O., Dunlosky, J., White, D. M., Steinberg, J., Townes, B. D., & Anderson, D. (1990). Cognition and metacognition at extreme altitudes on Mount Everest. *Journal of Experimental Psychology General, 119*(4), 367–374.

Nelson, T. O., & Narens, L. (1990). Metamemory: A theoretical framework and new findings. In *Vol. 26 Psychology of learning and motivation* (pp. 125–173). Elsevier. https://doi.org/10.1016/S0079-7421(08)60053-5.

Ohn, S. H., Park, C.-I., Yoo, W.-K., Ko, M.-H., Choi, K. P., Kim, G.-M., et al. (2008). Time-dependent effect of transcranial direct current stimulation on the enhancement of working memory. *Neuroreport, 19*(1), 43–47. https://doi.org/10.1097/WNR.0b013e3282f2adfd.

Parens, E. (1998). *Enhancing human traits: Ethical and social implications*. Georgetown University Press.

Parens, E. (2005). Authenticity and ambivalence: Toward understanding the enhancement debate. *The Hastings Center Report, 35*(3), 34. https://doi.org/10.2307/3528804.

Penolazzi, B., Di Domenico, A., Marzoli, D., Mammarella, N., Fairfield, B., Franciotti, R., et al. (2010). Effects of transcranial direct current stimulation on episodic memory related to emotional visual stimuli. *PLoS One*, *5*(5), e10623. https://doi.org/10.1371/journal.pone.0010623.

Savulescu, J., Sandberg, A., & Kahane, G. (2011). Well-being and enhancement. In J. Savulescu, R. ter Meulen, & G. Kahane (Eds.), *Enhancing human capacities* (pp. 1–18). Blackwell Publishing Ltd. https://doi.org/10.1002/9781444393552.ch1.

Schade, S., Moliadze, V., Paulus, W., & Antal, A. (2012). Modulating neuronal excitability in the motor cortex with tDCS shows moderate hemispheric asymmetry due to subjects' handedness: A pilot study. *Restorative Neurology and Neuroscience*, *3*, 191–198. https://doi.org/10.3233/RNN-2012-110175.

Schambra, H. M., Bikson, M., Wager, T. D., DosSantos, M. F., & DaSilva, A. F. (2014). It's all in your head: Reinforcing the placebo response with tDCS. *Brain Stimulation*, 7(4), 623–624. https://doi.org/10.1016/j.brs.2014.04.002.

Schwartz, B. L., & Díaz, F. (2014). Quantifying human metacognition for the neurosciences. In S. M. Fleming & C. D. Frith (Eds.), *The cognitive neuroscience of metacognition* (pp. 9–23). Springer-Verlag Publishing.

Schwartz, B. L., & Metcalfe, J. (1992). Cue familiarity but not target retrievability enhances feeling-of-knowing judgments. *Journal of Experimental Psychology: Learning, Memory, and Cognition*, *18*(5), 1074–1083. https://doi.org/10.1037/0278-7393.18.5.1074.

Slife, B. D., & Weaver, C. A. (1992). Depression, cognitive skill, and metacognitive skill in problem solving. *Cognition & Emotion*, *6*(1), 1–22. https://doi.org/10.1080/02699939208411055.

Smith, M. E., & Farah, M. J. (2011). Are prescription stimulants "smart pills"? The epidemiology and cognitive neuroscience of prescription stimulant use by normal healthy individuals. *Psychological Bulletin*, *137*(5), 717–741. https://doi.org/10.1037/a0023825.

Utz, K. S., Dimova, V., Oppenländer, K., & Kerkhoff, G. (2010). Electrified minds: Transcranial direct current stimulation (tDCS) and Galvanic Vestibular Stimulation (GVS) as methods of non-invasive brain stimulation in neuropsychology—A review of current data and future implications. *Neuropsychologia*, *48*(10), 2789–2810. https://doi.org/10.1016/j.neuropsychologia.2010.06.002.

von Hippel, W., & Trivers, R. (2011). The evolution and psychology of self-deception. *Behavioral and Brain Sciences*, *34*(1), 1–16. https://doi.org/10.1017/S0140525X10001354.

Wexler, A. (2016). The practices of do-it-yourself brain stimulation: Implications for ethical considerations and regulatory proposals. *Journal of Medical Ethics*, *42*(4), 211–215. https://doi.org/10.1136/medethics-2015-102704.

Wexler, A. (2017). The social context of "do-it-yourself" brain stimulation: Neurohackers, biohackers, and lifehackers. *Frontiers in Human Neuroscience*, *11*, 224. https://doi.org/10.3389/fnhum.2017.00224.

Wexler, A. (2018). Who uses direct-to-consumer brain stimulation products, and why? A study of home users of tDCS devices. *Journal of Cognitive Enhancement*, *2*(1), 114–134. https://doi.org/10.1007/s41465-017-0062-z.

Wexler, A., & Reiner, P. B. (2017). Home use of tdcs: From "do-it-yourself" to "direct-to-consumer". In S. M. Johnson & K. S. Rommelfanger (Eds.), *The Routledge handbook of neuroethics* (pp. 271–284). Routledge.

Wurzman, R., Hamilton, R. H., Pascual-Leone, A., & Fox, M. D. (2016). An open letter concerning do-it-yourself users of transcranial direct current stimulation: DIY-tDCS. *Annals of Neurology*, *80*(1), 1–4. https://doi.org/10.1002/ana.24689.

CHAPTER EIGHT

What is neurohacking? Defining the conceptual, ethical and legal boundaries

Marcello Ienca*, James Scheibner
Department of Health Sciences and Technology, ETH Zurich, Zurich, Switzerland
*Corresponding author: e-mail address: marcello.ienca@hest.ethz.ch

Contents

1. Introduction

The expression neurohacking (sometimes also referred to as "brain hacking" or "mind hacking") has been often used in recent years to denominate activities involving the human use of neurotechnology. References to neurohacking and brain hacking have appeared in scientific articles from disciplines as different as cognitive (Koch, 2010) and behavioral neuroscience (Bela et al., 2016; Varazzani, 2017), neurophysiology (Kaniusas et al., 2019), the social sciences (Gardner & Wray, 2013; Wexler, 2017),

Developments in Neuroethics and Bioethics, Volume 3
ISSN 2589-2959
https://doi.org/10.1016/bs.dnb.2020.03.008

philosophy (Spreeuwenberg, 2019; Tillas, 2018), law (Lee, 2016), computer science and cybersecurity (Bak, Pyo, & Jeong, 2019; Frank, Mabrey, & Yoshigoe, 2017; Sinha, Sharma, Kumari, & Vyas, 2015) and, most importantly for this book's readership, bioethics and neuroethics (DiEuliis & Giordano, 2017; Ienca, 2015; Ienca & Haselager, 2016; Jebari & Hansson, 2013; Kellmeyer, 2018; O'Brolchain & Gordijn, 2014; Steinert & Friedrich, 2020; Sullivan & Illes, 2018; Takagi, 2012). Furthermore, they have appeared in a number of mainstream publications for the general audience (Ackerman, 2018; Leggett, 2009; Marcus, 2015; Nutt, 2016; Sofge, 2014; Strickland, 2014; Talbot, 2009).

In spite of this growing scientific and journalistic interest, it is unclear whether these terms are being used to denote the same class of activities or phenomena. This chapter presents the findings of a scoping review and subsequent conceptual and normative analysis with a twofold aim. First, we intend to map and clarify the various meanings of "neurohacking" to ensure a more rigorous and semantically unambiguous use of this word in scholarly and science communication literature. Second, we intend to assess the various uses of the word "neurohacking" from an ethical and legal perspective. In particular, we will investigate which ethical principles and international regulations pertain to each of the activities described as "neurohacking" as to prepare the terrain for an ethico-legal taxonomy.

2. Methods

We conducted a scoping review of the literature by searching three databases (Web of Science, MEDLINE/Pubmed, IEEE Xplore) to retrieve relevant articles published until August 20, 2019. We searched title, abstract, and keywords for the terms: "neurohacking" OR "brain hacking" to explore how these notions are used in the scientific literature. This methodological strategy is grounded on the philosophical assumption that the use of a word in the language might offer valuable proxy information about the meaning of that word. As stated by Ludwig Wittgenstein in the Philosophical Investigations (Section 43): "For a large class of cases—though not for all—in which we employ the word 'meaning' it can be defined thus: the meaning of a word is its use in the language" (Wittgenstein, 1953). Query logic was modified to adapt to the language used by each engine or database. We complemented this systematic search with a purposive screening of the Google Scholar web search engine anonymously to retrieve additional entries that might not be indexed in academic databases. We terminated the review of Google Scholar's webpages after theoretical saturation was reached, which typically occurred after the twentieth webpage. Systematic screening identified a total of 76 entries, 37 of which were included into the final synthesis after duplicate removal and eligibility check. Twenty-eight further entries were retrieved through Google Scholar, 18 of which were included into the synthesis after duplicate removal.

All entries were imported automatically into the endnote literature manager software (version X8 for Mac) and analyzed using both a descriptive numerical summary and a thematic analysis. In the former analysis, both relative and cumulative frequencies were extracted and graphically represented (see Table 1). In several occurrences, the same article addressed multiple meanings of neurohacking, hence was associated with multiple themes. In the latter analysis, recurrent semantic patterns were identified through full-text screening and subsequent coding. Once conceptually stable semantic patterns emerged from the codes, these were grouped together into a system of themes and subthemes (see Table 1). A minority of articles (6) contain more than one theme.

All entries were checked anew through an automated text search for the presence of the emerging themes. Following Braun and Clarke (Braun & Clarke, 2006), codes that did not seem to fit into any main theme were temporarily housed in a group identified as miscellaneous. These codes were subsequently clustered into a new theme or reallocated to an existing thematic group after consultation. Internal consultation was performed between the authors to crosscheck and validate the findings.

3. Findings

The search revealed a total of 34 papers meeting our eligibility criteria. A majority of papers (28) were articles, 4 papers were conference papers, 1 paper was a book chapter and 1 was published on a website. These papers were published in fields including psychology and neuroscience, biotechnology, computer science, chemistry, law, ethics and general science. The majority of papers retrieved (29) were published between 2010 and 2019. By contrast, 4 were published between 2000 and 2009, and only one result was published before 2000.

The thematic analysis reveals that the terms "neurohacking" and "brain hacking" are being used to denote four distinct families of activities or phenomena related to the use of neurotechnology. These are, in order of frequency:

Table 1 The themes and relevant articles identified via the search string.

Theme	Subthemes	Frequency	References
Human enhancement	Cognitive enhancement Emotional enhancement Behavioral enhancement	13	Borrione and Brunoni (2017), Buchan (2018), Conner (2010), Dresler et al. (2019), Front (2017), Mehrali et al. (2018), Prpa, Quesnel, Vidyarthi, Kitson, and Riecke (2016), Plotogea and Zamfiroiu (2017), Schiller (2016), Siva (2006), Strickland (2014), Tillas (2018), Wade (2018)
Malicious hacking	Misuse of neurodevices Unauthorized access to data	12	Folgieri (2017), Ienca and Haselager (2016), Ienca, Jotterand, and Elger (2018), Ienca, Shaw, and Elger (2019), Ienca, Vayena, and Blasimme (2018), Kermani, Zhang, Raghunathan, and Jha (2013), Malin, Gudaitis, Holt, and Kilger (2017), Patel, Taghavi, Júnior, Latih, and Zin (2012), Plotogea and Zamfiroiu (2017), Pugh, Pycroft, Sandberg, Aziz, and Savulescu (2018), Pycroft and Aziz (2018), Pycroft et al. (2016), Sage and Hardell (2018), Szu and Mehmood (2006)
Causal understanding of brain processes	Decoding neural processes in the brain Understanding consciousness and other phenomena in the brain	10	Heaven (2014), Macknik and Martinez-Conde (2009), Mathalon, Ford, and Pfefferbaum (2000), Matos, Pinto-Gouveia, and Duarte (2012), McNaughton (1998), Nazari and Faez (2018), Schiller (2016), Siva (2006), Stark and Campbell (2018), Tillas (2018)
Non-malicious hacking and self-experimentation	Ethical hacking of neurodevices Open source neurotechnology DIY neurotechnology	5	Conner (2010), Li and Johnson (2015), McMahon and Schukat (2018), Strickland (2014), Wexler (2017)

1. The use of neurotechnology by healthy individuals for the purpose of cognitive, emotional or behavioral enhancement;
2. The malicious use or unauthorized access to neurotechnology systems or related databases by third parties;
3. The use of neuroscience and neurotechnology to unveil the causal mechanisms of brain processing; and,
4. The non-malicious hacking of neurodevices and their open development and self-experimentation.

Each of these different thematic families include various subthemes which will be further discussed in the following.

3.1 Enhancement

Most articles (38%) associate the idea of neurohacking with the use of neurotechnology, especially non-invasive brain stimulation, for the purpose of optimizing brain function, improving cognition, and achieving a better performance at various activities including work, sports, leisure activities and even the military (Borrione & Brunoni, 2017). In these articles, the verb "to hack" is interpreted as "to modify/improve." Authors hypothesize this form of neurohacking might result from the growing "demands for cognitive functioning" of our "complex information society" (Dresler et al., 2019, p. 1137). The same authors observe that neurohacking for cognitive enhancement is not "a monolithic phenomenon" but "a multifaceted concept." In turn, this multifaceted concept encompasses a "great variety of interventions that can be clustered into biochemical, physical, and behavioral enhancement strategies" (Dresler et al., 2019, p. 1137). Among these interventions, authors attribute a prominent role to non-invasive neuromodulation tools such as transcranial direct current stimulation (tDCS) (Folgieri, 2017). Some authors go further to speculate that neurohacking technologies "are capable of leading humanity forward into the age of cyborgs, a time in which human biology can be hacked at will to yield beings with abilities beyond their natural capabilities" (Mehrali et al., 2018). Some articles associate neurohacking with enhancing not only cognitive but also emotional traits. For example, (Schiller, 2016) applies the neurohacking label to describe a novel approach to fear reduction via unconscious processing based on the reinforcement of neural activity (Koizumi et al., 2016).

3.2 Malicious hacking

Another significant portion of the literature (35%) associates the idea of neurohacking with the misuse or malicious use of personal neurotechnologies, especially brain-computer interfaces (BCIs) (Ienca & Haselager, 2016) and invasive neurostimulation such as deep brain stimulation (DBS) (Pugh et al., 2018). In these articles, the verb "to hack" is interpreted as "to gain access to a computer illicitly." This segment of the literature, particularly common among ethics and cybersecurity researchers, sees neurohacking in continuity with malicious computer hacking (also known as "cracking" or "black hat hacking"). Articles of this type typically explore the cybersecurity vulnerabilities of neurodevices. Authors identify two main cybersecurity risks associated with malicious neurohacking. The first risk is the unauthorized access to brain-related information from either cloud-based data repositories or from the devices themselves (Dupont, 2013; Ienca, 2019; Ienca & Haselager, 2016). Cybersecurity researchers warn that the risk of cyberattacks and security breaches in personal neurotechnology devices could lead "to losses of private patient medical and financial data" emphasizing that these records are strictly protected under health information privacy laws such as the "US HIPAA [Health Information Portability and Accountability Act]" (Sage & Hardell, 2018, p. 95). A detailed analysis of the cyber-vulnerabilities of BCIs suggested that security breaches could occur at various phases of the BCI cycle including signal acquisition, feature extraction and decoding (Ienca & Haselager, 2016). A second risk highlighted in the literature is the so-called brainjacking, namely the risk that malevolent attackers might "exert malicious control over brain implants" without the users' consent (Pycroft et al., 2016, p. 454). While the first type of vulnerability regards the unauthorized extraction of information, attacks of the second type involve the hijacking of the device's control by third parties with consequent disempowerment of the user/victim. Attacks of this type include cessation of stimulation, draining implant batteries, and inducing tissue damage. They might cause impairment of motor function, alteration of impulse control, modification of emotions or affect, induction of pain, and modulation of the reward system of the victim (Pycroft et al., 2016).

3.3 Causal understanding of brain processes

A slightly smaller portion of documents (29%) uses the word "neurohacking" to refer to advances in neuroscience and neurotechnology which could unveil the causal mechanisms underlying brain processes in humans or other animals. In these documents, the word "hacking" is used in a metaphorical sense, as a synonym of "unlocking," "unveiling" or "revealing." Articles of this type focus on the explanatory and heuristic function of neuroscience research, i.e., its potential for providing causal explanations of brain functions and processes. Among these functions and processes, phenomena such as consciousness and personal identity seem to be of particular interest for researchers (Front, 2017; Heaven, 2014). According to this metaphor, the brain is seen as a complex system that needs to be deciphered or decrypted by advances in neuroscience and neurotechnology. Articles of this type do not refer to a specific application of neurotechnology but rather focus on models and approaches for developing explanations of brain function at the causal level.

3.4 Non-malicious hacking and self-experimentation

The smallest portion of articles (15%) associates neurohacking with the quest to democratize neuroscience and neurotechnology, especially the tools of scientific experimentation. This type of neurohacking activities are seen in parallel to the DIY Biology (or "biohacking") movement (Wexler, 2017). An example of this approach is Open BCI, a US-based incorporated company founded "to harness the power of the open source movement to accelerate ethical innovation of human-computer interface technologies."[a] Open BCI's mission statement explicitly refers to the values of open science as it proclaims the belief "that science advancements will only—and should only—be made through an open forum of shared knowledge and concerted effort, by people from a variety of backgrounds" (ivi). One common practice among these types of neurohackers is the organization of hackathons, otherwise known as hack days, hackfests or codefests. Hackathons are design sprint-like events in which computer programmers and others involved in software-hardware development collaborate intensively on software projects. For example, Li and Johnson describe the experience of a hackathon held in Brazil with the aim of developing novel technological solutions for stroke detection (Li & Johnson, 2015). According to the authors, the hackathon model is an effective way to foster innovation and raise awareness about stroke and other health or social problems. Within this thematic family, some documents use the language of neurohacking to refer to the self-experimentation of personal neurotechnology devices, especially DIY devices, by non-expert users (Strickland, 2014). While intimately intertwined with the previously described category of neurohacking as open development, these two

[a] See: https://openbci.com/ (last accessed: September 15, 2019).

categories do not supervene upon each other. In fact, open neurotechnology development can be pursued without involving self-experimentation and vice versa. Strickland (2014) reports the stories of several self-experimenters of non-invasive neuromodulation, especially tDCS. Most of these self-experimentation attempts are reported to be conducted in unsupervised environments, primarily at home. Individuals who engage in these activities include both professional engineers and amateurs without formal scientific training. By investigating the social context of this form of neurohacking, Wexler argued that it should be seen as "an offshoot of the 'life hacking' (or 'quantified self') movement" (Wexler, 2017, p. 1). In this movement, individuals self-track minute aspects of their daily lives in order to enhance productivity or performance. Neurohacking activities of this type are primarily aimed at modifying the experimenter's mental abilities for cognitive enhancement purposes. Reported desired effects include "faster learning, better math skills, improved memory, more creativity" (Strickland, 2014). This indicates an intimate link between self-experimentation and cognitive enhancement. However, the actual safety, effectiveness and success rate of DIY-experiments has been questioned by researchers (Wurzman, Hamilton, Pascual-Leone, & Fox, 2016).

4. Discussion and ethical implications

Our analysis reveals that the terms "neurohacking" and "brain hacking" have been used to describe very different activities related to the use and development of neurotechnology. To avoid semantic ambiguity in scientific research, ethics and science journalism, we suggest breaking down the neurohacking/brain-hacking label into more detailed semantic categories. In particular, we advise that future research articles adopting the terms "neurohacking" or "brain hacking" include a short disambiguation note clarifying which of the four meanings they are using. Such a disambiguation note would avoid possible misunderstandings and ensure that the reader can understand whichever of the possible meanings is used in the article. This need for disambiguation is important not only to enhance scientific rigor and conceptual clarity but also to ensure adequate ethical assessment and public discussion. The four categories of neurohacking-labeled activities described above appear to have different degrees of ethical permissibility. Consequently, a semantic disambiguation is necessary, not only to advance neuroscience and neurotechnology, but also and foremost for the benefit of neuroethics and technology governance. Without this disambiguation, ethical implications and policy options could be misattributed. For example, a critical stance against maliciously hacking neurodevices based on the principles of data privacy and personal autonomy could be misinterpreted as opposing open development in neurotechnology research. Similarly, a critical stance against self-experimentation and unsafe DIY-neuromodulation could be misinterpreted as a general opposition to neurotechnology-induced cognitive enhancement or even to advancing basic neuroscience research through causal models of brain function.

To illustrate this point, in the next section of this chapter we will provide a succinct ethical and legal analysis of the four main neurohacking-labeled activities. This analysis entails an overview of the ethical principles and international regulations that are relevant to each of the four categories. Based on this analysis, we suggest that the four distinct families of activities or phenomena commonly referred to by the terms neurohacking and brain hacking have different ethical and legal significance, hence require separate assessment and evaluation. In the remaining part of this chapter, we will describe in detail the main ethical and regulatory implications of each thematic family. While not exhaustive, this analysis can offer useful guidance on the normative boundaries of each of the four domains of activities described in this chapter. Finally, we will look at the problem of regulatory approval to exemplify the main regulatory challenges of each type of neurohacking-labeled activity.

4.1 Neurohacking for enhancement

To narrow the scope of the research questions considered in this chapter, it is necessary to focus on a particular type of neurotechnology and neuroenhancement. Specifically, this section will consider a single type of neurotechnology as a case study; namely non-invasive cognitive enhancement devices (CEDs) (Maslen, Douglas, Cohen Kadosh, Levy, & Savulescu, 2014; Schmitz-Luhn, Katzenmeier, & Woopen, 2012). CEDs can be used to influence the brain in specific ways, and include neurofeedback equipment (NE) and non-invasive electrical brain stimulation such as tDCS and transcranial alternate current stimulation (tACS). While the latter two deliver current via electrodes on the head (see chapter "Do-it-yourself and direct-to-consumer neurostimulation" by Wexler in this volume), neurofeedback equipment uses neuroimaging to provide real time displays of electroencephalographic (EEG) data. This display allows the user to attempt to regulate his or her brain activity by examining the display and controlling their brainwave activity accordingly. To approach the question of cognitive enhancement via CEDs, it is also helpful to consider this question through the lens of both positive and negative liberties (Ienca & Andorno, 2017). Positive liberties may include the right of individuals to enhance their cognitive abilities. By contrast, negative liberties relate to the right of individuals to be protected from intrusion by corporations or other users of neurotechnology, or not to be enhanced.

Authors have argued that the moral baseline for CEDs should permit the freedom to modify their cognition using neurotechnology (Ienca & Andorno, 2017). Nevertheless, exceptions might apply in certain context specific circumstances. One scenario where researchers have argued this positive liberty might be outweighed by other policy considerations concerns the use of neurostimulation in sport. Under the World Anti-Doping Authority (WADA) International Code, the use of performance enhancing substances and methods is strictly regulated. Athletes and teams facing severe consequences for anti-doping rule violations. The 2015 WADA Code prohibits substances or methods that enhance performance, represent a risk to the health of athletes, or are against the spirit of sport. However, WADA has not issued a statement as to whether CEDs should be considered performance enhancing substances or methods (Imperatori, Milbourn, & Garasic, 2018). Further, Mihailov and Savulescu, among others, argue the effects of CEDs are complicated and should not be proscribed from sporting contests (including mentally stimulating sport such as chess) (Mihailov & Savulescu, 2018). Imperator, Milbourn and Garasic consider this issue as well in distinguishing between tDCS, which requires a degree of training to achieve any benefit, and other CEDs technology. Imperator, Milbourn and Garasic argue that tDCS should be considered permissible for use in sport if there is ongoing evidence to demonstrate its safety. Other domains where exceptions to cognitive liberty might apply include the military setting. In this context, compelled enhancement-oriented interventions may be admitted if they contribute to make warfighters and other human operators fit for duty (Ienca, Jotterand, et al., 2018; Tennison & Moreno, 2012).

These examples demonstrate how different contexts may change whether CEDs are appropriate or not. An intimately related question is whether CEDs should be considered safe and effective for human use. On the one hand, CEDs can be a powerful tool to complement cognitive enhancement programs, particularly among older adults and people with cognitive disabilities (Ienca et al., 2019). On the other hand, there are also safety concerns associated with the use of CEDs. As discussed by Talbot, there has always been significant concern about the potential for neurotechnology and neurostimulants to harm users (Talbot, 2009). Further, Fitz and Reiner note that the absence of any studies demonstrating harm from CEDs is not proof of their safety (Fitz & Reiner, 2015). In particular, Stagg and Nitsche note that incorrectly applying the electrodes may result in long term neuronal damage (Stagg & Nitsche, 2011).

In addition to potential harms for users of neurotechnology, it is necessary to consider harms for third parties, as well as whether third parties can be held liable for harm to a person who uses CEDs. This question relates to the negative liberty described above; that is, the liberty for users of CEDs and other forms of neuroenhancement to be free from interference. However, as discussed above, there is considerable debate as to whether interference with a medical device constitutes tortious or criminal conduct. Although there is no precedent yet, potential torts or civil actions concerning the misuse of CEDs include invasion of privacy, damage to property, negligence, or battery. As discussed by Quigley and Ayihongbe, establishing whether there has been damage to property or a person will very much depend on the circumstances. In this regard, Quigley and Ayihongbe consider damage to cybernetic components deeply integrated into a person as an offense against a person (Quigley & Ayihongbe, 2018). Whether the same claims for offenses against a person could be made about CEDs attached to a person is significantly more uncertain.

4.2 Malicious neurohacking

The second category of neurohacking, which could be labeled malicious (Ienca & Haselager, 2016), is the most problematic one in ethical terms. The reason stems from the fact that activities such as third-party misuse of a neurodevice and unauthorized access to personal data represent violations of fundamental ethical principles—primarily: respect for personal autonomy, privacy and non-maleficence (Beauchamp & Childress, 2001; Ienca, 2015).

With respect to device misuse, there is no international legal standard on what is criminalized conduct in the context of neurotechnology misuse. However, a relevant international framework is the Convention ETS 185 *Convention on Cybercrime (The Budapest Convention)*, as criminal activities involving digital neurotechnologies can be seen as instances of cybercrime—i.e., criminal activities involving computer systems. It is therefore important to consider what types of crimes fall under the umbrella of a computer crime under the convention.

The *Budapest Convention* defines four categories of crime. The first category includes offenses against confidentiality, integrity, and availability of computer systems and data. These crimes can include illegal access (Article 2), illegal interception (Article 3), data interference (Article 4), system interference (Article 5), and misuse of devices (Article 6). The second category includes computer related offenses, such as computer related forgery (Article 7) and computer related fraud (Article 8). The third category includes content related offenses, such as access to child exploitation material (Article 9). The fourth category includes offenses related to infringements of copyright and related rights (Article 10).

It is also important to consider the types of devices that might be subject to access and interception. Article 2 requires signatories to criminalize access to the whole or part of any computer system without authorization. However, the literature uncovered during the scoping review above describes use cases such as machine learning software for examining EEG data (Kellmeyer, 2018). Accordingly, an attempt to use a device to examine or modify neurological processors is unlikely to constitute any crime prosecutable under Article 2. Nevertheless, the criminal requirements imposed by Articles 3–6 may provide greater avenues for the criminalization of neurotechnology misuse.

First, Article 3 requires signatories to criminalize illegal interception of non-public transmissions of computer data. Second, Article 4 requires signatories to criminalize damaging, deleting, deteriorating or altering computer data. Third, Article 5 requires each party to criminalize seriously impairing the functioning of a computer system by either inputting, transmitting, damaging, deleting, deteriorating, altering or suppressing computer data. In the cybersecurity context, systems carrying personal health information (PHI) are extremely vulnerable in this regard, as this data is considered highly sensitive and valuable (Mrcela & Vuletic, 2018). However, it is not clear from the *Budapest Convention* whether "computer data" encompasses all forms of computer data, or just content data. Likewise, it remains unclear whether illegal transmission would encompass just the digitized version of neurological data, or the acts of "brainjacking" to directly access data stored in invasive devices (Ienca & Haselager, 2016; Pycroft et al., 2016). Likewise, the use of adversarial images to fool human sight may fall outside the realm of illegal transmission (Ackerman, 2018).

Yet another possible avenue for criminalizing malicious neurohacking could exist under Article 6. This article requires signatories to prohibit the sale or distribution of devices and programs for committing the offenses in Articles 2–5. However, the extent to which signatories could prohibit the sale of these devices is questionable. By way of analogy, a study by Halperin and colleagues considered the vulnerability of implantable cardiac defibrillators (ICDs). Using an oscilloscope and a software radio (which are available from general hardware stores) Halperin and colleagues were able to intercept wireless data sent by an ICD, including private data. In addition, Halperin and colleagues were able to use their equipment to deliver electric shocks via the ICD (Halperin et al., 2008). As Kellmeyer notes, the wide availability of these devices demonstrates the difficulty in preventing the sale of equipment with which to carry out neurohacking (Kellmeyer, 2018).

However, it should be noted that, as observed by Tosoni, a number of crimes are missing from the list expounded upon under the *Budapest Convention*. These include identity theft, cyberterrorism, and crucially, breaches of data protection law. In particular, during the Fifteenth International Congress of Penal Law, there were serious disagreements as to whether criminal law should be involved in the protection of privacy (Tosoni, 2018). There are a number of reasons for opposition to introducing criminal requirements for breaches of privacy. The first is the principle of subsidiarity, which supplies that certain forms of behavior should only be criminalized as a last resort. The second is the fact that most data protection laws and treaties are classified as "framework" norms, designed to provide national legislatures with flexibility in derogating these treaties. However, these framework style laws conflict with the harm principle, which states that a person cannot face criminal punishment for an act criminalized after that act occurred. Third, data protection laws are targeted at data processors and controllers, whereas most jurisdictions limit criminal liability to individuals and not corporations (Tosoni, 2018). Therefore, the absence of data protection related crimes contained within the *Budapest Convention* could undermine attempts at developing a coherent international framework for criminalizing neurohacking.

Further, differing interpretations of public international law, as well as the conflict of laws between states remains an open question under the *Budapest Convention* (Tosoni, 2018). Chapter III of the Convention does provide that signatories are to cooperate with one another to the widest extent possible to investigate cybercrime and collect electronic evidence. Article 23 of the Convention also requires signatories to provide mutual assistance to the widest extent possible. The fact that the phrase "to the widest extent" possible is used here allows member states to compensate for variations in criminal law. In particular, parties to the convention may restrict their cooperation more narrowly in cases of extradition, real time collection of data or interception of content data. In addition, a common requirement for extradition is dual criminality, in that both jurisdictions need to recognize the crime. In the field of neurohacking, countries having divergent perspectives on what constitutes the elements of the crime, may impede the development of international standards for *sui generis* laws prohibiting such conduct.

Finally, Weber and Studer note that attacks on wearables and "Internet of Things" are not covered as criminal activity under the *Budapest Convention*. Further, Weber and Studer note that the *Budapest Convention* does not consider new attack avenues, such as botnets and ransomware (Weber & Studer, 2016). Therefore, drawing analogies between wearable and Internet of Things devices on one hand and neurotechnology on the other demonstrates the relatively outdated nature of the *Budapest Convention*. Specifically, it may be difficult to use the *Budapest Convention* as to develop a criminal framework prohibiting the misuse of neurotechnology. Accordingly, another mechanism, possibly at the UN level, is required to aid the development of neurotechnology crimes, particularly in fields of cyberterrorism and economic crime (Weber & Studer, 2016).

With respect to invasion of privacy, the interference or theft of neurofeedback or EEG data might also represent a breach of data protection legislation (Ienca & Haselager, 2016). This legislation includes General Data Protection Regulation (GDPR) in the European Union or Health Information Portability and Accountability Act (HIPAA) in the United States. Unlike liability under tort, which rests with the party accessing the data, liability for breach under data protection legislation rests with the party holding the data. Under the GDPR, data controllers, who decide how data should be used and processed, are responsible for the handling and security of personal data via appropriate technical measures (GDPR, Art. 24). Further, data processors, who process data on behalf of controllers, are also required to comply with the terms of the GDPR (Art. 28). Data controllers must report any breaches to affected data subjects as soon as possible (Art. 33). Further, data controllers must notify the relevant supervisory authority within 72 h of becoming aware of the breach (Art. 34). Failure to comply with this condition, or indeed any of the other terms of the GDPR, results in significant fines (Art. 84).

Unlike the GDPR, which has a significantly broader scope of liability for personal data held by data processors, HIPAA only applies to protected health information processed by covered entities (45 CFR § 160.103).

Protected health information includes any health information about an identifiable individual. However, this definition does not include personal information generated by an individual about their own health or information that might be indirectly related to health (Glenn & Monteith, 2014). Further, covered entities only include health care providers, health insurance providers or healthcare clearing houses, as well as their business associates, as opposed to all entities processing health related data (Price & Cohen, 2019). Similar breach notifications to the GDPR exist under HIPAA, with covered entities required to notify each individual affected by a breach within 60 days (45 CFR § 164.404). Further, all of these breach notifications must be sent to the Department of Health and Human Services. Despite this, the breach notification requirements under HIPAA are noticeably less strict than those under the GDPR.

Further, many companies providing direct to consumer CEDs may be responsible for processing information that is sensitive as protected health information (Kreitmair, 2019). In particular, current non-invasive CEDs appear to fall in the middle ground between covered medical devices and non-covered medical devices such as wearables. Accordingly, HIPAA may not provide the same expansive protection to neurological data as the GDPR provides. As the next section discusses, punishment under criminal law may provide a more appropriate course of action for malicious or negligent misuse of CEDs.

4.3 Neurohacking to understand brain processes

Unlike malicious neurohacking, the ethical implications of using neuroscience and neurotechnology to better understand brain processes are pronouncedly tending toward the positive side of the ethical spectrum. Developing models and causal explanations of brain functions can improve current preventive and predictive models in brain and mental health, cast light on the etiology of neurological and psychiatric disorders, enable earlier diagnosis, and possible deliver better treatments to patients (Ienca, Vayena, et al., 2018). This, in turn, can not only improve knowledge but also alleviate the global burden of neurological and psychiatric disorders, hence promote the ethical principle of beneficence.

Advancing the understanding of brain processes via neuroscience is intimately linked to the right of everyone to "share in scientific advancement and its benefits" which is enshrined in the UDHR Universal Declaration of Human Rights (UDHR) adopted in 1948 by the United Nations (UN) General Assembly. Since 1966, this right is also incorporated into the International Covenant on Economic, Social and Cultural Rights, a treaty signed by 169 countries. It has been observed, however, that the right to science, unlike most other human rights has never been legally defined and is often ignored in practice by national governments (Wyndham & Vitullo, 2018).

Furthermore, recent developments in the citizen science and biohacking movements might challenge the theoretical and normative boundaries of this right. This has implications also for neurohacking. As Wexler notes, in both developed and developing countries there is an increased interrelationship between open development in neurotechnology and other open science or citizen science movements (Wexler, 2017). This interrelationship raises the question of the relationship between open development of CEDs and patents on diagnostic methods (Edris-Roskams, Anderson-Redick, Kiss, & Illes, 2017). The question of open source CEDs in turn dovetails into the different forms of openness within biotechnology research and development. First, open science, most commonly found in forms of basic academic research, involves publications, data and research tools are made available on a gratis basis. Second, open source involves research tools such as software are released openly but on a conditional basis using open source licensing. These open source licenses may require subsequent authors to attribute the original author or creator in derivative works, or relicense these derivative works under the same license. Finally, open innovation involves using conventional intellectual property rights to increase collaboration and subsequent market exploitation (Caulfield & Ogbogu, 2012).

Since the early uses of neuroinformatics software for research, the academic community has recognized the potential for patents to simultaneously supplement research funding yet undermine open source development (Eckersley et al., 2003). In particular, there have been concerns that biotechnology patents, including neurotechnology patents, could hinder the free flow of knowledge otherwise required for subsequent research (the anti-commons effect) (Caulfield & Ogbogu, 2012). There is equivocal evidence of an anti-commons effect emerging for research patents. However, concerns about this effect, along with moral concerns regarding biotechnology patents, has led to calls to exclude neurotechnology as patentable subject matter (Caulfield & Ogbogu, 2012).

4.4 Non-malicious hacking and self-experimentation

The quest for democratizing neuroscience and neurotechnology, especially the tools of scientific experimentation, is not only rooted in the right to science (including the right to citizen science) but also in a general dissenting view of the current state of proprietary and closed-source neurotechnology. For decades, authors have discussed the admissibility of exceptions to intellectual property restrictions on biotechnology development. One notable exception is that for medical treatments. Strandburg and Bechtold have written separately on the intersection between the medical treatment exception for patents, physician innovation, and open innovation (Bechtold, 2014; Strandburg, 2014). Article 27(3)(a) of the Trade Related Aspects of Intellectual Property Rights (TRIPS) Agreement[b] permits signatories to exclude "diagnostic, therapeutic and surgical methods for the treatment of humans or animals from patentability." However, there is substantial divergence whether medical methods of treatment are to be regarded as patentable. On the one hand, countries such as the United States and Australia permit patents to be directed toward methods of medical treatment. By contrast, Article 53(c) of the European Patent convention holds that methods of medical treatment should be excluded from the ambit of patent eligibility. This exclusion inoculates medical practitioners from patent claims. However, over time the European Patent Office (EPO) permitted Swiss-type use or process patent claims to bypass this restriction (Sterckx & Cockbain, 2012).

Beyond patent claims, the manufacturers of CEDs may also claim intellectual property rights in the software used for these devices. Simultaneously, thriving communities have emerged, revolving around the *gratis* exchange of open source hardware and circuit designs for CEDs (Jwa, 2015). Further, software tools for processing and analyzing neurological data can be easily and widely distributed through open source licensing . However, outside of neurotechnology, and despite the equivocal evidence described above, the success of open source and open innovation initiatives depend on whether patents negatively impact such initiatives. Reichman and Simpson describe how a critical patent on high throughput sequencing (HTS) for drug discovery may have completely nullified attempts at open drug discovery if held to be valid (Reichman & Simpson, 2016). Accordingly, open source "neurohackers" who modify the source code of CEDs could potentially infringe upon intellectual property rights associated with these devices (Quigley & Ayihongbe, 2018).

The self-experimentation of neurotechnology raises similar legal issues as open development approaches to neurotechnology. DIY neurohackers and

[b] Agreement on Trade Related Aspects of Intellectual Property Rights (TRIPS), Annex 1C to the Agreement Establishing the World Trade Organisation, 1994.

self-experimenters who modify CEDs could potentially infringe upon intellectual property rights associated with these devices (Quigley & Ayihongbe, 2018). It is questionable, however, which intellectual property rights apply to neurohackers who assemble their own CEDs instead of modifying pre-existing proprietary devices. Finally, self-experimentation with neuromodulation devices raises safety concerns. Experts have observed the "improvised devices or practices" are often adopted by DIY neurohackers and self-experimenters "without reference to established protocols." These scholars argue that such practices raise greater safety risks compared to technologies developed in the research settings with prior scientific validation and using guidance from institutional ethics review boards (Bikson, Bestmann, & Edwards, 2013). More recent studies suggest that home users generally follow scientific protocols regarding current level and length of stimulation. However, the same study found that home users stimulate themselves far more frequently than recommended by scientific literature (Wexler, 2018).

Besides open development and criminal neurohacking, it is also necessary to consider the ethical-legal implications of non-malicious hacking (also referred to as ethical hacking). In security literature, hackers are usually divided into "black hat" and "white hat" hackers. Although hacking is driven by a wide range of motivations (including political, criminal, financial and egotistical reasons), black hat or "gray hat" hackers act without consent or for malicious reasons. By contrast, white hat hackers act with consent and non-maliciously. The ultimate aim of white hat hackers is to use software tools to improve the security of a system or organization. In addition, white hat may occur in a number of different forms. One scenario is where white hat hacking, although done with the consent of the organization, occurs without the organization's employees or members being aware of the testing as an alternative to traditional forms of internal testing (Votipka, Stevens, Redmiles, Hu, & Mazurek, 2018). As alluded to previously, the relevant sections of the *Budapest Convention* that relate to ethical neurohacking concern device related misuse (Article 6). Preventing the sale of such devices will carry with it the same challenges for ethical neurohacking as well as unethical neurohacking. In addition, offenses related to copyright and related rights offenses (Article 10) may also be included in this category. As discussed below, there are a wide variety of intellectual property rights that might apply to neurotechnology devices. It is therefore conceivable that white hat hackers may (incidentally or deliberately) infringe upon these rights.

5. The various meanings of neurohacking and the problem of regulatory approval

The analysis above indicates that different neurohacking-labeled activities exhibit distinctive ethical and legal implications. However, our review revealed that at least one issue, namely regulatory approval, applies to the entire spectrum of neurotechnology uses. Therefore, in addition to the category-specific issues discussed above, it is necessary to consider the relevant international regulatory framework in order to assess the ethico-legal status of neurohacking-related activities. In this final section, we will examine both the EU and the US regulatory frameworks.

Directive 93/42/EEC of June 14, 1993 established the initial approval process for medical devices in the EU. However, Directive 93/42/EEC has been replaced by a new regulation for medical devices (Regulation 2017/745), which will come into force in 2020. Unlike Directive 93/42/EEC, which was not directly enforceable in EU member states without national legislation, Regulation 2017/745 can be directly enforced in national courts. Regulation 2017/745 has also expanded the definition of medical devices to remove the requirement that a medical device must be intended to be used for a medical purpose (Article 2). Another question that has frequently featured in Court of Justice of the European Union (CJEU) case law regarding Directive 93/42/EEC is whether a device in question is a medical device or not, as the device must be explicitly used for a medicinal purpose. In particular, devices used for research subjects and for experimental research have been held not to fall within the purview of Directive 93/42/EEC (Case C-219/11 *Brain Products GmbH v BioSemi VOF and Others*). This definition includes some (but not all) software that may be used for medical purposes, confirming case law regarding Directive 93/42/EC's applicability to medical software (Case C-329/16 *Snitem and Phillips France v Ministre des Afairs sociales et de la Santé*). In addition, software that can be used for prognosis and prediction may fall within the category of medical devices under the new regulation. Therefore, software NE or CED devices may be considered medical devices if used to predict the likelihood of a condition improving or worsening (Lang, 2017).

The European medical device regulatory system classifies devices based on their risk level to patients into Class I, Class IIa, IIb and III, with Class III as the highest risk class (Article 9, Directive 93/42/EEC).

However, Regulation 2017/745 modifies the classification schema and approval mechanism for medical devices (Article 51). The changes in risk level between Directive 93/42/EEC and Regulation 2017/745, as well as the relevant article and procedure for approval, are displayed in Table 2.

Table 2 Comparison between device classification under Medical Devices Directive 93/42/EEC and Regulation 2017/745 (De Maria et al., 2018; Mehrali et al., 2018).

	Directive 93/42/EEC	Regulation 2017/745
Custom	Procedure in Annex VIII (Art. 4, 11)	Annex XIII (Art. 10, 21, 52, 86)
Class I	Declaration of conformity based on Technical Documentation (Annex VII)	Declaration according to Article 19 after drafting the documentation referred to in Annex I and II (Art. 52)
Class Im (measurement)	Declaration of conformity based on the Technical Documentation (Annex VII) + one of procedures referred to in Annex II, IV, V or VI (Annex VII, Section 5)	Part A of Annex IX (without chapter II) OR Annex XI (product conformity) (Art. 52)
Class Is (sterile)	Declaration of conformity based on the Technical Documentation (Annex VII) + (only for sterility) one of the procedures referred to in Annex II, IV, V or VI (Annex VII, Section 5)	Part A of Annex IX (without chapter 2) OR Annex XI (product conformity) (Art. 52)
Class IIa	Full declaration of conformity and quality assurance (Annex II), without examination of design (Section 4) AND Conformity assessment procedure (Art. 11) via: 1. Verification of conformity (Annex IV) 2. Production quality assurance (Annex V) 3. Product quality assurance (Annex VI)	Part A of Annex IX without chapter 2 OR General Safety and Performance Requirements + Design and Manufacture Requirements + Labeling (Annex I), Technical Documentation (Annex II)

Table 2 Comparison between device classification under Medical Devices Directive 93/42/EEC and Regulation 2017/745 (De Maria et al., 2018; Mehrali et al., 2018).—Cont'd

	Directive 93/42/EEC	Regulation 2017/745
Class IIb	Full declaration of conformity and quality assurance (Annex II) without examination of design (Section 4) OR EC Type Examination (Annex III) AND Conformity assessment procedure (Art. 11) via: 1. Verification of conformity (Annex IV) 2. Production quality assurance (Annex V) 3. Product quality assurance (Annex VI)	Part A of Annex IX without chapter 2 if not implanted (quality management system assessment, auditing, surveillance assessment (Annex IX, ss2-3)), otherwise all of Annex IX OR General Safety and Performance Requirements + Design and Manufacture Requirements + Labeling (Annex I), Technical Documentation (Annex II)
Class III	Annex II (complete) OR EC Type Examination (Annex III) AND 1. Verification of conformity (Annex IV) 2. Production quality assurance (Annex V)	Annex IX (complete) OR Annex X (type) and Annex XI (product)

In the EU the approval and classification processes for medical devices are conducted by notified bodies that ensure devices conform to appropriate regulations and can freely circulate within the EU. Medical devices may only be put on the market if they bear the appropriate CE marking and have undergone the appropriate conformity assessment to ensure safety for human usage (Case C-288/08 *Kemikalieinspektionen v Nordiska Dental AB*). This requirement must be balanced against the need for consistent standards for the exchange of devices within the European Single Market (Case C-662/15 *Lohman & Rauscher International GmbH & Co KG v BIOS Medical Services GmbH*). Previously, the CJEU held the responsibility to ensure the safety of CE marked devices depends on national law

(Case C-219/15 *Elisabeth Schmitt v TÜV Rheinland LGA Products GmbH*). In this case, the CJEU examined what notified bodies were required to do when assessing CE marked devices, namely negligently designed and manufactured breast implants. The CJEU rejected the plaintiff's argument that notified bodies were obliged to carry out unannounced inspections of premises and check business records under Directive 93/42/EEC. However, all of the responsibilities that the plaintiff argued were imposed on notified bodies have been imposed by Regulation 745/2017. These include the requirement to inspect premises and business records unannounced every 5 years, take samples of the CE marked devices and inspect the raw materials (Annex IX, Arts. 3.4 and 3.5). These reforms therefore appear to ameliorate many of the concerns expressed previously regarding the effectiveness of Directive 93/42/EEC (Kuersten & Hamilton, 2014; Maslen, 2014). In particular, Kuersten and Hamilton note that the self-regulating nature of the direct to consumer CED industry will be severely restricted by the new EU regulations.

In the United States, the Federal *Food Drug and Cosmetic Act* 1976 regulates the sale and distribution of medical devices and medical products. Section 201(h) of the Federal *Food Drug and Cosmetic Act* 1976 defines a medical device as an instrument intended for diagnosing disease or other conditions, or curing, mitigating, treating or preventing disease. "Intended use" is defined within Title 21, Part 801, subpart A, Section 801.4 as the objective intent of the person legally responsible for labeling the devices. Unlike in the EU, a single regulatory body, the Food and Drug Administration (FDA), is responsible for classification and approval. Nevertheless, in a similar fashion to the European regime, medical devices are classified in three categories; low (Class I), moderate (Class II) and high-risk (Class III) devices, according to the risks the devices present to patients. Class I devices are simple in design and manufacture, and usually include devices that have a history of safe use. Class II devices are more complicated than Class I devices, whereas Class III devices support human life, can prevent the impairment of human health or present a risk of injury to the patient. For Class II devices, the Section 510 (k) approval process provides a less burdensome route to enable newer versions of existing devices to enter the market. To seek approval, the sponsor does not require clinical trials or manufacturing inspections to demonstrate efficacy or safety. The only requirement was to demonstrate the device was *substantially equivalent* to an already marketed device. Only in rare cases are clinical trials (or some other special control, such as labeling requirements or mandatory post-market surveillance) required for Section 510(k) approval

(Sorenson & Drummond, 2014). By contrast, Class III devices generally must go through an FDA premarket approval process (PMA). This PMA requires extensive testing including "valid scientific evidence" that provides reasonable assurance that the device is safe and effective for its intended use (McDonald, Lau, Coates McCall, & Illes, 2019).

Initially, Section 510(k) reviews were mandated for a very small subset of devices that already had equivalents on the market (such as surgical gloves or hearing aids). However, from a historical perspective Zuckerman, Brown and Nissen note that the FDA did not have the resources to manage the increasing number of device applications through the PMA process. Instead, the vast majority of these approvals were sent through the 510 (k) approach (Zuckerman, Brown, & Nissen, 2011). These requirements were again loosened with the passage of the *Medical Device User Fee and Modernization Act* 2002 (MDUFMA). Crucially, this reform shifted the regulatory standard to the "least burdensome approach in all areas of medical device regulation" (Zuckerman et al., 2011, p. 1007). Subsequently, the definition of *substantially equivalent* was modified to include products made from different materials and using a different mechanism of action. Further, no clinical trials would be required, as clinical trials are not required for 510 (k) clearance. Therefore, the FDA has used the 510(k) provision as the primary mechanism for approving devices. To reinforce this point, Zuckerman et al. examine the number of recalled FDA devices that are subject to either PMA or a 510(k) clearance between 2005 and 2009. 71% of these devices were subject to a 510(k) filing (Zuckerman et al., 2011). These cases demonstrate the importance of guaranteeing the physical reliability of neurotechnological devices more broadly.

While it appears that US FDA regulations on tDCS are more flexible than Regulation 745/2017, the FDA has a number of strategies to prevent the release of defective devices. First, as Wexler notes, US case law has considered other factors, including the circumstances of distribution and "oral and written" statements made about the device. Further, the FDA has the capacity to rebrand a device if it considers it incorrectly labeled (Wexler, 2015). It should be noted, however, that the FDA and other agencies are currently making efforts to close the regulatory gap on direct to consumer tDCS. To illustrate this, Wexler draws reference to the California Department of Public Health (CDPH), which forced a tDCS firm to withdraw consumer tDCS devices (Wexler, 2015). Nevertheless, there still remain regulatory challenges. Wexler and Reiner note that the FDA has recently clarified that direct to consumer tDCS devices fall within the agency's jurisdiction. However, in the same document the FDA indicated that it would not be enforcing medical device regulations against low risk devices marketed for wellness purposes (Wexler & Reiner, 2019).

6. Conclusions

This review has shown that the word neurohacking is being increasingly used in the popular and scientific literature. Specifically, this term denominates a broad spectrum of activities related to the development, use and misuse of neurotechnology. However, conceptual analysis has shown that these activities are qualitatively heterogeneous and might raise different ethico-legal implications. Based on our analysis, we have identified four distinct yet overlapping families of activities or phenomena related to the use of neurotechnology that are denoted by the neurohacking label.

To avoid semantic ambiguity in scientific research, ethics and science journalism, we propose future academic research and media coverage should break down the neurohacking label into more detailed semantic categories. In particular, it would be advisable to include a short disambiguation note and clarify which of the four meanings they are referring to. Such semantic disambiguation is necessary to make neurohacking a viable terminology for rigorous scientific or normative purposes. An alternative conclusion, which would be equally consistent with our analysis, is that the term neurohacking is not suitable for use in rigorous scientific or normative discussions, and the phenomena discussed under its banner are already more appropriately captured by the labels we used in this chapter. While we remain agnostic between which of these two solutions should be preferred, we emphasize the importance of avoiding conceptually ambiguous uses of the term.

Furthermore, our ethical and legal analysis highlights that the four distinct activities falling under the neurohacking label raise different ethical issues and are governed by distinct regulatory mechanisms. The neurohacking spectrum encompasses activities with very different degrees of ethical permissibility and regulatory standards. Accordingly, a semantic disambiguation is necessary not only to advance neuroscience and neurotechnology but also and foremost for the benefit of neuroethics and technology governance. This disambiguiation would also prevent normative considerations regarding a certain neurohacking-labeled activity are misattributed to another type of activity or illegitimately generalized to the entire neurohacking spectrum.

While different neurohacking-labeled activities typically exhibit distinctive ethical and legal implications, we observed that the issue of regulatory approval applies to the entire spectrum of neurotechnology uses.

References

Ackerman, E. (2018). *Hacking the brain with adversarial images*. IEEE Spectrum: Technology, Engineering, and Science News.

Bak, S., Pyo, Y., & Jeong, J. (2019). Protection of EEG data using blockchain platform. In Paper presented at the 2019 7th international winter conference on brain-computer interface (BCI).

Beauchamp, T. L., & Childress, J. F. (2001). *Principles of biomedical ethics*. USA: Oxford University Press.

Bechtold, S. (2014). Physicians as user innovators. In R. C. Dreyfuss & J. Ginsburg (Eds.), *Intellectual property at the edge: The contested contours of IP* (pp. 343–357). Cambridge University Press.

Bela, G., Peltola, T., Young, J. C., Balázs, B., Arpin, I., Pataki, G., et al. (2016). Learning and the transformative potential of citizen science. *Conservation biology: The Journal of the Society for Conservation Biology*, *30*(5), 990–999. https://doi.org/10.1111/cobi.12762.

Bikson, M., Bestmann, S., & Edwards, D. (2013). Transcranial devices are not playthings. *Nature*, *501*(7466), 167. https://doi.org/10.1038/501167b.

Borrione, L., & Brunoni, A. R. (2017). Primum non nocere or primum facere meliorem? Hacking the brain in the 21st century. *Trends in Psychiatry and Psychotherapy*, *39*(4), 232–238. https://doi.org/10.1590/2237-6089-2017-0075.

Braun, V., & Clark, V. (2006). Using thematic analysis in psychology. *Qualitative Research in Psychology*, *3*(2), 77–101. https://doi.org/10.1191/1478088706qp063oa.

Buchan, M. J. (2018). Book review: The genius within: Smart pills brain hacks and adventures in intelligence. *Frontiers in Psychology*, *9*.

Caulfield, T., & Ogbogu, U. (2012). Stem cell research, scientific freedom and the commodification concern: Vague concerns about possible commodification should not serve as justification to limit the freedom of research. *EMBO Reports*, *13*(1), 12–16. https://doi.org/10.1038/embor.2011.232.

Conner, M. (2010). Hacking the brain: Brain-to-computer-interface hardware moves from the realm of research. https://www.edn.com/hacking-the-brain-brain-to-computer-interface-hardware-moves-from-the-realm-of-research/.

De Maria, C., Di Pietro, L., Díaz Landata, A., Madete, J., Makobore, P. N., Mridha, M., et al. (2018). Safe innovation: On medical device legislation in Europe and Africa. *Health Policy and Technology*, 7(2), 156–165. https://doi.org/10.1016/j.hlpt.2018.01.012.

DiEuliis, D., & Giordano, J. (2017). Why gene editors like CRISPR/Cas may be a game-changer for neuroweapons. *Health Security*, *15*(3), 296–302.

Dresler, M., Sandberg, A., Bublitz, C., Ohla, K., Trenado, C., Mroczko-Wasowicz, A., et al. (2019). Hacking the brain: Dimensions of cognitive enhancement. *ACS Chemical Neuroscience*, *10*(3), 1137–1148. https://doi.org/10.1021/acschemneuro.8b00571.

Dupont, B. (2013). Cybersecurity futures: How can we regulate emergent risks? *Technology Innovation Management Review*, *3*(7), 6–11.

Eckersley, P., Egan, G. F., De Schutter, E., Yivuan, T., Novak, M., Sebesta, V., et al. (2003). Neuroscience data and tool sharing. *Neuroinformatics*, *1*(1), 149–165. https://doi.org/10.1007/s12021-003-0002-1.

Edris-Roskams, D., Anderson-Redick, S., Kiss, Z. H. T., & Illes, J. (2017). Situating brain regions among patent rights and moral risks. *Nature Biotechnology*, *35*(2), 119–121. [Patents].

Fitz, N. S., & Reiner, P. B. (2015). The challenge of crafting policy for do-it-yourself brain stimulation. *Journal of Medical Ethics, 41*(5), 410–412. https://doi.org/10.1136/medethics-2013-101458.

Folgieri, R. (2017). Brain computer interface and transcranical stimulation: Frontiers, reliability, safety and threats. *Biolaw Journal-Rivista Di Biodiritto, 2017*(3), 103–110.

Frank, D., Mabrey, J., & Yoshigoe, K. (2017). Personalizable neurological user authentication framework. In Paper presented at the 2017 international conference on computing, networking and communications (ICNC).

Front, S. (2017). Hacking the brain: Duncan Jones's source code. *Swiat I Slowo, 1*(28), 11–26.

Gardner, P., & Wray, B. (2013). From lab to living room: Transhumanist imaginaries of consumer brain wave monitors. *Ada: A Journal of Gender, New Media, and Technology,* (3).

Glenn, T., & Monteith, S. (2014). Privacy in the digital world: Medical and health data outside of HIPAA protections. *Current Psychiatry Reports, 16*(11), 494. https://doi.org/10.1007/s11920-014-0494-4.

Haplerin, D., Heydt-Benjamin, T., Ransford, B., Clark, S. S., Defend, B., Morgan, W., et al. (2008). Pacemakers and implantable cardiac defibrillators: Software radio attacks and zero-power defenses. In *2008 IEEE symposium on security and privacy,* (pp. 129–142).

Heaven, D. (2014). The great brain hack. *New Scientist, 223*(2979), 8–9. https://doi.org/10.1016/s0262-4079(14)61420-2.

Ienca, M. (2015). Neuroprivacy, neurosecurity and brain-hacking: Emerging issues in neural engineering. In *Paper presented at the Bioethica Forum.*

Ienca, M. (2019). Democratizing cognitive technology: A proactive approach. *Ethics and Information Technology, 21*(4), 267–280.

Ienca, M., & Andorno, R. (2017). Towards new human rights in the age of neuroscience and neurotechnology. *Life Sciences, Society and Policy, 13*(1), 5.

Ienca, M., & Haselager, P. (2016). Hacking the brain: Brain-computer interfacing technology and the ethics of neurosecurity. *Ethics and Information Technology, 18*(2), 117–129.

Ienca, M., Jotterand, F., & Elger, B. S. (2018). From healthcare to warfare and reverse: How should we regulate dual-use neurotechnology? *Neuron, 97*(2), 269–274. https://doi.org/10.1016/j.neuron.2017.12.017.

Ienca, M., Shaw, D. M., & Elger, B. (2019). Cognitive enhancement for the ageing world: Opportunities and challenges. *Ageing & Society, 39*(10), 2308–2321.

Ienca, M., Vayena, E., & Blasimme, A. (2018). Big data and dementia: Charting the route ahead for research, ethics, and policy. *Frontiers in Medicine, 5,* 13. https://doi.org/10.3389/fmed.2018.00013.

Imperatori, L. S., Milbourn, L., & Garasic, M. D. (2018). Would the use of safe, cost-effective TDCS tackle rather than cause unfairness in sports? *Journal of Cognitive Enhancement, 2*(4), 377–387. https://doi.org/10.1007/s41465-018-0113-0.

Jebari, K., & Hansson, S.-O. (2013). European public deliberation on brain machine interface technology: Five convergence seminars. *Science and Engineering Ethics, 19*(3), 1071–1086.

Jwa, A. (2015). Early adopters of the magical thinking cap: A study on do-it-yourself (DIY) transcranial direct current stimulation (tDCS) user community. *Journal of Law and the Biosciences, 2*(2), 292–335.

Kaniusas, E., Kampusch, S., Tittgemeyer, M., Panetsos, F., Gines, R. F., Papa, M., et al. (2019). Current directions in the auricular vagus nerve stimulation I—A physiological perspective. *Frontiers in Neuroscience, 13,* 854. https://doi.org/10.3389/fnins.2019.00854.

Kellmeyer, P. (2018). Big brain data: On the responsible use of brain data from clinical and consumer-directed neurotechnological devices. *Neuroethics,* 1–16.

Kermani, M. M., Zhang, M., Raghunathan, A., & Jha, N. K. (2013). Emerging frontiers in embedded security. In *2013 26th international conference on VLSI design and 2013 12th international conference on embedded systems* (pp. 203–208).

Koch, C. (2010). A smart vision of brain hacking. *Nature*, *467*(7311), 32.
Koizumi, A., Amano, K., Cortese, A., Shibata, K., Yoshida, W., Seymour, B., et al. (2016). Fear reduction without fear through reinforcement of neural activity that bypasses conscious exposure. *Nature Human Behaviour1*, 0006. https://doi.org/10.1038/s41562-016-0006 https://www.nature.com/articles/s41562-016-0006-supplementary-information.
Kreitmair, K. V. (2019). Dimensions of ethical direct-to-consumer neurotechnologies. *AJOB Neuroscience*, *10*(4), 152–166. https://doi.org/10.1080/21507740.2019.1665120.
Kuersten, A., & Hamilton, R.H. (2014). The brain, cognitive enhancement devices, and European regulation. *Journal of Law and the Biosciences*, *1*(3), 340–347. https://doi.org/10.1093/jlb/lsu019.
Lang, M. (2017). Heart rate monitoring apps: Information for engineers and researchers about the new European medical devices regulation 2017/745. *JMIR Biomedical Engineering*, *2*(1), e2. https://doi.org/10.2196/biomedeng.8179.
Lee, K. J. (2016). A call for law schools to link the curricular trends of legal tech and mindfulness. *University of Toledo Law Review*, *48*, 55.
Leggett, H. (2009). *The next hacking frontier: Your brain*. Wired Magazine.
Li, L. M., & Johnson, S. (2015). Hackathon as a way to raise awareness and foster innovation for stroke. *Arquivos de Neuro-Psiquiatria*, *73*(12), 1002–1004.
Macknik, S. L., & Martinez-Conde, S. (2009). Real magic: Future studies of magic should be grounded in neuroscience. *Nature Reviews Neuroscience*, *10*(3), 241.
Marcus, A. (2015). *The weird world of brain hacking*. WSJ Online. *(Nov. 9, 2015)(4 pages)*.
Malin, C. H., Gudaitis, T., Holt, T., & Kilger, M. (2017). Deception in the digital age: Exploiting and defending human targets through computer-mediated communications. Elsevier.
Maslen, H., Douglas, T., Cohen Kadosh, R., Levy, N., & Savulescu, J. (2014). The regulation of cognitive enhancement devices. *Journal of Law and the Biosciences*, *1*(1), 68–93. https://doi.org/10.1093/jlb/lst003.
Mathalon, D. H., Ford, J. M., & Pfefferbaum, A. (2000). Trait and state aspects of P300 amplitude reduction in schizophrenia: A retrospective longitudinal study. *Biological Psychiatry*, *47*(5), 434–449.
Matos, M., Pinto-Gouveia, J., & Duarte, C. (2012). Above and beyond emotional valence: The unique contribution of central and traumatic shame memories to psychopathology vulnerability. *Memory*, *20*(5), 461–477.
McDonald, P. J., Lau, C., Coates McCall, I., & Illes, J. (2019). Regulatory oversights for implantable neurodevices. *The Lancet Neurology*, *18*, 913.
McMahon, M., & Schukat, M. (2018, August). A low-cost, open-source, BCI-VR game control development environment prototype for game based neurorehabilitation. In *2018 IEEE games, entertainment, media conference (GEM)* (pp. 1–9).
McNaughton, B. L. (1998). The neurophysiology of reminiscence. *Neurobiology of Learning and Memory*, *70*(1–2), 252–267.
Mehrali, M., Bagherifard, S., Akbari, M., Thakur, A., Mirani, B., Mehrali, M., et al. (2018). Blending electronics with the human body: A pathway toward a cybernetic future. *Advanced Science*, *5*(10), 1700931. https://doi.org/10.1002/advs.201700931.
Mihailov, E., & Savulescu, J. (2018). Social policy and cognitive enhancement: Lessons from chess. *Neuroethics*, *11*(2), 115–127. https://doi.org/10.1007/s12152-018-9354-y.
Mrcela, M., & Vuletic, I. (2018). Healthcare, privacy, big data and cybercrime: Which one is the weakest link? *Annals of Health Law*, *27*(2), [viii]–280.
Nazari, S., & Faez, K. (2018). Empowering the impaired astrocytes in the tripartite synapses to improve accuracy of pattern recognition. *Soft Computing*, *23*(17), 1–13.
Nutt, A. (2016). Brain hacking: The mind's biology. *The Washington Post, 2016*.
O'Brolchain, F., & Gordijn, B. (2014). Brain-computer interfaces and user responsibility. In G. Grübler & E. Hildt (Eds.), *Brain-computer-interfaces in their ethical, social and cultural contexts* (pp. 163–182). Springer.

Patel, A., Taghavi, M., Júnior, J. C., Latih, R., & Zin, A. M. (2012). Safety measures for social computing in wiki learning environment. *International Journal of Information Security and Privacy (IJISP)*, *6*(2), 1–15.

Plotogea, R., & Zamfiroiu, A. (2017). Use of IoT technology in the medical field. *Romanian Journal of Information Technology and Automatic Control-Revisata Romana De Informatica Si Automatica*, *27*(3), 19–28.

Price, W. N., & Cohen, G. I. (2019). Privacy in the age of medical big data. *Nature Medicine*, *25*(1), 37. https://doi.org/10.1038/s41591-018-0272-7.

Prpa, M., Quesnel, D., Vidyarthi, J., Kitson, A., & Riecke, B. E. (2016). Sonic cradle—Immersive interaction design combining breathing–and neurofeedback to foster focused attention meditation on breath. In Poster presented at the 2nd international conference on mindfulness (Rome).

Pugh, J., Pycroft, L., Sandberg, A., Aziz, T., & Savulescu, J. (2018). Brainjacking in deep brain stimulation and autonomy. *Ethics and Information Technology*, *20*(3), 219–232. https://doi.org/10.1007/s10676-018-9466-4.

Pycroft, L., & Aziz, T. Z. (2018). Security of implantable medical devices with wireless connections: The dangers of cyber-attacks. *Expert Review of Medical Devices*, *15*(6), 403–406.

Pycroft, L., Boccard, S. G., Owen, S. L., Stein, J. F., Fitzgerald, J. J., Green, A. L., et al. (2016). Brainjacking: Implant security issues in invasive neuromodulation. *World Neurosurgery*, *92*, 454–462.

Quigley, M., & Avihongbe, S. (2018). Everyday cyborgs: On integrated persons and integrated goods. *Medical Law Review*, *26*(2), 276–308. https://doi.org/10.1093/medlaw/fwy003.

Reichman, M., & Simpson, P. B. (2016). Open innovation in early drug discovery: Roadmaps and roadblocks. *Drug Discovery Today*, *21*(5), 779–788. https://doi.org/10.1016/j.drudis.2015.12.008.

Sage, C., & Hardell, L. (2018). Fatal collision? Are wireless headsets a risk in treating patients? *Electromagnetic Biology and Medicine*, *37*(2), 95–99. https://doi.org/10.1080/15368378.2017.1422261.

Schiller, D. (2016). Neuroscience: Hacking the brain to overcome fear. *Nature Human Behaviour*, *1*(1), 1–2.

Schmitz-Luhn, B., Katzenmeier, C., & Woopen, C. (2012). Law and ethics of deep brain stimulation. *International Journal of Law and Psychiatry*, *35*(2), 130–136. https://doi.org/10.1016/j.ijlp.2011.12.007.

Sinha, Y., Sharma, K., Kumari, M., & Vyas, V. (2015). Brain fingerprinting: A brain hacking tool. *International Journal of Engineering and Management Research (IJEMR)*, *5*(1), 48–50.

Siva, N. (2006). Hacking the brain with computer chips. *The Lancet Neurology*, *5*(12), 1006–1007.

Sofge, E. (2014). *You have already complied: RoboCop and the all-too-feasible horror of brain hacking*. Popular Science.

Sorenson, C., & Drummond, M. (2014). Improving medical device regulation: The United States and Europe in perspective. *The Milbank Quarterly*, *92*(1), 114–150. https://doi.org/10.1111/1468-0009.12043.

Spreeuwenberg, L. (2019). Taking the love pill: A reply to Naar and Nyholm. *Journal of Applied Philosophy*, *36*(2), 248–256.

Stagg, C. J., & Nitsche, M. A. (2011). Physiological basis of transcranial direct current stimulation. *The Neuroscientist*, *17*(1), 37–53. https://doi.org/10.1177/1073858410386614.

Stark, L., & Campbell, N. D. (2018). The ineffable: A framework for the study of methods through the case of mid-century mind-brain sciences. *Social Studies of Science*, *48*(6), 789–820.

Steinert, S., & Friedrich, O. (2020). Wired emotions: Ethical issues of affective brain–computer interfaces. *Science and Engineering Ethics*, *26*(1), 351–367.

Sterckx, S., & Cockbain, J. (2012). Exclusions from patentability: How far has the European patent office eroded boundaries? In *Cambridge University Press*.
Strandburg, K. J. (2014). Legal but unacceptable: Pallin v. Singer and physician patenting norms. In R. C. Dreyfuss & J. Ginsburg (Eds.), *Intellectual property at the edge: The contested contours of IP,* (pp. 14–42). Cambridge University Press.
Strickland, E. (2014). *The latest DIY craze: Brain hacking home experimenters are building rigs to send currents through their heads.* IEEE Spectrum.
Sullivan, L. S., & Illes, J. (2018). Ethics in published brain-computer interface research. *Journal of Neural Engineering, 15*(1), 013001.
Szu, H., & Mehmood, A. (2006). Authenticated, private, and secured smart cards (APS-SC). In H. Szu (Ed.), In *Independent component analyses, wavelets, unsupervised smart sensors, and neural networks IV* (Vol. 6247, p. 62470L). International Society for Optics and Photonics.
Takagi, M. (2012). Safety and neuroethical consideration of deep brain stimulation as psychiatric or dementia treatment. *Asian Bioethics Review, 4*(1), 48–64.
Talbot, M. (2009). *Brain gain.* The New Yorker.
Tennison, M. N., & Moreno, J. D. (2012). Neuroscience, ethics, and national security: The state of the art. *PLoS Biology, 10*(3), e1001289.
Tillas, A. (2018). Hacking our brains for learning. *Humana Mente Journal of Philosophical Studies, 11*(33), 83–102.
Tosoni, L. (2018). Rethinking privacy in the council of Europe's convention on cybercrime. *Computer Law & Security Review, 34*(6), 1197–1214.
Varazzani, C. (2017). *The risks of ignoring the brain.* Behavioral Scientist.
Votipka, D., Stevens, R., Redmiles, E., Hu, J., & Mazurek, M. (2018). Hackers vs. testers: A comparison of software vulnerability discovery processes. In *Paper presented at the 2018 IEEE symposium on security and privacy (SP).*
Wade, M. (2018). Virtuous play: The ethics, pleasures, and burdens of brain training. *Science as Culture, 27*(3), 296–321.
Weber, R. H., & Studer, E. (2016). Cybersecurity in the Internet of Things: Legal aspects. *Computer Law & Security Review, 32*(5), 715–728. https://doi.org/10.1016/j.clsr.2016.07.002.
Wexler, A. (2015). A pragmatic analysis of the regulation of consumer transcranial direct current stimulation (TDCS) devices in the United States. *Journal of Law and the Biosciences, 2*(3), 669–696. https://doi.org/10.1093/jlb/lsv039.
Wexler, A. (2017). The social context of "do-it-yourself" brain stimulation: Neurohackers, biohackers, and lifehackers. *Frontiers in Human Neuroscience, 11*, 224.
Wexler, A. (2018). Who uses direct-to-consumer brain stimulation products, and why? A study of home users of tDCS devices. *Journal of Cognitive Enhancement, 2*(1), 114–134. https://doi.org/10.1007/s41465-017-0062-z.
Wexler, A., & Reiner, P. B. (2019). Oversight of direct-to-consumer neurotechnologies. *Science, 363*(6424), 234. https://doi.org/10.1126/science.aav0223.
Wittgenstein, L. (1953). *Philosophical investigations* (3rd ed.). 31–36. GEM Anscombe (Nueva York, 1953).
Wurzman, R., Hamilton, R. H., Pascual-Leone, A., & Fox, M. D. (2016). An open letter concerning do-it-yourself users of transcranial direct current stimulation. *Annals of Neurology, 80*(1), 1–4. https://doi.org/10.1002/ana.24689.
Wyndham, J. M., & Vitullo, M. W. (2018). Define the human right to science. *Science, 362*(6418), 975. https://doi.org/10.1126/science.aaw1467.
Zuckerman, D. M., Brown, P., & Nissen, S. E. (2011). Medical device recalls and the FDA approval process. *Annals of Internal Medicine, 171*(11), 1006–1011. https://doi.org/10.1001/archinternmed.2011.30.

CHAPTER NINE

Assessing current mechanisms for the regulation of direct-to-consumer neurotechnology

Ishan Dasgupta*
Department of Philosophy, University of Washington, Seattle, WA, United States
Center for Neurotechnology, University of Washington, Seattle, WA, United States
*Corresponding author: e-mail address: idasg@uw.edu

Contents

1. Introduction

The brain is having its moment. Like the genetics revolution of the 1990s, neuroscience has captured the attention of both scientists and consumers. Neuropsychiatric disorders are one of the leading causes of disease burden and have been rapidly increasing over the last 20 years (Sun et al., 2019). In order to combat this epidemic, governments around the world have invested heavily in research aimed at developing next-generation

Developments in Neuroethics and Bioethics, Volume 3
ISSN 2589-2959
https://doi.org/10.1016/bs.dnb.2020.03.009

neurotechnologies to treat disease (Richardson, 2017). In parallel, the market for direct-to-consumer (DTC) neurotechnologies, ready-to-use products available without approval from a physician, has skyrocketed (Wexler & Reiner, 2019). Unlike products approved for clinical use, DTC neurotechnologies often have unknown efficacy and are not subject to the same forms of regulation. Due to these differences, neuroethicists have raised concerns about whether consumers are adequately protected from a range of potential harms (see Table 1). These worries are exacerbated by the fact that most DTC neurotechnologies seek to enhance capacities beyond normal levels rather than treat an illness or condition, thereby making any risks associated less acceptable. While the negative impacts from therapies may be offset by the potential for benefit to a patient, any potential harms may be less justified when consumers are aiming to be "better than well" (Elliott, 2011).

Despite these worries, there is no consensus about what regulatory mechanisms should be utilized to adequately protect consumers (see Table 1). Although there is a growing literature on this topic, there is a lack of scholarship examining why some types of oversight are better suited to protect consumers from particular harms. The aim of this paper is to fill that gap by reviewing the main legal mechanisms for regulation of DTC neurotechnologies in the United States and the relative pros and cons of each of these options. When taken alone, the current regulatory mechanisms are insufficient to protect consumers from all the potential harms that are likely to result from widespread DTC neurotechnology use due to the patchwork of consumer protection laws in the United States. However, as Anna Wexler has argued in the context of neurostimulation devices, if we take a multifaceted approach to regulation it is more likely that associated harms can be minimized (Wexler, 2015).

In Section 2, I highlight two conceptual issues that have plagued the conversation around appropriate regulation and argue that effective regulation will depend on considering the specific nature of the harms implicated and the ways in which regulatory mechanisms are built to address those harms. This discussion foregrounds the breadth and complexity of the different harms introduced by DTC neurotechnologies in order to motivate the need for utilizing different regulatory mechanisms. Despite the diversity, I argue that five types of harm are common to most current DTC neurotechnologies: safety, transparency, privacy, data security, and justice. In Section 3, I review the existing legal mechanisms for regulation and weigh the relative merits and limitations of different approaches. The purpose of this discussion is to illustrate how each mechanism has inherent advantages and limitations that can either assist or hinder effective regulation. In Section 4, I briefly consider two alternate modes of regulation that do not involve formal legal institutions. In this section, I explore ways in which self-regulation and expert working groups may be the best approach for addressing harms such as data security and justice. Finally, in Section 5, I discuss why a multi-faceted approach is necessary and address limitations to my approach.

Table 1 Types of DTC neurotechnologies, harms, and regulatory mechanisms.

Source	DTC neurotech discussed	Potential harms mentioned	Regulatory mechanisms mentioned
Armontrout, Torous, Cohen, McNiel, and Binder (2018)	– Mental health apps	– Safety – Privacy – Malfunctions – Hacking	– Food and Drug Administration (FDA) Federal Trade Commission (FTC) – Health Insurance Portability and Accountability Act (HIPAA) – Tort liability
Fitz and Reiner (2015)	– Neurostimulation	– Safety	– FDA – Open communication – Education – Self-regulation
Ienca, Haselager, and Emanuel (2018)	– Neurostimulation – Neuromonitoring	– Privacy – Data security – Informed consent – Reliability – Safety	– FDA – HIPAA
Kreitmair (2019)	– Neurostimulation – Neuromonitoring – Mental health apps – Brain training apps	– Safety – Transparency – Privacy – Epistemic – Authenticity – Unjust distribution – Lack of oversight	– FDA – Working group

Continued

Table 1 Types of DTC neurotechnologies, harms, and regulatory mechanisms.—cont'd

Source	DTC neurotech discussed	Potential harms mentioned	Regulatory mechanisms mentioned
Kuersten and Hamilton (2016)	– Neurostimulation	– Safety – Informed consent	– FDA – FTC – CPSC – Tort liability
Coates McCall, Lau, Minielly, and Illes (2019)	– Neurostimulation – Neuromonitoring	– Unsubstantiated claims – Safety – Inadequate warnings	– FDA – FTC
Martinez-Martin and Kreitmair (2018)	– Mental health apps	– Privacy – Data protection – Safety – Accountability – Liability – Informed consent	– HIPAA – FDA
Purcell and Rommelfanger (2017)	– Mental health apps – Brain training apps	– Aggregation of user data for research – Validating efficacy – Privacy	– FTC – Family Educational Rights and Privacy Act (FERPA) – US Personal Data Privacy and Security Act of 2014
Terry and Gunter (2018)	– Mental health apps	– Quality – Efficacy – Safety – Data protection	– FDA – FTC – Tort liability – HIPAA
Wexler and Reiner (2019)	– Neurostimulation – Neuromonitoring – Mental health apps – Brain training apps	– Safety – Reliability – Misleading claims – Distributive justice	– FDA – FTC – Working group
Wexler and Thibault (2019)	– Neuromonitoring	– Misleading claims – Unsubstantiated claims	– FDA – FTC

2. Types of DTC neurotechnology and harms

2.1 What is a DTC neurotechnology?

The last decade has seen a proliferation in DTC neurotechnologies. Neuroethicists, however, have failed to agree on a common typology for DTC neurotechnologies (Kreitmair, 2019). The term DTC neurotechnology has no consistent definition in the literature and generally refers to devices, tools, and applications that attempt to read, modulate, or enhance part of the human brain. While devices that stimulate the brain or monitor neural data are universally categorized as DTC neurotechnologies, others such as wearables that track non-neural data have been included by some (Kreitmair, 2019) and excluded by others (Ienca & Vayena, 2019). I will not attempt to resolve this debate. Rather, for the purpose of this paper, I use the categories articulated in a recent commentary on oversight of DTC neurotechnologies: neurostimulation, neuromonitoring, mental health apps, and cognitive training applications (Wexler & Reiner, 2019). What is important to note, however, is that the different types of DTC neurotechnologies raise slightly different concerns about potential harms.

2.1.1 Neurostimulation

Neurostimulation devices utilize a battery and electrodes to provide electrical or magnetic forces through the scalp to certain regions of the brain (Fitz & Reiner, 2015). Consumer headsets generally provide stimulation at lower than clinical levels using technologies such as transcranial direction stimulation (tDCS) and cranial electrotherapy stimulation (CES) (Ienca et al., 2018). Neurostimulation companies claim that their devices can improve concentration, help with brain training, improve sleep efficiency, reduce stress, and treat conditions such as depression, migraines, anxiety, and insomnia (Ienca et al., 2018). Although the stimulation is considered "non-invasive" the ability to modulate the brain without surgery is both what makes such devices promising and potentially unsafe.

Neurostimulation devices introduce potential consumer harms related to safety, unsubstantiated and oversold claims, and reliability (Coates McCall et al., 2019). Claims about neurostimulation devices that are not backed up by rigorous evidence can cause harm by undermining trust in the medical system and diverting consumers from proven therapies that could improve their health (Coates McCall et al., 2019). Neurostimulation devices, such as tDCS, can also pose a threat to physical and psychological health and have been reported to cause itching or burning sensations, skin lesions from electrode placement, temporary blindness, mood changes, impaired cognition and memory, and, in one case, a seizure (Zettler, 2017). The technology is considered generally safe when used for short periods of time, although there is evidence that longer use can lead to more adverse effects (Bikson, Paneri, & Giordano, 2016). Finally, there are concerns that these devices could exacerbate distributive justice problems if they become widely adopted and actually provide users with cognitive gains (Kreitmair, 2019).

2.1.2 Neuromonitoring

Neuromonitoring devices are designed to detect, analyze, and report brain activity (Wexler & Thibault, 2019). Devices utilize electroencephalography (EEG) to measure brain signals through the scalp. In the near future, advancements in magnetoencephalography (MEG) technology may allow for wearable devices in the consumer market (Boto et al., 2018; Ienca et al., 2018). According to device manufacturers, neuromonitoring headsets are capable of elevating a user's meditation experience, assisting with self-assessment, improving cognitive training, controlling external devices, improving sleep efficiency, and more (Ienca et al., 2018; Wexler & Thibault, 2019).

Privacy and data security related harms are mentioned in the context of neuromonitoring devices (Ienca et al., 2018; Kreitmair, 2019). There is disagreement, however, about whether current technologies are developed enough to collect and store the type of neural data we would be worried about being stolen (Wexler, 2019; Wexler & Thibault, 2019). More likely, in the short term, consumer harms will be related to receiving information that is misleading, confusing, or potentially dangerous (Coates McCall et al., 2019; Wexler & Thibault, 2019). In the future, if devices do become more sophisticated and integrated to other mobile devices in our lives, privacy and data security harms may be more likely (Ienca & Haselager, 2016). What happens to this data after it is entered into the system is often unclear to consumers, and data can be sold or utilized for purposes that would be against the user's morality (Martinez-Martin & Kreitmair, 2018). This is less of a concern with neuromodulation products, because they are unlikely to contain as much brain data as monitoring devices. Privacy harms can also arise from the use of personal data in unexpected ways, even if they are legal, because "[n]onprofessional

DTC digital psychotherapy providers do not have a duty to keep consumer data confidential and generally state so in the terms and conditions" (Martinez-Martin & Kreitmair, 2018, p. 3).

2.1.3 Mental health apps

Mental health apps are software programs that are designed to operate on smart phones, tablets, or mobile devices in order to assist the user do a variety of tasks related to improving their mood and behavior (Armontrout et al., 2018). Mental health apps aim to address the gap between the high demand for clinical therapies for psychiatric disorders and the limited supply of available resources (Torous, Wisniewski, Liu, & Keshavan, 2018). There has been an explosion of mental health apps with as many as 10,000 currently available for download. There are several distinct types of mental health apps—self-help or self-assessment apps, psychotherapy apps, and diagnostic or predictive apps (Terry & Gunter, 2018). Apps can also be divided by whether they are provider-facing or consumer-facing. I will focus on consumer-facing apps due to the fact that provider-facing apps are largely regulated as medical devices and are an extension of traditional practice as opposed to a replacement.

Mental health apps raise concerns about safety, protection of user data, privacy, accountability, liability, informed consent, and hacking (Armontrout et al., 2018; Martinez-Martin & Kreitmair, 2018; Terry & Gunter, 2018). If a user picks the wrong therapy app, this can lead to harm by preventing her from receiving better treatment or by directly causing harm to her mental health by suggesting inappropriate therapies (Martinez-Martin & Kreitmair, 2018). The storage of sensitive neural data is also a concern given that most consumer-facing mental health apps are not regulated by the FDA and do not have consistent protections to ensure that private consumer information is not inappropriately shared (Terry, 2019). Mental health apps also introduce issues of accountability since they can replace traditional relationships between doctors and patients (Terry & Gunter, 2018). Without a clinician involved in the treatment, there are questions as to who is accountable and eventually liable for the patient's mental health if there are problems with the app that lead to negative consequences (Martinez-Martin & Kreitmair, 2018). Concerns about informed consent are motivated by the lack of reliable evidence for the effectiveness of some mental health apps (Martinez-Martin & Kreitmair, 2018). This uncertainty makes it difficult for the consumer to assess which apps are of high quality and which are better to be avoided. Similar to neuromonitoring devices, there is a concern that unauthorized users may try to gain access to private data (Armontrout et al., 2018). This is especially worrying in the mental health context, since a hacker could theoretically change treatment parameters without the consumer knowing.

2.1.4 Brain training apps

Cognitive training or brain training apps are similar to mental health apps but instead of helping the user mange or treat their mental illness, the focus is on improving cognitive abilities such as problem solving and memory (Simons et al., 2016). Brain training apps represent a billion dollar industry, with companies such as Lumos, creator of the popular Lumosity, having as many as 70 million active users (Torous, Staples, Fenstermacher, Dean, & Keshavan, 2016). Lumosity purportedly enhances a consumer's cognitive skills using a game based interface that allows the user to complete basic mental tasks intended to improve abilities like memory, attention, and problem solving (Simons et al., 2016).

Brain training apps raise concerns about consumer harms related to aggregation of user data for research purposes, validation of efficacy, privacy, and data security (Purcell & Rommelfanger, 2017). These worries about data aggregation and use are different from the other three types of DTC neurotechnologies given the sheer number of users that are using brain training apps (Purcell & Rommelfanger, 2017). Concerns about efficacy, privacy, and data security are similar to those raised in the mental health app and neuromonitoring setting. Given that these three technologies have the ability to read and store some sort of neural data, there are related concerns with how this information is shared and protected.

2.2 What are the harms caused by DTC neurotechnologies?

The preceding discussion highlights a number of different ethical concerns raised by neuroethicists and a number of potential harms faced by users. The breadth of ethical concerns is a by-product of the diversity of types of DTC neurotechnologies and uncertainty regarding their safety and efficacy. Part of the difficulty in arriving at a coherent list of harms caused by DTC neurotechnologies is determining how forward looking our analysis should be. For example, some have claimed that concerns about brain hacking are especially urgent due to the ability for DTC neurotechnologies to interface wirelessly with devices in the "Internet of Things" (Ienca et al., 2018). For others, this represents, at best, a projection of not yet achieved technology and, at worst, a misstatement of the current science and unjustified worry that distracts from more immediate consumer harms (Wexler, 2019). While it is vital for us to understand the precise nature, magnitude, and proximity of harms related to different types of DTC neurotechnologies, such an analysis is both beyond the scope of this paper and covered elsewhere in this volume. Rather, in what follows, I discuss some of the harms that are frequently raised in the neuroethics literature.

2.2.1 Safety

Consumer products contain an inherent warranty of safety that ensures that regular use as defined by instructions will not lead to any harm (Owen, 2007). Safety is especially paramount in the consumer setting because there is no third-party, such as a physician, who oversees use. As one might expect, neurostimulation devices, such as tDCS, pose the biggest threat to both physical and psychological health (Bikson, Bestmann, & Edwards, 2013). Although neuromonitoring devices have no overt safety risks to physical health, they may be psychologically impactful if they reveal information that is either upsetting or confusing to the consumer (Coates McCall et al., 2019). Similarly, mental health apps raise issues about psychological well-being if a consumer is given information that leads to anxiety or stress (Martinez-Martin & Kreitmair, 2018). Furthermore, receiving a mental health diagnosis can be a life-changing experience for a person in any scenario but can be especially problematic if the consumer is not properly connected to appropriate resources (Torous et al., 2018). By comparison, brain training apps are the least concerning when assessing safety related harms.

2.2.2 Transparency

There are important transparency related harms that are implicated by DTC neurotechnologies. At present, there is little scientific data to evaluate the efficacy of many DTC neurotechnologies (Kreitmair, 2019). This means that consumers are almost entirely dependent on the claims made by the companies marketing the products. If companies fail to maintain transparency about the effects of the device, consumers are at risk of being misled (Wexler & Thibault, 2019). Neurotechnologies are often seen by the public through a lens of awe which jeopardizes their ability to rationally decide whether use is beneficial given the associated risks (Giattino, Kwong, Rafetto, & Farahany, 2019). Consumers may be even more vulnerable if companies utilize exaggerated or misleading claims about the technology (Wexler & Reiner, 2019). Transparency is a concern for all types of DTC neurotechnologies. At minimum, companies should be honest about what their product can and cannot do. They should also communicate with consumers how their neural data is being utilized and protected. This can help foster a trusting relationship between developers of DTC neurotechnologies and consumers.

2.2.3 Privacy

Privacy harms are one of the most discussed concerns related to DTC neurotechnologies (see Table 1). Our brain data can reveal personal information that may not be intended for access by others. Opening channels that directly gather information about you without conscious processing introduces worries that one may inadvertently share data that would otherwise remain private. Privacy harms are most relevant for neuromonitoring devices, mental health apps, and brain training games. This is because these technologies will be collecting our neural data in one form or another. Although privacy concerns are not unique to neural data there are a few ways in which information collected directly from a person's brain raises unique worries. First, neural data, like genetic information, is unique to each person and has been proposed as a possible metric for identification and authentication (Bonaci, Calo, & Chizeck, 2014). Second, at least one study has shown that highly personal information such as passwords and month of birth can be obtained using only visual stimuli and EEG data (Martinovic et al., 2012). What happens to neural data after it is entered into a system is often unclear to consumers due to a lack of transparency in company data sharing policies (Martinez-Martin & Kreitmair, 2018). Finally, there is a way which neural data seems more private than other forms of information. This could be due to the cultural importance of mental privacy in western society. Recently, there have been increasing concerns about the selling of personal data, prompting the passage of the General Data Protection Regulation in Europe to establish new civil rights around data privacy (Mulder, 2019).

2.2.4 Data security

Related to privacy is the notion that certain information about you can be the target of hacking. Security harms are distinct concerns that focus on how DTC neurotechnologies create a new avenue for unauthorized access to personal data (Ienca & Haselager, 2016). This worry is not unique to the DTC neurotechnology setting. Rather data security harms are likely to occur as a result of these products becoming yet another access point for our personal data (Ienca et al., 2018). For neuromodulation devices, there is a concern that connectivity with the Internet of Things, via Bluetooth or cellular networks, will create vulnerabilities that can be exploited by hackers to send unwanted stimulation to a consumer's brain (Dorri, Kanhere, Jurdak, & Gauravaram, 2017). For neuromonitoring devices, the concern is more about what personal information a hacker might be able to figure out based on brain data. Mental health apps raise security concerns related to the storage of sensitive medical information (Terry & Gunter, 2018). Unlike medical record companies, smaller DTC companies may not have sophisticated protection for information about consumer's psychiatric conditions (Purcell & Rommelfanger, 2017). Security concerns around brain training apps are less apparent since there is unlikely to be much sensitive or personal information contained in these apps. Nevertheless, unauthorized access to results of certain games or tests could lead to embarrassment or stigma. Some of the data security concerns may be overhyped given the current state of the technology (Wexler, 2019). Nevertheless, these concerns could become more problematic in the future as DTC neurotechnologies interface with other household devices (Ienca et al., 2018).

2.2.5 Justice

If the promise of DTC neurotechnologies is fully realized, they may have the potential to exacerbate existing inequities in a variety of ways. First, the most well-off in society will have greater access to these technologies as they do to most innovations that impact health (Wexler & Reiner, 2019). We can expect the gap to be similar to the digital divide, in which lower socioeconomic groups are severely hampered in their access to information (Gonzales, 2016). Second, there is a potential that many of these devices will be designed by the majority and neglect those who are worst off. For example, devices that aim to correct "certain behaviors" may unfairly target minorities. Third, it is likely that the lack of diversity among developers of DTC neurotechnologies will introduce bias into the design of the products (Gianfrancesco, Tamang, Yazdany, & Schmajuk, 2018). For example, algorithms will likely be trained on a white population and may not always be effective when used by people of color. Despite these concerns, it is worth nothing that some have argued that DTC neurotechnologies have the ability to *mitigate* concerns about justice by providing relatively cheap access to products that can help the least well-off in society (Fitz & Reiner, 2015; Wexler & Reiner, 2017). If this is true, we must be careful that any regulation we choose is not overly burdensome toward developers in a way that may prevent technologies from ever reaching the market or being made available at prices that are affordable for the entire population.

3. Legal mechanisms for DTC neurotechnology oversight

Given the range of potential harms introduced by DTC neurotechnologies, there has been a robust discussion in the neuroethics literature about the appropriate form of regulation to employ (see Table 1 for diversity of regulatory mechanisms mentioned). Is there a regulatory gap in the existing mechanisms for oversight or do DTC neurotechnologies introduce salient harms that go beyond the scope of extant approaches? Some have argued that additional federal regulation should bring DTC products under the definition of medical devices (Maslen, Douglas, Cohen Kadosh, Levy, & Savulescu, 2014). Others, however, have rightfully noted that regulation at the federal level comes with considerable hurdles, and that DTC neurotechnologies are better suited to oversight mechanisms like expert working groups (Wexler & Reiner, 2019). The reality is that no one mechanism is likely to be sufficient to protect against the multitude of harms presented by DTC neurotechnologies. Therefore, it is important for policy makers to understand the ways in which existing approaches are suited to regulating products in this space and how the different mechanisms implicate each other. In what follows, I review the major legal approaches to consumer protection in the United States and whether they are well suited for addressing the harms introduced by DTC neurotechnologies.

Before I begin my discussion of current regulatory mechanisms for DTC neurotechnologies, I want to briefly explain and justify the focus on the US setting. First, many of the developers of DTC neurotechnologies are based in the United States. Companies such as Kernel, Emotiv, Neurosky, Neuralink, and others mentioned in the neuroethics literature were founded in the United States and list it as their primary place of business. This makes it likely that the technologies they design will be developed with American regulation in mind. Even if they are later subject to international regulation, it is important for readers outside the United States to understand the context in which these technologies were originally introduced. Furthermore, given these companies are based in the United States, it may be difficult for other jurisdictions to enforce regulations efficiently. For example, Apple has recently been involved in litigation around the issue of whether they are required to pay taxes to Ireland (Barrera & Bustamante, 2018). This dispute illustrates how American companies may try to skirt

European regulation—requiring costly and time-consuming litigation to enforce rules.

Second, there are important lessons that can be learned from how the United States approaches regulations. On one hand, there are examples from history where agencies like the Food and Drug Administration (FDA) were ahead of European regulators in terms of protecting the public. The history of thalidomide demonstrates this point. The drug, which did not receive FDA approval due to safety concerns, was widely available in the UK, Germany, and other European markets (Krapohl, 2007). The adverse effects from using the drug were catastrophic. Newborn babies whose mothers took Thalidomide during pregnancy developed dangerous deformities that lead to death or severe disability. The incident lead to an overhaul of the way European agencies regulated pharmaceutical drugs (Krapohl, 2007). There are enough similarities between the US regulatory system and other jurisdictions where lessons from the United States could easily be transferrable. On the other hand, the United States has been a more permissive jurisdiction when it comes to regulating issues surrounding food additives (Rabin, 2019). There are undoubtedly differences in regulatory approaches that will lead to different results outside the United States. Nevertheless, researchers outside the United States should understand the implications of having more restrictive policies toward DTC neurotechnologies. For example, if the United States allows an easier pathway to get products to the markets, the European market for these technologies may suffer. The effects of globalization have made it so regional markets are not immune to the activities of foreign governments. An understanding of the US regulatory systems may help international scholars determine the impact of US regulations in their own jurisdiction.

Finally, readers from outside the United States are likely to have some familiarity with their local regulatory systems. A detailed comparison of the US regulatory system to other jurisdictions is beyond the scope of this paper. Therefore, my focus on the US setting prioritizes depth over breadth. My point here is that it is hard to deny the influence the US regulatory system has on the rest of the world. By devoting space to exploring the specifics of the US system instead of offering a more comparative look, at the very least, readers will gain an understanding of the nuances involved in the United States and will be able to use it as a point of reference when evaluating their own country's regulatory framework. Nevertheless, there are limitations to this approach that must be highlighted. These are discussed in Section 5.

3.1 Federal agencies

Federal administrative agencies are responsible for providing oversight and regulations for a wide range of human activities. As opposed to legislation, which can be overly broad and is created by policymakers that lack specific knowledge or expertise, agencies offer the advantage of subject matter specialization. For example, while Congress is charged with passing all kinds of laws, an agency like the Environmental Protection Agency is confined in scope by the enabling statute that created it. Since administrative agencies are part of the executive branch, they do not require bipartisan support to enact new policy. Furthermore, regulations created by federal agencies have the force of law and can be enforced if the agency has been given authority to conduct investigations or levy punishments. Agency rules are also less susceptible to judicial review by political opponents because the Supreme Court has ruled that courts must give deference to agency interpretations of enabling statutes (Walker, 2018). Finally, agency regulation can be prospective in nature and, when done successfully, can prevent harms rather than being reactionary.

Federal agencies, however, can also be a poor option for regulating DTC neurotechnologies for several reasons. Writing new regulations can be time consuming, often taking years to implement. This is problematic when addressing a fast-evolving technology, since regulation will be one step behind the science. Additionally, agency regulations can be overturned by subsequent administrations making them sensitive to political turnover. For example, federal funding for stem cell research was halted during the majority of the Bush administration, only to be reversed within the first year of the Obama administration (Murugan, 2009). Another problem is that agency enforcement is heavily dependent on available resources and inevitably requires prioritization. This means enforcement is sometimes driven by political pressure and public outrage more than sound science. Due to the limited resources and lack of efficiency, although agency regulation can be prospective in theory, it is often reactive in practice.

3.1.1 Food and Drug Administration (FDA)

In theory, the FDA is an excellent choice to regulate DTC neurotechnologies due to its subject matter expertise and wide authority to create new regulations with the force of law and enforce them through both civil and criminal means. The FDA is well positioned to address safety and transparency harms given its mission of "protecting the public health by ensuring the safety, efficacy, and security of human and veterinary drugs, biological products, and medical devices" (FDA, 2018). At present, however, the FDA does not regulate most DTC neurotechnologies, with the exception of some types of mental health apps. Under its enabling statute, the Food, Drug, and Cosmetic Act (2017), the FDA has authority to regulate any "device intended for use in the diagnoses of disease or other conditions, or in the cure, mitigation, treatment, or prevention of disease" or "intended to affect the structure or any function of the body of man." Despite this authority, the FDA has chosen to use its enforcement discretion to avoid regulating many DTC neurotechnologies that may fit the agency's definition for a medical device (Wexler, 2015). The FDA has stated that it does not intend to regulate "general wellness products" that are considered to be "low risk" (FDA, 2019a).

Recent consumer medical products, like DTC genetic tests, may provide an example of how the FDA might enter the neurotechnology space in the future. The FDA initially chose to take a permissive approach to regulating the quickly developing DTC genetic testing market (Green & Farahany, 2014). It was not until a report by the US Government Accountability Office (GAO) demonstrated that the four major DTC genetic testing companies had tests that were "misleading and of little or no practical use to consumers" and had "egregious examples of deceptive marketing" that the FDA decided to regulate genetic tests as medical devices (Kutz, 2010). In 2013, this culminated in the FDA sending cease and desist letters to several major DTC genetic testing companies commanding them to remove their health-related products from the market (Lutins & Terry, 2015). Eventually, 23andMe cooperated with the FDA by submitting numerous studies to indicate that its products were analytically valid and that results could be comprehended by consumers in order to get approval for genetic testing kits for Alzheimer's disease and breast cancer (Sharkey, 2018).

Following these enforcement actions, however, the FDA appears to have made the decision to reduce the number of consumer products it regulates (Sharkey, 2018). Other scholars have noted that the FDA has backed away from regulating products that would previously have fallen within its jurisdiction (Terry, 2019). This cautionary approach to regulation was made more explicit with the passage of the 21st Century Cures Act. That law specifically exempted a wide range of app-based DTC neurotechnologies from FDA regulation (FDA, 2019b). While provider-facing mental health apps or apps that interface in some way with traditional medical care are still considered medical devices and are under FDA jurisdiction, user-facing apps like ones that allow users to monitor their mood are specifically exempt under the new rules (Terry, 2019).

I agree with Wexler (2015) and Kuersten and Hamilton (2016) that the FDA's unwillingness to regulate does not indicate a gap in oversight. Rather, there is a strong argument to be made that should safety related harms from DTC neurotechnologies increase the FDA is well situated to regulate the majority of products in the space. The example of DTC genetic testing provides support for this view. Once the FDA was made aware of the potential harms related to the product, they used their authority to step in. As some have noted, as DTC neurotechnologies advance it is likely that some devices, especially those that stimulate the brain, will become regulated by the FDA due to increasing safety risks (Kreitmair, 2019). This would still leave a large majority of low risk DTC neurotechnologies that are not subject to regulation by the FDA yet could cause harms related to transparency, privacy, data security, and justice. In that sense, there is a legitimate regulatory gap when it comes to companies making misleading claims about the efficacy of their products and marketing them using ambiguous claims about wellness. For medical devices, the FDA is responsible for oversight of false or misleading marketing claims (Anderson et al., 2016). However, since the FDA does not consider most DTC neurotechnologies to be medical devices, any regulation for misleading or inappropriate claims will likely fall to the Federal Trade Commission (FTC) (Wexler & Reiner, 2019). Additionally, even if the FDA did choose to regulate in this space, it would not be well suited to address harms related to privacy and data security.

Of note, is the FDA's handling of the technology MicroSort. The device was designed to help split up sperm bearing an X chromosome from those bearing a Y chromosome (Macklin, 2010). The FDA initially approved clinical trials but later prohibited the company from pursuing further studies using the device (Dahl, 2011). Since documents related to the FDA's decision to withdraw authorization are not public it is difficult to know what factors influenced their decision. At least one commentator, however, has argued that the FDA overreached beyond assessing safety and efficacy and instead pulled the product from clinical trials because of ethical issues related to the potential for couples to choose the sex of their child (Dahl, 2011). The FDA's effective ban of MicroSort from the US market indicates it may be willing to regulate a DTC neurotechnology that it feels raises ethical concerns related to justice.

3.1.2 Federal Trade Commission (FTC)

The FTC is tasked with protecting consumers from "unfair, deceptive or fraudulent practices in the marketplace" (FTC, 2013). The agency has the authority under the Federal Trade Commission Act of 1914 to bring civil claims against companies on behalf of consumers who have been harmed due to a violation of the law. It also develops regulations to increase competition between companies and educate consumers and businesses about their rights (Federal Trade Commission Act of 1914). Given the FDA's decision not to regulate DTC neurotechnologies, the FTC is the agency best positioned to address harms related to transparency and privacy. In fact, the FTC is the only federal agency that has taken enforcement action against a DTC neurotechnology company (FTC, 2016). Lumos Lab, the creator of Lumosity, a set of brain training games, claimed that consumers could sharpen their minds based on the "science of neuroplasticity" (Underwood, 2016). The ad claimed that by playing just 10–15 min per day several days a week one could improve work and school performance and potentially prevent Alzheimer's and other conditions (Underwood, 2016). The FTC fined Lumos Labs $2 million for false advertising and required the company to display a popup that informed consumers about the FTC enforcement and how to unsubscribe (Wexler, 2017). The enforcement was one of the first of its kind in this market and indicates the FTC's authority and willingness to regulate certain kinds of DTC neurotechnologies.

Another enforcement action taken by the FTC was not as uncontroversial. In 2015, the FTC fined two employees of Carrot Neurotechnology a total of $150,000 for the claim that its training program improved binocular visual acuity by an average of 31% (Underwood, 2016). This figure was based on a study of 19 baseball players at the University of California Riverside which argued that using the company's learning program improved vision, decreased strikeouts, and created more runs. The agency said that such claims had to be backed up by statistically significant results from randomized, blinded, placebo-controlled clinical trials as opposed to the small sample psychologist Aaron Seitz pulled his 31% figure from (FTC, 2015). This enforcement action was seen by some as a clear attempt by the FTC to regulate an untamed market (Underwood, 2016). Since 2016, however, the FTC has not taken any action against DTC neurotechnology companies. It is difficult to say whether the agency's change in approach is due to companies being wary of overselling their products or a policy decision to step back from regulation of DTC neurotechnologies. Nevertheless, the FTC's prior enforcement actions illustrate the ability and willingness to go after companies that are making false claims in the neurotechnology market. This should help to mitigate transparency related harms.

The FTC is also leading the way in terms of regulation of consumer privacy and data. Since the late 1990s, the FTC has been regulating company privacy policies through its authority to police unfair and deceptive trade practices (Solove & Hartzog, 2014). Through its enforcement decisions, the FTC has created a set of de facto regulations by letting companies know the types of behaviors that will lead to agency action. The FTC's privacy jurisprudence centers around the deception theory of broken promises of privacy (Solove & Hartzog, 2014). The FTC will seek enforcement in cases where the company promises to do the following types of things but later violates their agreement: "promises to maintain confidentiality or to refrain from disclosing to third parties; promises to only collect data consistent with the company's privacy policy; promises to provide adequate security of personal data; promises to maintain anonymity; and promises not to disclose personal data to third parties by selling in bankruptcy proceedings" (Solove & Hartzog, 2014). Recent enforcement actions during the Trump administration illustrate that the FTC is continuing to regulate in the privacy area through enforcement actions (Krawiec, Schlager, Mehta, Griffith, & Ryan, 2018). The FTC's regulation in this area is not sufficient to address all the privacy harms related to DTC neurotechnologies. Yet, it is evidence that the regulatory mechanisms to deal with violations of privacy and data security exist and can be functional.

3.1.3 Consumer Product Safety Commission (CPSC)

While some DTC neurotechnologies will be regulated by the FDA, when it comes to ensuring the safety of the majority of DTC products the responsibility falls to the CPSC. The CPSC has jurisdiction over all injuries related to consumer products except food, drugs, devices, cosmetics, motor vehicles, tobacco, aircrafts, and pesticides (15 U.S.C. § 2052(a)(5)). The CPSC has the authority to create safety standards for products and in the past has regulated products like bicycles, children's toys, and garage door openers (Crane, Eichenseer, & Glazer, 2011). The CPSC also has the power to recall or ban products it deems to not meet set safety standards or finds otherwise dangerous to consumers. At first glance, the CPSC seems like an ideal candidate to cover any remaining concerns about the safety of DTC neurotechnologies. However, a recent review of non-fatal injuries caused by consumer products found that the majority of harms were caused by items in the categories of sports and recreation, home structures and construction materials, and home furnishings and fixtures (Lawrence, Spicer, & Miller, 2015). In contrast, the authors found that home electronics and hobby, where DTC neurotechnologies would likely be classified, accounted for only one perfect of total incidents. This indicates that while the CPSC would have broad authority to regulate DTC neurotechnologies that fell through the cracks of FDA regulation, they are likely to be given a low priority since they are generally safe when compared to other consumer products.

3.2 Legislation

If regulatory agencies are unable to address all the harms introduced by DTC neurotechnologies, legislation may be a potential option. There is an emerging need for federal legislation to address harms related to privacy and data security (Terry, 2019). Federal legislation offers a powerful tool for introducing new regulation for DTC neurotechnologies. Legislation is advantageous because it can address broad and expansive problems that are beyond the scope of specific agencies. For example, whereas the FDA, FTC, and CPSC are all limited by the language of their enabling statute, federal legislation can be expansive enough to address multiple types of DTC neurotechnologies and harms. Additionally, new bills have the ability to create new rights, causes of action, and allocate funding for additional oversight measures. Recently, Anna Eshoo and Zoe Lofgren, members of the US House of Representatives, proposed federal legislation that would create a new regulatory agency and more robust consumer rights. The proposed bill would create a Digital Privacy Agency that would enforce privacy rights and grant users the ability to "access, correct, delete and transfer data about them" (Feiner, 2019). Representative Eshoo stated that "the FTC lacks the staff, the expertise and the culture to take [this on]. This is a monumental task of protecting privacy" (Feiner, 2019). In addition to providing new rights such as requiring opt-in consent for users' data to be used in machine-learning algorithms, it creates new causes of action for consumers to file lawsuits for declaratory or injunctive relief and damages when not filing as part of a class action (Feiner, 2019). This type of legislation would mirror privacy legislation passed in Europe called the General Data Protection Regulation (GDPR) and present a way out of the regulatory patchwork that currently exists in US consumer protection law (Mulder, 2019).

Federal legislation, however, comes with challenges. At times, it can stifle innovation by increasing the costs of entering the market (Goldschlag & Tabarrok, 2018). Although it is important to minimize the harms posed by innovative technologies, regulation should not be so burdensome as to prevent developments of novel products. State legislation offers the possibility of introducing experimental policies to test the effectiveness of certain regulations. In theory, different states can utilize their regional expertise to craft more specific regulations that target their constituents. In reality, however, state regulations are not as flexible as one might hope since companies are forced to comply with the regulations in the strictest state so that they do not have to change their operations in every state. State legislation also introduces the issue of pre-emption. Recently, California passed the Consumer Privacy Act of 2018. As one scholar notes, introduction of this state-level legislation has galvanized industry support around federal legislation protecting privacy (Terry, 2019). This is because federal legislation introduces a question of whether the latter legislation will overrule or supplement state-level bills. If a federal law does pre-empt state law it creates a situation where a more robust state law protecting privacy could be replaced by a weaker federal law (Terry & Gunter, 2018). Perhaps the biggest drawback to seeking federal legislation is that it requires bipartisan support or control of both houses of Congress. In the past few years, Congress has consistently been rated as one of the least approved government institutions in the United States, in part, due to its inability to find middle ground and pass necessary legislation (Jones, 2015). A federal law with as many political implications as a data privacy bill faces a lengthier process to codification than regulations through agencies.

3.3 Tort liability

Litigation has historically been a successful method for providing industry oversight and preventing consumer harm when there has been an inability or unwillingness to regulate on the part of the executive and legislative branches of government. Litigation against big tobacco predated and inspired later regulation by Congress and the FDA (Viscusi & Hersch, 2010). In the consumer DTC neurotechnology space, consumers could file lawsuits to recover damages based on harms resulting from product defects and failure to warn of risks (Terry & Wiley, 2016). There are potentially other causes of action related to privacy but the complexity of the tort law framework for privacy related harms makes it beyond the scope of the current discussion. Regulation via litigation has several advantages over other types of oversight. First, regulation through litigation does not require other policy makers and can seek legal rights through judicial precedent. Furthermore, litigation provides consumers with monetary damages for harms they suffered because of DTC neurotechnologies. Punitive damages against companies can also discourage them from acting poorly in the future.

Nevertheless, there are some inherent limitations to using litigation as a regulatory mechanism. First, litigation is always reactionary and therefore will only be useful after consumers have been harmed. To establish a cause of action for most torts, the consumer must provide evidence of harm. Given the uncertainty around whether DTC neurotechnologies are actually efficacious it may be difficult to prove a causal chain of events that illustrates that harms suffered were connected to problems with the technology. Second, litigation is expensive and will be difficult to pursue on an individual level for most consumers. Class action lawsuits are an attractive solution but necessitates many consumers suffering harms before claims can be filed. Class action cases also require assembling plaintiffs who have suffered similar harms from the same product. Given the diversity of DTC neurotechnologies and the harms they implicate it may be difficult to find a consistent group of consumers in order to form a class for litigation.

4. Alternative governance models

The prior discussion has illustrated that although there are extant regulatory mechanisms that can mitigate most harms caused by DTC neurotechnologies, concerns regarding justice remain largely beyond the scope of US consumer protection law in its current form. This is because the regulatory framework in the United States is not designed to control the pricing of consumer products. In fact, even the price of medical products is largely determined by the relative negotiating powers of device manufacturers, private insurers, and large government programs like Medicare and Medicaid. The recent controversy surrounding EpiPen pricing illustrates this point. Although the FDA can consider expediting review of generic drugs to compete with medicines that have seen price surges, it does not consider the pricing of drugs when deciding whether to approve them (Costales, 2018). Given the government's reluctance to regulate pricing of medical products, a common feature in the European context, it is unlikely that there would be effective legal mechanisms to control the cost of DTC neurotechnologies. If DTC neurotechnologies do genuinely provide benefit to users, they could join other technologies, such as the internet or smartphones, in creating a "digital divide" between communities that have access to these innovations and those who are left out.

In addition, due to the limitations of the legal mechanisms discussed above (see Table 2), there are reasons to believe harms related to transparency and data security will not be comprehensively addressed through agency regulation, legislation, and litigation. The prior discussion should have illustrated that an efficient and well-functioning regulatory system in the United States has the mechanisms to address most DTC neurotechnology related harms. The reality, however, is that these systems will not always function effectively. Necessary agency action could be stalled due to lack of resources to fund enforcement. Federal legislation can take years to pass, and then more years to implement. Litigation, though effective

Table 2 Pros and cons of different regulatory mechanisms.

Regulatory mechanism	Pros	Cons
Federal agencies	– Subject matter expertise – Easier to create new regulations than legislation – Less political hurdles than legislation – Has force of law – Can be prospective (although in practice is often reactionary)	– Limited to enabling statute – Time consuming – Less permanent than legislation – Lag between regulation and science – Limited by available resources – Enforcement decisions can be political or based on public outrage – Open to judicial challenge
Legislation	– Can be broad and expansive in nature – Can create new rights, agencies, causes of actions, funding sources – Has force of law – More difficult to create new legislation than regulations	– Overly broad statutes can lead to judicial challenge about scope – More difficult to overturn existing legislation than regulations – Lack subject matter expertise – Issues with pre-emption – Lag between legislation and science – Often only possible after harms have been felt by many and public outrage is high

Table 2 Pros and cons of different regulatory mechanisms.—cont'd

Regulatory mechanism	Pros	Cons
Tort liability	– Does not require policy or decision makers – Can establish legal rights and precedent without involving political branches – Can seek damages for harm – Punitive damages provide economic incentive to change behavior – Can aggregate harms suffered through class action lawsuits	– Costly – Time consuming – Reactionary – Depends on existence of rights, causes of actions, pre-emption issues – Requires aligning harms and consumers against a common opponent
Self-regulation	– Familiarity with subject matter can lead to narrowly tailored regulation – Can be more prospective than other regulation – Can adapt to changes quickly	– Lacks force of law – Regulations could be ignored if it hurts economic interests too much – Companies could be wary of extra cost
Independent expert group	– Diverse stakeholders can share different perspectives – Multi-subject expertise – Can focus on very specific issue – Can adapt to changes quickly	– Lacks force of law – Can be ignored if it hurts economic interests too much – Unfunded mandate – Long-term uncertainty

at making harmed users whole, is inherently reactive in nature. Therefore, a multi-faceted approach to regulation of DTC neurotechnologies must include alternative governance models to be effective.

4.1 Self-regulation

To address these shortcomings, self-regulation has been offered as a potential solution (Cohen & Sundararajan, 2015). By putting the burden of regulation on stakeholders of DTC neurotechnologies, the hope is that they will be intimately familiar with the unique problems faced by the group, be able to flexibly and efficiently respond to new changes, and avoid the pitfall of having to be overly broad in order to cover a wide range of interests. Congress can certainly pass legislation requiring companies to invest in better hardware in order to prevent hacking, but it might be more effective to rely on the company to regulate itself as a means of competing in the marketplace. One would expect a company to be interested in producing a technology that sells well and has few negative impacts. Furthermore, companies have an inherent incentive to protect consumer interests since it reflects back on their brand as a whole and can ultimately impact sales.

However, there are reasons to be skeptical about a regulatory framework that relies *only* on methods of self-regulation. If companies are large enough to dominate their industry, they may lack sufficient incentives to regulate their own behavior. Furthermore, if there are no established outcome standards in the field, there may be a lack of guidance about what it is companies should actually regulate internally. Overreliance on industry self-regulation has led to several recent examples of corporate negligence that harmed consumers. One need only look at companies like Facebook and Equifax to see examples of large corporations who failed to prevent consumer harm despite calls for reform (Downes, 2018; Gellman & Dixon, 2016). It is overly optimistic to think that because a company is creating a product that intends to help people that it will be sufficiently motivated to protect consumers when faced with drastic financial constraints or unbridled greed. One could argue that these companies suffered massive reputational and financial harm because of their actions and that additional government regulation would have been unlikely to prevent misbehavior. On the other hand, these are just examples of companies that have been found guilty of misconduct—there are presumably countless similar companies that continue to behave unethically but will never be discovered. This illustrates that federal regulation may be preferable in order to mitigate privacy harms. While we may be able to rely on companies to respect the privacy of one individual, can we rely on them to go against data harvesting trends and fall behind competitors in order to protect consumer privacy on the population level?

Nonetheless, self-regulation seems like a particularly good candidate for addressing justice and data security related harms. Although it may be difficult for federal agencies or even policy makers to enact legislation that controls the price of DTC neurotechnologies, there are reasons to think companies would voluntarily commit themselves to social causes in order to improve their public image and increase the value of their brand. Nike's recent embracing of controversial athlete Colin Kaepernick's fight against racial injustice and Apple's marketing of their data security functions illustrates a growing awareness by corporations that consumers care about issues of social justice and data security.

4.2 Independent working groups

Another potential solution that has been proposed is to form an independent body of experts and interested parties to deliberate about potential concerns and democratically arrive at a regulatory plan. This improves upon self-regulation by a single group of stakeholders by providing diverse perspectives on a particular issue. For example, Wexler and Reiner (2019) argue that federal oversight is ill-suited for the task: "[g]iven the sheer number of products, the dynamic nature of software applications that can change with each update, the flexibility required to oversee them, and the potential ethical issues involved, current regulatory oversight leaves much to be desired" (Wexler & Reiner, 2019, p. 235). They instead propose the creation of an independent working group that provides consumer guidance on which DTC neurotechnologies to use. This group would also be responsible for educating the public about the risks involved with certain DTC products and act as a clearing house for complaints that may be of interest to administrative agencies (Wexler & Reiner, 2019).

Independent working groups avoid some of the pitfalls of industry self-regulation but lack sufficient force to compel companies to change behavior. Who would fund such a group? Current groups of that nature are sometimes funded by federal agencies as a way of gaining insight into whether or not an area requires additional oversight. However, relying on government funding means resources could be limited, especially for areas of study that are not priorities for current leaders. It is also possible that these groups could be funded by industry themselves. Yet, there is an inherent conflict of interest in a company funding the oversight committee that is created to regulate its actions. Furthermore, since any recommendations would lack the force of law, companies have the option of simply ignoring the advice and moving on with business as usual.

This is not to say that these mechanisms lack value. They serve important functions at different stages of the policy development process. For example, these groups can act as a means to address issues of transparency by aggregating information and disseminating it effectively to the public. Federal agencies may be limited in terms of the amount of advocacy work they can do to raise awareness of an issue. Independent working groups, however, can be useful in addressing things like harms caused by unreliable or misleading information by providing purchasers of DTC mental health apps with objective reviews of different companies so that the consumer is aware of the relevant risks and benefits associated with each app.

5. Discussion and limitations

I have argued that a multi-faceted approach to regulation is required to address all the harms raised by DTC neurotechnologies. In order to organize the above discussion, it is useful to review the ways in which the different harms implicated by the use of DTC neurotechnologies can be best mitigated using specific regulatory mechanisms. This is not meant to be a definitive discussion about which mechanisms are necessarily best to address certain harms, but rather a way to illustrate how different approaches need to be married for effective regulation.

5.1 Discussion

For safety concerns raised by DTC neurotechnologies, a combination of agency action, litigation, and independent working groups should address most worries. The FDA's history suggests that it would step in to regulate the DTC neurotechnology market if there were reports about adverse safety issues with products. If these products caused physical or psychological harm, it is likely that consumers could pursue legal action to seek damages from companies. This litigation, if successful, could also be a powerful way to ensure companies take safety concerns seriously. Nonetheless, part of the problem in the DTC neurotechnology space is that we do not have consistent data about who uses DTC neurotechnologies and whether they are being harmed. An independent working group would be a good candidate to devise and create a reporting system that could aggregate user data related to DTC neurotechnologies. This aggregation could help to create the type of dataset required to get the attention of federal agencies like the FDA and to ensure that they are taking enforcement actions seriously.

For transparency related concerns, a mixture of agency enforcement by the FTC and the creation of a database for evaluating the safety and efficacy of consumer DTC neurotechnologies is necessary to mitigate harms. The FTC, rather than the FDA, is the agency that is best positioned to tackle the problem of unreliable or misleading products that enter the market

(Dasgupta, 2019). Given the FTC's limited resources, however, there is concern about whether the FTC will prioritize DTC neurotechnologies over other consumer products. If the FTC neglects to use its authority, claims about products could still be evaluated by other mechanisms. Wexler and Reiner (2019) suggest utilizing an expert working group to assess the different harms and efficacy of DTC neurotechnologies broadly and make this information available to the consumers. Self-regulation could also play an important part in improving transparency related issues. In a growing field, companies may innovate to create safer and more effective products in order to outcompete their rivals. To aid self-regulation efforts, agencies or independent working groups could create a set of standards that if adopted could guide company behavior prospectively. If companies adopt these standards, it also opens up the door to litigation if they violate their own terms and commitments.

Privacy and data security related harms pose an interesting dilemma for regulators in the DTC neurotechnology space. This is because these concerns are not entirely unique to devices that interact with the brain—although neural data can be more sensitive than other types. Broader societal challenges that are implicated by many different types of technologies may require legislation to address the multitude of issues that need to be tackled. In this sense, while agency action by the FTC or litigation may help address some of these concerns, legislation like the GDPR would create new mechanisms for regulation. It will also be important for neuroethicists to clearly delineate the specific types of harms within the domains of privacy and data security. For example, it is important to draw a distinction between companies improving their systems in order to prevent hacking and the willful sharing and monetization of user data. As noted in Section 4.1, there are reasons to think that companies would be successful at creating their own norms for data security. On the other hand, given that many companies derive much of their revenue from sharing user data with third parties, it may require federal legislation to force the hands of tech giants like Facebook and Google.

Finally, for justice related issues, it may be up to the companies themselves to address issues related to cost and access to new devices. Certainly, it is possible that agencies or legislation could create new regulations that require companies creating DTC neurotechnologies that have the capacity to enhance human capabilities to make them be affordable. It is a different question, however, whether such an approach would ever be politically feasible or desired. Attempting to control the price of consumer products may discourage companies from entering the market and innovating to create new technologies because of fear of diminishing profits. If DTC neurotechnologies are to reach their promise and offer benefits to the least well-off in society, one might argue that overly restrictive regulations could stifle the innovation that is so desperately needed. In that sense, regulation could actually exacerbate justice issues by creating an environment where only large companies with enough capital could invest in creating DTC neurotechnologies. If there are fewer products on the market, there would conceivably be less competition and companies could charge higher prices and target only high-income clientele. For these reasons, justice related harms may be best addressed by the companies themselves.

5.2 Limitations

Nevertheless, there are limitations to the approach I have chosen and ways to improve upon the analysis that would advance our understanding of the regulatory options for DTC neurotechnologies. First, my review of types of DTC neurotechnologies was not systematic and risks overlooking products that may well be worthy of including under the term DTC neurotechnology. This is part of a larger conceptual issue of what counts as DTC neurotechnology? Without taking a side in the debate, my paper risks being overly broad or overly narrow in its consideration of technologies. The risk of being overly broad is that too many technologies are identified making it more difficult to identify common concerns and propose efficient regulatory solutions. The risk of being overly narrow, on the other hand, is failing to see how different types of technologies may be culturally or socially connected and therefore have implications for the kind of harms they introduce. As the field grows, the typology of DTC neurotechnologies used in this paper will be insufficient conceptually to address future harms. Future work should be undertaken to resolve these problems.

Second, I did not evaluate the justifications for the potential harms that were identified by neuroethicists. By taking the harms mentioned at face value without assessing whether or not they are justified or overly alarmist, I risk overhyping minor concerns over more pressing issues. Despite this shortcoming, my approach has the advantage of highlighting regulatory gaps in addressing future harms even if it gives too much attention to speculative technologies. A systematic review of the state of the art of current DTC neurotechnologies and their efficacy would greatly aid neuroethicists in focusing their attention on the products that are currently on the market or will be in the near future rather than those that are never likely to be a reality.

Third, as discussed earlier, since my analysis focuses on the US setting, it may be difficult to apply the lessons to other jurisdictions. For example, there are different pressure points that influence policy change in the United States as opposed to Europe. Regulatory agencies like the FDA, make assessments related to safety and efficacy using different calculations and different levels of evidence than in other settings. Future scholarship should focus on making these differences more explicit so that neuroethicists can identify which jurisdictions do a better job addressing potential consumer harms. Nevertheless, this paper contains a wealth of information for scholars outside the United States who are interested in understanding the nuance of how the American regulatory system operates. Furthermore, there are higher level takeaways about types of regulation that are consistent across jurisdictions. Agency regulation, for example, will almost always offer a more specific set of rules than legislation at the national level. In that sense, even though readers outside the United States should gain important insight from my analysis, they should be careful in how they utilize the takeaways when they apply it to DTC neurotechnologies in their home jurisdictions.

Finally, my analysis of the current regulatory mechanisms lacked the depth required for effective policy development. In assessing adequacy, my analysis assumed that regulatory mechanisms would be functioning efficiently, consistently, and similarly to how they have in the past. This neglects questions related to cost effectiveness, political feasibility, and rate of technological advancement. Future research should develop models and collect data in order to answer questions such as whether it would be more cost effective to pursue self-regulation methods as opposed to federal legislation when trying to address privacy harms related to companies using data collected from DTC neurotechnologies for social media advertising. Expert working groups could be a useful tool for bringing together diverse stakeholders in order to facilitate policy analysis with the goal of developing effective regulations to address lingering worries about harms.

6. Conclusion

In this chapter, I have attempted to provide an overview of the ways in which existing regulatory mechanisms are sufficient to provide adequate oversight of DTC neurotechnologies. I began by explaining how a lack of clarity regarding what counts as a DTC neurotechnology and which harms they may cause has made it difficult to create a consistent framework for regulation in this space. Despite these issues, I reviewed the main types of DTC neurotechnologies and the harms raised in the literature to provide the necessary context for evaluating the relative strengths and weaknesses of different types of regulatory mechanisms. This discussion illustrated that while agency regulation, legislation, and tort liability are effective ways of mitigating most harms related to safety, transparency, privacy, and data security, alternative governance models may be necessary to address justice related harms and to resolve thus far refractory conceptual problems. Taken together, these mechanisms provide an adequate framework for regulating most DTC neurotechnologies we are likely to encounter. The aim of this discussion has been to argue that certain mechanisms are better suited to protect certain harms. My hope is that this chapter aids ethicists, scientists, and policy makers in determining the next steps toward creating a comprehensive regulatory framework for oversight of DTC neurotechnologies.

References

Anderson, L., Antkowiak, P., Asefa, A., Ballard, A., Bansal, T., Bello, A., et al. (2016). FDA regulation of neurological and physical medicine devices: Access to safe and effective neurotechnologies for all Americans. *Neuron, 92*(5), 943–948. https://doi.org/10.1016/j.neuron.2016.10.036.

Armontrout, J. A., Torous, J., Cohen, M., McNiel, D. E., & Binder, R. (2018). Current regulation of mobile mental health applications. *The Journal of the American Academy of Psychiatry and the Law, 46*(2), 8.

Barrera, R., & Bustamante, J. (2018). The rotten apple: Tax avoidance in Ireland. *The International Trade Journal, 32*(1), 150–161.

Bikson, M., Bestmann, S., & Edwards, D. (2013). Transcranial devices are not playthings. *Nature, 501*(7466), 167. https://doi.org/10.1038/501167b.

Bikson, M., Paneri, B., & Giordano, J. (2016). The off-label use, utility and potential value of tDCS in the clinical care of particular neuropsychiatric conditions. *Journal of Law and the Biosciences, 3*(3), 642–646. https://doi.org/10.1093/jlb/lsw044.

Bonaci, T., Calo, R., & Chizeck, H. J. (2014, May). App stores for the brain: Privacy & security in brain-computer interfaces. In *Proceedings of the IEEE 2014 international symposium on ethics in engineering, science, and technology* (p. 47). IEEE Press.

Boto, E., Holmes, N., Leggett, J., Roberts, G., Shah, V., Meyer, S. S., et al. (2018). Moving magnetoencephalography towards real-world applications with a wearable system. *Nature, 555*(7698), 657–661. https://doi.org/10.1038/nature26147.

Coates McCall, I., Lau, C., Minielly, N., & Illes, J. (2019). Owning ethical innovation: Claims about commercial wearable brain technologies. *Neuron, 102*(4), 728–731. https://doi.org/10.1016/j.neuron.2019.03.026.

Cohen, M., & Sundararajan, A. (2015). Self-regulation and innovation in the peer-to-peer sharing economy. *University of Chicago Law Review Dialogue, 82*, 116.

Costales, S. S. (2018). EpiPen® Pricing—Evidence of a systemic problem with drug pricing in the USA. *Journal of Generic Medicines, 14*(1), 29–34.

Crane, E. M., Eichenseer, N. J., & Glazer, E. S. (2011). U.S. Consumer Protection Law: A federalist patchwork. *Defense Counsel Journal, 78*(3), 305–330.

Dahl, E. (2011). FDA bans gender selection procedure. Institute for Ethics and Emerging Technologies.
Dasgupta, I. (2019). Ethical oversight of direct-to-consumer neurotechnologies: The FDA, the FTC, or Self-Regulation? *AJOB Neuroscience*, *10*(4), 200.
Dorri, A., Kanhere, S. S., Jurdak, R., & Gauravaram, P. (2017). Blockchain for IoT security and privacy: The case study of a smart home. In *2017 IEEE international conference on pervasive computing and communications workshops (PerCom workshops)* (pp. 618–623), IEEE.
Downes, L. (2018). GDPR and the end of the Internet's grand bargain. *Harvard Business Review,* April 9, Retrieved from https://hbr.org/2018/04/gdpr-and-the-end-of-the-internets-grand-bargain.
Elliott, C. (2011). Enhancement technologies and the modern self. *Journal of Medicine and Philosophy*, *36*(4), 364–374. https://doi.org/10.1093/jmp/jhr031.
FDA, What we do, Retrieved October 22, 2019, from FDA website: http://www.fda.gov/about-fda/what-we-do, 2018, November 3.
FDA. (2019a). General wellness: Policy for low risk devices—Guidance for Industry and Food and Drug Administration Staff. U.S. Food and Drug Administration 12.
FDA. (2019b). Policy for device software functions and mobile medical applications—Guidance for Industry and Food and Drug Administration Staff. U.S. Food and Drug Administration 45.
Feiner, L. (2019, November 5). Two Silicon Valley congresswomen propose a new federal agency to enforce online privacy rights. Retrieved November 20, 2019, from CNBC website. https://www.cnbc.com/2019/11/05/reps-lofgren-and-eshoo-propose-online-privacy-act.html.
Fitz, N. S., & Reiner, P. B. (2015). The challenge of crafting policy for do-it-yourself brain stimulation. *Journal of Medical Ethics*, *41*(5), 410–412. https://doi.org/10.1136/medethics-2013-101458.
FTC. (2013, June 7). What we do. Retrieved October 22, 2019, from Federal Trade Commission website. https://www.ftc.gov/about-ftc/what-we-do.
FTC. (2015). FTC charges marketers of'vision improvement' app with deceptive claims. Sep 17. https://Www.Ftc.Gov/News-Events/Press-Releases/2015/09/Ftc-Charges-Marketers-Vision-Improvement-App-Deceptive-Claims, [Accessed 2019-03-08] [WebCite Cache ID 76ipKE4WS].
FTC. (2016). Lumosity to pay 2$ million to settle FTC deceptive advertising charges for its "brain training" program. Federal Trade Commission.
Gellman, R., & Dixon, P. (2016). Failures of privacy self-regulation in the United States. In D. Wright & P. De Hert (Eds.), *Enforcing privacy* (pp. 53–77). Cham: Springer.
Gianfrancesco, M. A., Tamang, S., Yazdany, J., & Schmajuk, G. (2018). Potential biases in machine learning algorithms using electronic health record data. *JAMA Internal Medicine*, *178*(11), 1544–1547.
Giattino, C. M., Kwong, L., Rafetto, C., & Farahany, N. A. (2019). The seductive allure of artificial intelligence-powered neurotechnology. In *Proceedings of the 2019 AAAI/ACM conference on AI, ethics, and society* (pp. 397–402). https://doi.org/10.1145/3306618.3314269.
Goldschlag, N., & Tabarrok, A. (2018). Is regulation to blame for the decline in American entrepreneurship? *Economic Policy*, *33*(93), 5–44. https://doi.org/10.1093/epolic/eix019.
Gonzales, A. (2016). The contemporary US digital divide: From initial access to technology maintenance. *Information, Communication & Society*, *19*(2), 234–248.
Green, R. C., & Farahany, N. A. (2014). Regulation: The FDA is overcautious on consumer genomics. *Nature News*, *505*(7483), 286. https://doi.org/10.1038/505286a.
Ienca, M., & Haselager, P. (2016). Hacking the brain: Brain–computer interfacing technology and the ethics of neurosecurity. *Ethics and Information Technology*, *18*(2), 117–129. https://doi.org/10.1007/s10676-016-9398-9.

Ienca, M., Haselager, P., & Emanuel, E. J. (2018). Brain leaks and consumer neurotechnology. *Nature Biotechnology, 36*(9), 805–810. https://doi.org/10.1038/nbt.4240.

Ienca, M., & Vayena, E. (2019). Direct-to-consumer neurotechnology: What is it and what is it for? *AJOB Neuroscience, 10*(4), 149–151. https://doi.org/10.1080/21507740.2019.1668493.

Jones, D. R. (2015). Declining trust in congress: Effects of polarization and consequences for democracy. *The Forum, 13*, 375–394, De Gruyter.

Krapohl, S. (2007). Thalidomide, BSE and the single market: An historical-institutionalist approach to regulatory regimes in the European Union. *European Journal of Political Research, 46*(1), 25–46.

Krawiec, M., Schlager, I., Mehta, N., Griffith, K., & Ryan, L. (2018). FTC trends in consumer protection. *Loyola Consumer Law Review, 31*(2), 225–258.

Kreitmair, K. (2019). Dimensions of ethical direct-to-consumer neurotechnologies. *AJOB Neuroscience, 10*(4), 152–166.

Kuersten, A., & Hamilton, R. H. (2016). Minding the 'gaps' in the federal regulation of transcranial direct current stimulation devices. *Journal of Law and the Biosciences, 3*(2), 309–317. https://doi.org/10.1093/jlb/lsw015.

Kutz, G. (2010). *Direct-to-consumer genetic tests: Misleading test results are further complicated by deceptive marketing and other questionable practices: Congressional testimony*. DIANE publishing.

Lawrence, B. A., Spicer, R. S., & Miller, T. R. (2015). A fresh look at the costs of non-fatal consumer product injuries. *Injury Prevention: Journal of the International Society for Child and Adolescent Injury Prevention, 21*(1), 23–29. https://doi.org/10.1136/injuryprev-2014-041220.

Lutins, E., & Terry, S. F. (2015). Will the wall come tumbling down? *Genetic Testing and Molecular Biomarkers, 19*(5), 226–227. https://doi.org/10.1089/gtmb.2015.1566.

Macklin, R. (2010). The ethics of sex selection and family balancing. *Seminars in Reproductive Medicine, 28*, 315–321, © Thieme Medical Publishers.

Martinez-Martin, N., & Kreitmair, K. (2018). Ethical issues for direct-to-consumer digital psychotherapy apps: Addressing accountability, data protection, and consent. *JMIR Mental Health, 5*(2) e32. https://doi.org/10.2196/mental.9423.

Martinovic, I., Davies, D., Frank, M., Perito, D., Ros, T., & Song, D. (2012). On the feasibility of side-channel attacks with brain-computer interfaces. In *Presented as part of the 21st {USENIX} security symposium ({USENIX} Security 12)* (pp. 143–158).

Maslen, H., Douglas, T., Cohen Kadosh, R., Levy, N., & Savulescu, J. (2014). The regulation of cognitive enhancement devices: Extending the medical model. *Journal of Law and the Biosciences, 1*(1), 68–93. https://doi.org/10.1093/jlb/lst003.

Mulder, T. (2019). Health apps, their privacy policies and the GDPR. *European Journal of Law and Technology, 10*(1). Retrieved from http://ejlt.org/article/view/667/898.

Murugan, V. (2009). Embryonic stem cell research: A decade of debate from Bush to Obama. *The Yale Journal of Biology and Medicine, 82*(3), 101–103.

Owen, D. G. (2007). The evolution of products liability law. *The Review of Litigation, 26*(4), 955–990.

Purcell, R. H., & Rommelfanger, K. S. (2017). *Neuroscience online: Real ethical issues in virtual realms*. Taylor and Francis.

Rabin, R. (2019). *What foods are banned in Europe but not banned in the U.S.?* The New York Times, D4.

Richardson, R. M. (2017). Global brain initiatives. *Neurosurgery, 80*(5), N21–N22. https://doi.org/10.1093/neuros/nyx118.

Sharkey, C. M. (2018). Direct-to-consumer genetic testing: The FDA's dual role as safety and health information regulator twenty-fourth annual Clifford symposium on tort law and social policy: Patient safety: How might the law help. *DePaul Law Review, 68*(2), 343–384.

Simons, D. J., Boot, W. R., Charness, N., Gathercole, S. E., Chabris, C. F., Hambrick, D. Z., et al. (2016). Do "brain-training" programs work? *Psychological Science in the Public Interest*, *17*(3), 103–186. https://doi.org/10.1177/1529100616661983.
Solove, D. J., & Hartzog, W. (2014). The FTC and the new common law of privacy. *Columbia Law Review*, *114*(3), 583–676.
Sun, Y., Yao, Y., Bao, Y., Sun, Y., Shi, L., Shi, J., et al. (2019). The effect of changes in society, economy and public health on the disease burden of neurological, mental and substance use disorders: A global longitudinal study. *The Lancet*, *394*, S54. https://doi.org/10.1016/S0140-6736(19)32390-6.
Terry, N. P. (2019). Assessing the thin regulation of consumer-facing health technologies. *SSRN Electronic Journal*. https://doi.org/10.2139/ssrn.3468480. https://papers.ssrn.com/sol3/papers.cfm?abstract_id=3468480.
Terry, N. P., & Gunter, T. D. (2018). Regulating mobile mental health apps. *Behavioral Sciences & the Law*, *36*(2), 136–144. https://doi.org/10.1002/bsl.2339.
Terry, N. P., & Wiley, L. F. (2016). Liability for mobile health and wearable technologies. *Annals of Health Law*, *25*, 62.
Torous, J., Staples, P., Fenstermacher, E., Dean, J., & Keshavan, M. (2016). Barriers, benefits, and beliefs of brain training smartphone apps: An Internet survey of younger US consumers. *Frontiers in Human Neuroscience*, *10*, 180. https://doi.org/10.3389/fnhum.2016.00180.
Torous, J., Wisniewski, H., Liu, G., & Keshavan, M. (2018). Mental health mobile phone app usage, concerns, and benefits among psychiatric outpatients: Comparative survey study. *JMIR Mental Health*, *5*(4), e11715. https://doi.org/10.2196/11715.
Underwood, E. (2016). Regulators seek to tame brain training's 'Wild West'. *Science*, *351*(6270), 212–213. https://doi.org/10.1126/science.351.6270.212.
Viscusi, W. K., & Hersch, J. (2010). Tobacco regulation through litigation: The master settlement agreement. In D. P. Kessler (Ed.), *Regulation vs. litigation: Perspectives from economics and law* (pp. 71–101). University of Chicago Press.
Walker, C. J. (2018). Attacking Auer and Chevron deference: A literature review. *Georgetown Journal of Law & Public Policy*, *16*, 103.
Wexler, A. (2015). A pragmatic analysis of the regulation of consumer transcranial direct current stimulation (TDCS) devices in the United States. *Journal of Law and the Biosciences*, *2*(3), 669–696. https://doi.org/10.1093/jlb/lsv039.
Wexler, A. (2017). The social context of "do-it-yourself" brain stimulation: Neurohackers, biohackers, and lifehackers. *Frontiers in Human Neuroscience*, *11*, 224. https://doi.org/10.3389/fnhum.2017.00224.
Wexler, A. (2019). Separating neuroethics from neurohype. *Nature Biotechnology*, *37*(9), 988–990. https://doi.org/10.1038/s41587-019-0230-z.
Wexler, A., & Reiner, P. B. (2017). Home use of tDCS: From "do-it-yourself" to "direct-to-consumer". In L. S. M Johnson & K. S. Rommelfanger (Eds.), *The Routledge handbook of neuroethics* (pp. 271–284). Routledge.
Wexler, A., & Reiner, P. B. (2019). Oversight of direct-to-consumer neurotechnologies. *Science*, *363*(6424), 234–235. https://doi.org/10.1126/science.aav0223.
Wexler, A., & Thibault, R. (2019). Mind-reading or misleading? Assessing direct-to-consumer electroencephalography (EEG) devices marketed for wellness and their ethical and regulatory implications. *Journal of Cognitive Enhancement*, *3*(1), 131–137. https://doi.org/10.1007/s41465-018-0091-2.
Zettler, P. J. (2017). TDCS research in a world with FDA regulation. *AJOB Neuroscience*, *8*(1), 1–2. https://doi.org/10.1080/21507740.2017.1293192.

CHAPTER TEN

A view on incidental findings and adverse events associated with neurowearables in the consumer marketplace

Nicole Minielly, Viorica Hrincu, Judy Illes*
Neuroethics Canada, Division of Neurology, Department of Medicine, University of British Columbia, Vancouver, BC, Canada
*Corresponding author: e-mail address: jilles@mail.ubc.ca

Contents

> [...] we were very much encouraged to wear the devices at work in order for them to basically collect as much data and to refine their models as much as possible. There was a situation where one of the head data scientists came to me and asked to see my device because some unusual data was flowing through it. [...] they wanted to double-check that the hardware and the sensors were all working appropriately. After he double-checked it, he said, "Well the device is fine, but I don't think you are...." (Interview alpha)

1. Introduction

Wearable technologies that record from and stimulate the brain are predicted to be a $31 billion-dollar industry by 2021 and are becoming ubiquitous in daily life (Kaul & Wheelock, 2016). For the home, classroom, and

Developments in Neuroethics and Bioethics, Volume 3
ISSN 2589-2959
https://doi.org/10.1016/bs.dnb.2020.03.010

workplace, wearables that detect body or brain signals, or that introduce low voltage electrical current to the brain, are being marketed to boost cognitive function such as the speed of information processing and memory, and emotional regulation such as the reduction of stress and anxiety (Bard & Hildt, 2020). The growing presence of devices is exemplified by the excerpt above, taken from a case study interview we conducted with an employee mandated to use a body wearable in the workplace.

The FDA (U.S. Food and Drug Administration, 2019) views direct-to-consumer (DTC) neurowearable devices as low risk, general wellness products. The technologies or products do not pose a safety risk to the user or others, and are neither invasive nor implanted. Their utilization for wellness applies to a general state of health and the reduction of impact of certain chronic diseases or conditions through healthy lifestyle choices. However, the open market availability of these devices draws upon the medical and research uses for which they were originally developed, dating as far back as 1920 for neurorecording with electroencephalography (EEG) (Berger, 1929), and more recently for non-invasive neurostimulation with transcranial direct current stimulation (tDCS) (Nitsche & Paulus, 2000). Wexler (2018) has noted that some consumers use these devices to self-treat mood disorders such as depression.

Within this context, incidental findings of brain anomalies that are inherent to a user and exposed by the use of a device, and adverse events that are caused by a device, would seem to be inevitable. Accordingly, given more than a two-decade long history of related discourse both in the broad research ethics and regulatory systems (Kohane, Masys, & Altman, 2006; Wolf et al., 2008) and more specifically for incidental findings in neuroimaging (Illes, Desmond, Huang, Raffin, & Atlas, 2002; Illes et al., 2008, 2006; Katzman, Dagher, & Patronas, 1999; Sadatsafavi, Marra, Li, & Illes, 2010; Scott, Murphy, & Illes, 2012) we expected that companies would have strategies in place for their management. In this short chapter, we describe our findings and how we were both satisfied and disappointed.

2. Approach

Using an existing database built to examine industry claims about DTC neurowearable devices (Coates McCall, Lau, Minielly, & Illes, 2019) and snowball sampling, we identified and contacted industry leaders or representatives from 41 companies that sell, or intend to sell, neurowearables in the open marketplace. Through semi-structured interviews with participants, we sought to determine company policies about incidental findings and adverse events as they may occur through the data collection systems each company has in place. We used conventional qualitative methods to develop a coding strategy, tested our coding for reproducibility, and then performed the analysis to characterize responses. We integrated the information obtained from one email that we received in lieu of an interview into the results. We use descriptive data and illustrative quotes to report our findings.

3. Results and observations

We interviewed 12 company representatives (chief executive officers or their designates): 8 from neurowearable companies that produce recording devices (e.g., encephalography [EEG]); 3 from companies that produce stimulating device (e.g., transcranial direct current stimulation [tDCS]); one from a company whose device does both (e.g., EEG + tDCS).

3.1 Direct to consumer space

3.1.1 Data collection

Five participants from recording device companies reported that they collect brain data from users; six participants from recording device companies reported other forms of data collection, including demographic data and output from device algorithms. None of the participants from stimulating device companies reported that they collect data from users.

3.1.2 Instances of incidental findings

Across the interviews and email, actual instances of incidental findings were reported once and, in fact, only presumed:

> There are some examples from social media where people... have claimed that they've diagnosed themselves [with] something which our device doesn't claim to do...the only way of describing that is as an incidental finding. (Interview #09, recording device company participant)

While one participant suggested that the data may be crude:

> I think that four [dry] EEG sensors is [sic] unlikely to pick up anything that would fall under this kind of tricky category, so not really a priority for us. (Interview #01, recording device company participant)

another reported that:

> signal quality…is equivalent to clinical grade. (Interview #02, recording device company participant)

Some participants suggested that, even when high quality data are collected, incidental findings may go unnoticed in the DTC neurowearable industry because the data are not being analyzed for abnormalities and employees are not trained to detect incidental findings:

> […] the questions you are asking are legitimate ones, right, like can you detect certain mental illnesses from our data. Well, technically we don't know, it could be in there. (Interview #01, recording device company participant)
>
> […] we wouldn't know what we saw if we saw it. (Interview #06, recording device company participant)

3.1.3 *Instances of adverse events*

Previous literature has discussed different types of risks between recording and stimulating neurowearables (Wexler & Reiner, 2019). Notwithstanding these reports, we learned of only seven consumer-reported adverse events. Most of the adverse events discussed were minor, such as a skin rash, or occurred from improper use of the neurowearable, such as wearing the device too tightly:

> In some particular use cases, users might, against their training, choose to wear the device as a headband or a cap too tightly which can lead to a short-term external compression headache. (Interview #12, recording device company participant)
>
> Most of the reports we have with one of our headsets, the fact that after about an hour for our 5 electrode headsets inside, people find it to be uncomfortable to keep it on the head. (Interview #02, recording device company participant)

3.1.4 *Management strategies*

None of the participants reported that their companies had a formal strategy specifically for managing incidental findings. Ten participants, however, discussed customer service processes generally used to respond to questions about product complaints, including adverse events. These processes included the use of call centers, website chat windows, apps, and social media. None of the participants reported the use of an automated customer service interface on their company websites.

Four participants discussed an *a priori* formal strategy that their companies use to manage consumer reports. Strategies ranged from the documentation of adverse events to standardized follow-up with consumers and case consultation with a physician as needed:

> Every customer interaction in which someone reached out to us and reported any problem ... And anything that...has a technical issue that may have posed a risk to the consumer or to the end-user, we escalate that up to the hardware and the research team and we really dig into it. (Interview #06, recording device company participant)
>
> If it [an adverse event of clinical consequence] were to occur, we would put the person in touch with a physician. (Interview #02, recording device company participant)

However, frontline management is not necessarily carried out by a person with health training:

> Interviewer: Can you tell me ... is that a scientific person, a medical person ... what is the level and characteristics –
> Respondent: It's an administrative person.
> Interviewer: Okay. Not with a health or medical degree?
> Respondent: That's correct. No. (Interview #10, stimulating device company participant)

Among the eight participants who reported that their companies did not have a formal strategy for managing adverse events, we heard that:

> I have no legal policy. So, there's nothing that I can implement, put in place to protect me or the company or even the user. So, if someone had an adverse event, I can't even fabricate a policy that would be legitimate. So, if someone contacts me and they say, you know, "This is making me feel worse" I encourage them to return the device and stop using tDCS but in an indirect way. So that's yes, there is no policy. (Interview #11, stimulating device company participant)

One participant indicated that their company had already undertaken the initiative to create policies for incidental findings where they did not yet exist:

> Should we, for example, put some message to suggest they go see their doctor because there's some abnormal patterns in their sleep disturbance which might suggest they have a condition that should be assessed by a sleep physician? We're actually in the process of trying to make that decision, so in that sense your call is timely. (Interview #09, recording device company participant)

Another participant indicated that this study generated heated conversations regarding incidental findings within their company:

> It sparks particularly strong debate I found, more than I expected it to...one of our employees, became very strongly liberal on it... we have to own our data

> and we have to have right to information and any company that knows something that doesn't tell us about it is an evil company. (Interview #01, recording device company participant)

One participant, whose company sells neurowearables business-to-business, indicated that the company encourages the development of processes to manage adverse events and abnormal alarm signaling from devices worn by employees in the workplace. This participant further highlighted the importance of adhering to privacy legislation in different regions of the world:

> The privacy legislation in different jurisdictions has never kept up with technology. So, we've very much had to take it upon ourselves to implement very strict privacy controls. (Interview #12, recording device company participant)
>
> Europe stands out as having the strictest privacy laws with respect to personal data. And so therefore a very clear review of what data is being recorded and whether it falls under the definition of personal or identifiable information would really need to be looked at. (Interview #12, recording device company participant)

Four participants discounted the need for policies altogether, citing a number of reasons including the limitations of the data collected by recording neurowearables:

> I think that there's not really a strong utility for developing that policy for incidental findings of individual health criteria. (Interview #06, recording device company participant)
>
> Interviewer: Do you expect to put policies in place for the detection of an unexpected health finding in your device data?
> Respondent: Not before this call, no.
> Interviewer: After this call is that something you'll consider?
> Respondent: Probably not, reason being I don't think that the technology we're working with is likely to lead to those kinds of health findings. (Interview #01, recording device company participant)

3.2 Research studies and clinical trials

Seven participants volunteered that their DTC neurowearable device is also being used for research purposes. Two participants representing companies who manufacture recording devices and one stimulating device company reported that incidental findings can arise from the data collected:

> [...] when reviewing some of the EEG patterns there was the suspicion from the researchers that that particular individual might suffer from a mild form of epilepsy. (Interview #12, recording device company participant)

One participant from a recording device company and two from stimulating device companies also discussed adverse events:

> He wanted to start at very low frequency like 1 hz. Okay. So, he did. And we see him walking out of this like a drunk almost. (Interview #04, stimulating device company participant)

Strategies for handling incidental findings and adverse events involve oversight from external regulatory bodies and conform to requirements by institutional research ethics boards where research on the devices is conducted. From four participants, we learned that companies largely defer to these external resources about the ethical obligation to disclose or to not disclose incidental findings:

> [...] they, under the ethical approval that they had for that study, were obligated to communicate those findings with the individual. (Interview #12, recording device company participant)

Five participants discussed their responsibility to report adverse events to external regulatory bodies such as the FDA. As one participant described in detail:

> There is a procedure for adverse events. So, every time a client or a research project or a hospital reports on an adverse event, we do have our own procedure to report the event to the authorities as a manufacturer. (Interview #05, recording + stimulating device company participant)
>
> I have responsibility as a manufacturer towards the authorities like the FDA or the medical authorities to report any adverse events of the use of my device. But I don't have the relationship with the patient. (Interview #05, recording + stimulating device company participant)

4. Strategies forward

Information collected from neurowearables may be inherently consequential to users, yet there is no discernible consensus from the interviews we conducted with company representatives on the ability to gather and interpret these data from devices currently in the marketplace. Our findings also reveal little standardization across the DTC neurowearable industry for the reporting of incidental findings and adverse events.

In our view, there is an imperative for industry to offer a greater level of protection for consumers than currently exists. Furthermore, while as authors we recognize and have even written about the limitations of current neurowearable capabilities (Coates McCall et al., 2019), we see the transformational nature of the algorithms on which they rely, and anticipate a growing imperative in this regard. Drawing upon a comment from one the participants:

> [...] we don't know what pattern recognition can find if trained on a good enough data set. (Interview #01, recording device company participant)

This growth in algorithmic capacity is exemplified by pattern recognition improvements with other wearable technologies. Apple, for example, was recently able to detect atrial fibrillation using an algorithm on a large set of Apple Watch user data (Perez et al., 2019). According to the Recommendation of the Council on Responsible Innovation in Neurotechnology (OECD, 2019), anticipating the future potential uses of these technologies and proactively working toward mitigating their misuse is of vital importance.

Finally, there is a murky line between general wellness and health products on the DTC landscape (Kreitmair, 2019). Indeed, whereas a neurowearable may be considered a wellness product in one country it is considered a medical device in another. As a participant noted:

> It's over the counter everywhere except in the US. In the US it requires a healthcare authorization...The device is so mild that we actually can market it as an OTC (over the counter) product without healthcare imprimaturs or healthcare claims. Scientifically—you know, without medical claims. So instead of marketing it for insomnia, depression, anxiety, the over the counter increment...is marked with, you know, for sleep and mood, not for insomnia, depression and anxiety. (Interview #10, stimulating device company participant)

It is no surprise then, that consumers themselves might conflate clinical health with general wellness (Wexler, 2018). As one participant expressed:

> The only thing that will keep me awake at night is my concern that someone gets a hold of the technology and puts all their eggs in one basket and is very hopeful that it's going to help them and then it just becomes another thing that doesn't help them. (Interview #11, stimulating device company participant)

4.1 Recommendations

As the DTC neurowearable industry draws upon medical innovations and contemporary procedures for research and clinical trials, we offer three recommendations (Table 1):

1. *Data collection*: Minimize data collection, when possible. Standardize data collection when minimization is not possible. When data collection is

Table 1 Recommendations for the DTC neurowearable industry.

1. Minimize data collection where possible. Increase data security when data minimization is not possible
2. Standardize and make explicit incidental finding and adverse event management procedures
3. Encourage better regulatory and industry clarity about neurotechnologies for health and wellness

necessary to, e.g., *help…improve … product and post-market surveillance, (Interview #09, recording device company participant)* anonymization, and aggregation of collected data will mitigate the management burden of incidental findings.

When data are collected from individuals and the possibility exists to develop or improve record-keeping about incidental findings and adverse events, it should be pursued. Even small numbers can provide meaningful information for future neurotechnological developments.

2. *Data management pathways*: When the choice and possibility exist to detect incidental findings in the consumer space, there should be a company pathway from first data acquisition, triage, and contact with consumer. Elements of the pathway such as means of acquiring or utilizing personal information for contact, timing and mode of contact should be well defined and standardized. As highlighted by one of the participants, such findings should never be disclosed to a third party:

> The much bigger problem is the privacy problem, you know, how do you reassure your users that their data is protected, and they have—that the rights are theirs and not someone else's to access the data. (Interview #08, recording device company participant)

Consumers should have autonomy over their data including the opportunity to access, amend and delete data (OECD, 2019).

When a consumer has an adverse event to report, a customer service pipeline to receive this information should involve a standardized and secure reporting intake form and persons trained to triage it. Companies currently use call-in centers; chatbots are a real possibility for this context in the future. Therefore:

- Triage by a human may involve referral upward to a manager and/or health expert. This process is modeled by the company whose pipeline includes consulting with a physician retained to manage and contact a consumer when needed.

- When the customer service system is carried out by artificial intelligence, the website can direct a user to an online reporting form for the intake of information and have built-in features that trigger human review when needed.

Management *strategies* should be explicit online, on packaging information, or both.

3. *Definitions of health and wellness*: Akin to the review that has been undertaken for the regulatory classification of natural health products (Health Canada, 2013), we urge industry leaders and regulatory bodies to clarify definitions ofhealth and wellness vis à vis neurotechnology to protect consumer understanding and choice about products for the brain. On a global level, further collaborative work with the OECD and new efforts under the International Brain Initiative (International Brain Initiative, 2020) may provide opportunity to promote and co-ordinate such an effort.

4.2 Concluding thoughts

One of the participants interviewed shared that:

> It's important for us to do it...because we have [...] a sense of the ethical responsibility of marketing a product like this and we don't want people to feel that we're not being responsive. (Interview #10, stimulating device company participant)

If the goal of these DTC neurowearables is to give users power over their brain data thereby, as another participant suggested, for:

> democratizing access to brain health, (Interview #02, recording device company participant)

then they have a right to the brain-related information that may be collected from them, or to be proactively protected from harm.

We thank participants for sharing their views with us.

Acknowledgments

This work was supported by Technical Safety BC. The funding agency had no role in the preparation of this chapter. J.I. is the Canada Research Chair in Neuroethics.

References

Bard, I., & Hildt, E. (Eds.), (2020). *Ethical dimensions of commercial and DIY neurotechnologies*. Academic Press: Elsevier.

Berger, H. (1929). Über das Elektrenkephalogramm des Menschen. *Archiv für Psychiatrie und Nervenkrankheiten*, *87*, 527–570. https://doi.org/10.1055/s-0028-1130334.

Coates McCall, I., Lau, C., Minielly, N., & Illes, J. (2019). Owning ethical innovation: Claims about commercial wearable brain technologies. *Neuron*, *102*, P728–P731. https://doi.org/10.1016/j.neuron.2019.03.026.

Health Canada. (2013). *ARCHIVED—Charting a course: Refining Canada's approach to regulating natural health products*. Government of Canada.

Illes, J., Desmond, J. E., Huang, L. F., Raffin, T. A., & Atlas, S. W. (2002). Ethical and practical considerations in managing incidental findings in functional magnetic resonance imaging. *Brain and Cognition*, *50*, 358–365. https://doi.org/10.1016/S0278-2626(02)00532-8.

Illes, J., Kirschen, M. P., Edwards, E., Bandettini, P., Cho, M. K., Ford, P. J., et al. (2008). Practical approaches to incidental findings in brain imaging research. *Neurology*, *70*, 384–390. https://doi.org/10.1212/01.wnl.0000280469.17461.94.

Illes, J., Kirschen, M. P., Edwards, E., Stanford, L. R., Bandettini, P., Cho, M. K., et al. (2006). Incidental findings in brain imaging research. *Science*, *311*, 783–784. https://doi.org/10.1126/science.1124665.

International Brain Initiative. (2020). International Brain Initiative: An innovative framework for coordinated global brain research efforts. NeuroView*Neuron*, *105*, 212–216Yuste 2017.

Katzman, G. L., Dagher, A. P., & Patronas, N. J. (1999). Incidental findings on brain magnetic resonance imaging from 1000 asymptomatic volunteers. *Journal of the American Medical Association*, *282*, 36–39. https://doi.org/10.1001/jama.282.1.36.

Kaul, A., & Wheelock, C. (2016). *Enterprise wearable technology case studies*. Tratica White Paper.

Kohane, I. S., Masys, D. R., & Altman, R. B. (2006). The incidentalome: A threat to genomic medicine. *Journal of the American Medical Association*, *296*, 212–215. https://doi.org/10.1001/jama.296.2.212.

Kreitmair, K. V. (2019). Dimensions of ethical direct-to-consumer neurotechnologies. *AJOB Neuroscience*, *10*(4), 152–166. https://doi.org/10.1080/21507740.2019.1665120.

Nitsche, M. A., & Paulus, W. (2000). Excitability changes induced in the human motor cortex by weak transcranial direct current stimulation. *Journal of Physiology*, *527*, 633–639. https://doi.org/10.1111/j.1469-7793.2000.t01-1-00633.x.

OECD. (2019). *Recommendation of the council on responsible innovation in neurotechnology*. OECD/LEGAL/0457.

Perez, M. V., Mahaffey, K. W., Hedlin, H., Rumsfeld, J. S., Garcia, A., Ferris, T., et al. (2019). Large-scale assessment of a smartwatch to identify atrial fibrillation. *New England Journal of Medicine*, *381*, 1909–1917. https://doi.org/10.1056/NEJMoa1901183.

Sadatsafavi, M., Marra, C., Li, D., & Illes, J. (2010). An ounce of prevention is worth a pound of cure: A cost-effectiveness analysis of incidentally detected aneurysms in functional MRI research. *Value in Health*, *13*, 761–769. https://doi.org/10.1111/j.1524-4733.2010.00755.x.

Scott, N. A., Murphy, T. H., & Illes, J. (2012). Incidental findings in neuroimaging research: A framework for anticipating the next frontier. *Journal of Empirical Research on Human Research Ethics*, 7, 53–57. https://doi.org/10.1525/jer.2012.7.1.53.

U.S. Food and Drug Administration. (2019). *General wellness: Policy for low risk devices guidance for industry and food and drug administration staff*. U.S. Department of Health and Human Services Food.

Wexler, A. (2018). Who uses direct-to-consumer brain stimulation products, and why? A study of home users of tDCS devices. *Journal of Cognitive Enhancement*, *2*, 114–134. https://doi.org/10.1007/s41465-017-0062-z.

Wexler, A., & Reiner, P. B. (2019). Oversight of direct-to-consumer neurotechnologies. *Science*, *363*, 234–235. https://doi.org/10.1126/science.aav0223.

Wolf, S. M., Lawrenz, F. P., Nelson, C. A., Kahn, J. P., Cho, M. K., Clayton, E. W., et al. (2008). Managing incidental findings in human subjects research: Analysis and recommendations. *The Journal of Law, Medicine & Ethics*, *36*, 219–248. https://doi.org/10.1111/j.1748-720X.2008.00266.x.

FSC
www.fsc.org
MIX
Papier | Fördert
gute Waldnutzung
FSC® C083411

Kolibri 360 GmbH
Ferdinand-Jühlke-Straße 7
99095 Erfurt, Deutschland
produktsicherheit@kolibri360.de